EAT AND LIVESTOCK COMMISSION

THE L

PROGRESS IN SHEEP AND GOAT RESEARCH

Progress in Sheep and Goat Research

Edited by

A.W. Speedy

C·A·B International

C·A·B International   Tel: Wallingford (0491) 32111
Wallingford   Telex: 847964 (COMAGG G)
Oxon OX10 8DE   Telecom Gold/Dialcom: 84: CAU001
UK   Fax: (0491) 33508

A catalogue record for this book is available from the British Library

ISBN 0 85198 772 9

Printed and bound in the UK by Redwood Press Ltd., Melksham

Contents

Preface

The sheep is perhaps the most primitive of domestic animals, yet one of the most comprehensively studied by scientists. The biochemistry and physiology of this species has been studied in greater detail than any other, at least as much as the mouse or rat. Genetic studies of sheep populations have also been conducted by some eminent scientists, because of their importance in countries like Australia, New Zealand, parts of the USA and the UK.

New techniques that have emerged in recent years - not least the powerful armoury of molecular biology - have extended the research and produced fascinating new perspectives and possibilities.

The interest of scientists has been partly due to the size (50-80 kg) of the sheep in relation to the human, making it a useful animal "model". This has produced "spin-off" for the sheep production industry, producing many technical developments.

These observations are not true of the goat. Despite the title of this collection, there is noticeably less to report and describe on this related species. Nevertheless, there is increasing interest, particularly in the production of fibre from goats and, where possible, authors have contributed information on the physiology, nutrition, reproduction and management. A paper on fibre production from sheep and goats is included.

It is an appropriate time for a new collection of papers by some important members of the current sheep and goat research fraternity. Besides important new developments in the control of reproduction and genetic selection based on *in vivo* techniques of carcass assessment, there are new contributions on maternal/fetal endocrinology, nutrition, grazing management and parasite control. In particular, these apply new concepts - for example, optimal foraging theory - to some familiar subjects and consider some of the emerging problems such as anthelmintic resistance. In the area of disease, the more difficult problems posed by lentiviruses and sub-viral organisms are considered. Here we see the power of some new laboratory techniques to elucidate the complex aspects of the diseases. Finally, there are papers which consider not only physiological manipulation, but also the possibilities of transgenic animals to increase meat and wool production. The sheep appears to be becoming a "high-tech" species!

Yet, I am reminded of a paper presented by an eminent sheep researcher who promised that, within 10 years, sheep would be producing 3 lambs, twice a year. That was in the mid 1960s and it hasn't happened. The technology has existed for a long time but the application has not taken place, except in some isolated farms run by "sheep enthusiasts". Even less has existing scientific knowledge been applied in the extensive areas of the world where sheep are herded in traditional ways.

vii

The material presented here represents the current state of scientific research on these species and illustrates the considerable biological and technical potential that exists. In most cases, it still remains for the methodology for the application of this research to be developed, to produce real economic benefits and sustainable systems in the future.

Not all the information here will ever be fully applied. Yet the importance of the scientific methods is beyond question. This book makes a modest contribution to the "portfolio". With it, we can consider new possibilities for the increased production of meat, milk and fibre for the needs of people.

Andrew W. Speedy
Oxford, 1992

Chapter 1

Artificial Insemination and Embryo Transfer

Jacqueline M. Wallace

Rowett Research Institute, Greenburn Road, Bucksburn, Aberdeen AB2 9SB, UK

Introduction

Although the techniques of artificial insemination and embryo transfer have been practised commercially for over 50 and 30 years respectively, they have had little impact on the long term genetic improvement of sheep and goats. However, during the last decade, the advent of laparoscopic procedures for intrauterine insemination and the recovery and transfer of embryos have provided breeders with potentially powerful new tools to rapidly increase selection for genetically desirable traits. In sheep these traits include superior lean tissue growth, wool quality and growth, milk production and disease resistance, while in goats the predominant emphasis is on improving the yield of high quality fibre.

The aim of this chapter is to review the advantages offered by the new technologies while highlighting problem areas still to be improved.

Artificial Insemination

Semen collection

Semen collection methods have changed little with the passage of time. The collection of semen by artificial vagina remains the method of choice for both sheep and goats (Memon and Ott, 1981). Rams and bucks are trained to mount an intact oestrous female or alternatively an ovariectomized female in which oestrous behaviour has been induced by injection of oestradiol benzoate. The male mounts the female and the penis is deflected into the artificial vagina. The artificial vagina is filled with water to a pressure of 40-60 mm Hg and an inside temperature of 40-45°C.

Following ejaculation the semen is collected directly into an insulated sterile collection vessel (30-37°C) to prevent temperature shock. Alternatively semen may be obtained by electro-ejaculation using a stimulus (10-

15 volt output) applied via a rectal probe. The probe is inserted into the rectum to a depth of 15-20 cm and pressed towards the floor of the pelvis. Short stimuli (3-8 seconds) are applied at 15-20 second intervals until ejaculation occurs. However, this method is uncomfortable for the animal and results in a sample which contains fewer spermatozoa and more accessory secretions (Quinn *et al.*, 1968; Hernandez *et al.*, 1976). While useful for obtaining semen from valuable genotypes which for some reason may be unable to mount, it can result in animals being incorrectly diagnosed infertile.

Semen evaluation

Semen volume, spermatozoa motility, concentration and morphology should all be considered when assessing the quality of semen. Semen from rams and bucks is similar. An average ejaculate of 0.7-2 ml contains 2-6.5 x 10^9 spermatozoa per ml (Evans and Maxwell, 1987). Visual assessment of the motility of spermatozoa is one of the most rapid and widely used tests of semen quality. A drop of semen is examined under a microscope at low power. The wave motion is assessed on a 0-5 scoring system where 0 indicates motionless and 5 signifies dense, rapidly moving waves with 90% or more spermatozoa being active. However, this method of visual motility assessment does not always correlate with fertility in subsequent insemination trials (Eppleston *et al.*, 1986).

The use of time-lapse photography and computer-assisted analysis provides more accurate motility assessment but the equipment is expensive and rarely used in the field. For rams of normal fertility, the number of abnormal spermatozoa in a sample varies from 5 to 15%. Generally, a reduction in fertility is not observed until the number of abnormal spermatozoa exceeds 20% (Foote, 1974). The type and number of abnormal sperm are routinely and easily determined in smears using eosin-nigrosin stain. In recent years, the technique of *in vitro* fertilization of zona-free hamster oocytes has been widely used to assess the fertility of human spermatozoa (Yanagimachi, 1984). The technique has also been used in a variety of domestic species including sheep (Pavlok *et al.*, 1983; Pavlok and Fléchon, 1985) but its usefulness is confined to experimental studies of gamete fusion rather than assessing fertilizing ability *in vivo*. A simple and reliable *in vitro* test for standardizing semen evaluation across the world would be a valuable technological advance.

Semen preservation

A variety of extenders have been utilized to dilute ram and buck semen for use in the short (liquid-chilled storage) or long term (frozen storage). Most semen diluents utilize egg yolk or milk as the basic ingredient. Protection of the spermatozoa against cold shock is provided by yolk lipoprotein and lecithin (Foote, 1974) and milk casein respectively (O'Shea and Wales, 1966). Other diluent ingredients include buffer solutions such as tris (hydroxymethyl) aminomethane or citric acid and sugars such as fructose, glucose, lactose and sucrose as an energy source. Both egg yolk and milk are widely used in sheep semen diluents but goat semen tends to be extended with skimmed or ultra heat treated milk. While some groups report acceptable conception rates with egg yolk-based diluents others believe that egg yolk is toxic to goat spermatozoa (Roy, 1957). The presence of phosphotidase in the seminal plasma of goats catalyses the hydrolysis of lecithins in egg yolk to fatty acids and lysolecithins which are toxic to the spermatozoa. Indeed, diluents developed in Russia for sheep and goats contain antioxidants to decrease the oxidation of phospholipids and unsaturated fatty acids in the cell membrane (Milavanov and Sokolov-skaya, 1980; Grigoryan and Nazaryan, 1981; Milavanov *et al.*, 1981). Alternatively, for goats, diluents with a low concentration of egg yolk (1.5%, v/v) can be used (Ritar *et al.*, 1990b) or the seminal plasma can be removed by centrifugation (Ritar and Salaman, 1982).

Short-term liquid storage of semen is dependent on the reversible reduction of motility and metabolic activity of spermatozoa at reduced temperatures. The semen is diluted at 30°C and held in sterile straws or vials. It is then cooled in a vacuum flask or refrigerator at a rate of 0.25 °C per minute to 15 or 5°C, and maintained at that temperature until use. Using cervical insemination the viability of spermatozoa held at 15°C is 6-12 h for both ram and buck semen. When stored at 5°C, post-chilling motility can be maintained for several days but fertilization declines at a rate of 10-35% per day of storage. Acceptable fertility using cervical insemination is obtained using semen stored at 5°C for up to 24 h in sheep and 48 h in goats (Evans and Maxwell, 1987). This preservation method offers considerable mobility within a wide geographical area. To date, there are no data published on the use of liquid-chilled semen for laparoscopic intrauterine insemination but the technique could considerably extend the lifespan of chilled semen.

Long-term storage of semen in liquid nitrogen at -196°C completely stops spermatozoa metabolism hence allowing its transport worldwide and use for many years. The diluents used are essentially equivalent to those for liquid storage except that a cryoprotective agent, normally glycerol, is added to prevent membrane damage during freezing. In addition, when frozen semen is being deposited directly into the uterus, antibiotics (usually penicillin and streptomycin) are added to prevent bacterial growth. Both goat and sheep semen can be frozen in pellets or straws. The standard method involves diluting the semen, cooling it to 5°C over 1-2 h and then either dropping the semen on to dry ice or suspending straws on a rack in liquid nitrogen vapour. Freezing semen in pellets is generally easier and more popular than the straw method although few direct comparisons have been made. In goats the survival of spermatozoa after freezing as pellets was higher than that obtained using straws (Ritar *et al.*, 1990b). Similarly in sheep, Maxwell *et al.* (1980) reported that fertility tended to be higher when semen was frozen in the pellet form. The method used however may depend on whether or not the semen is being exported. Semen frozen as pellets is not suitable for export within Europe because legislation requires the easy identification offered by straws.

Type of Insemination

Cervical insemination

To date the most commonly used insemination technique involves deposition of semen into the first fold of the os-cervix, viewed with the aid of a speculum with an internal light source. This is a cheap and relatively easy method of insemination which utilizes fresh semen which may or may not be diluted and chilled. In sheep, the use of frozen-thawed semen for cervical insemination usually results in unacceptably low fertilization rates of 10-30% (Dzuik *et al.*, 1972; Maxwell *et al.*, 1980; Maxwell and Hewitt, 1986) due to its impaired transport through the cervix (Mattner *et al.*, 1969; Lightfoot and Salamon, 1970) and reduced retention/viability in the uterus (Hawk *et al.*, 1987). In goats, the use of frozen semen for cervical insemination results in higher conception rates of between 30 and 67% (Moore *et al.*, 1989; Ritar *et al.*, 1990a). This probably reflects the difference in the depth of insemination achieved in the two species. The anatomy of the goat cervix allows complete penetration into the body of the uterus in up to 60% of adult does (Ritar and Salamon, 1983; Ritar *et al.*, 1990a). Conception rates are positively correlated with the depth of insemination into the cervix, increasing by approximately 10% for each 1 cm advance. Nevertheless, insemination with frozen-thawed semen into the uterus via the cervix results in pregnancy rates which are lower than those achieved by direct laparoscopic insemination (45 versus 64%) (Ritar *et al.*, 1990a).

Vaginal insemination

Often termed "shot in the dark" or SID, this method simply involves depositing fresh diluted semen into the vagina of the female without using a speculum or attempting to locate the cervix. While some authors report results comparable with those obtained using cervical insemination (Maxwell and Hewitt, 1986; Tervit *et al.*, 1984), others suggest that conception rates are lower using this method (Kerton *et al.*, 1984; Rival *et al.*, 1984).

Transcervical insemination

Recently, a modified transcervical insemination method has been evaluated (Halbert *et al.*, 1990a) for sheep. This involves grasping the cervix and retracting it into the vagina with a pair of forceps to allow an inseminating instrument to be introduced into the cervical canal. Uterine penetration was achieved in 73 of 89 ewes studied and the time from the probe being introduced until it was assessed to be within the uterus was 2.6 minutes. To date, field trial evaluation of this technique is limited but suggests that acceptable pregnancy rates can be achieved when using this technique in conjunction with frozen-thawed semen (52 of 90 ewes; Halbert *et al.*, 1990b). However, the procedure does involve a high degree of manipulation and any accidental vaginal bleeding could cause vaginal wall adhesions to form and compromise the ewe's future ability to conceive naturally. As yet, no data are available on the efficacy of repeatedly using this technique.

Laparoscopic intrauterine insemination

The deposition of semen directly into the uterine lumen, thus circumventing the natural cervical barrier, has radically improved fertilization rates in a variety of physiological states in both sheep and goats. Initially, intrauterine insemination was carried out via mid-ventral laparotomy (Killeen and Moore, 1970). However, as well as being impractical for anything other than research purposes, it was associated with low embryo recovery and survival rates (Mattner *et al.*, 1969; Trounson and Moore, 1974; Boland and Gordon, 1978). In 1982, Killeen and Caffery first reported intrauterine insemination carried out by laparoscopy.

The instrumentation has been refined during the last 10 years but the basic technique remains unchanged. Ewes or does which have been deprived of food and water for 12-16 hours are suspended in a laparoscopy cradle at an angle of 40° to the horizontal. The area anterior to the udder is sheared and the skin sterilized. Local anaesthetic is injected subcutaneously 5-7 cm anterior to the udder, 3-4 cm on each side of the mid-ventral line. Two small incisions are made using a trocar and cannula to permit entry of the 5 mm diameter laparoscope, connected via a fibre optic cable to a powerful light source. The peritoneal cavity is inflated with CO_2 and the uterus is located anterior to the bladder. The glass inseminating pipette is introduced via the second cannula and stabbed through the uterine wall into the lumen. Normally both uterine horns are inseminated and then the apparatus is withdrawn. Current instrumentation is so fine that it does not require the entry point to be sutured and the whole procedure takes 1-2 minutes depending on the skill of the operator.

Advantages of laparoscopic intrauterine insemination

(1) *High fertilization rates*

Laparoscopic intrauterine insemination with fresh diluted semen in non-superovulated ewes results in high fertilization and pregnancy rates (>80%). However, the improvement beyond that obtained using cervical insemination is not vast (McKelvey *et al.*, 1985). In contrast, laparoscopic intrauterine insemination radically improves fertilization rates obtained when using frozen-thawed semen. Conception rates range from 50 to 80% in sheep (Maxwell, 1986a and b; Maxwell and Hewitt, 1986; Haresign, 1990) and goats (Moore *et al.*, 1988; Moore *et al.*, 1989; Ritar *et al.*, 1990a).

Arguably one of the major advantages of intrauterine insemination has been in guaranteeing high fertilization rates in ewes and does superovulated for embryo transfer programs. Superovulation is known to impair both sperm transport (Evans and Armstrong, 1984) and viability (Hawk *et al.*, 1987) following natural mating or cervical insemination. While high fertilization rates have been recorded in specific groups of cervically inseminated superovulated ewes, the technique is unreliable and often results in a high proportion of ewes with zero or partial fertilization. In studies at the Rowett Research Institute, fertilization rates in cervically inseminated ewes with a mean ovulation rate of 12.9 ± 0.98 varied from 0 to 100% (mean $\pm$ s.e.m., $70 \pm 7.5\%$, n=27). In contrast, fertilization rates following intrauterine insemination with fresh-diluted semen were consistently high (44 to 100%, mean $\pm$ s.e.m., $96 \pm 2.2\%$, n=34) (Robinson *et al.*, 1989). Similarly, high fertilization rates using intrauterine insemination (87%) have been achieved in pedigree ewes of various breeds submitted for embryo transfer (Scudamore *et al.*, 1991b). While the use of fresh-diluted semen in multiple ovulation and embryo transfer (MOET) schemes is feasible in the UK where valuable males can be transported around the country, the wider dissemination of valuable genotypes requires the use of frozen-thawed semen. Comparison of cervical *versus* intrauterine insemination of superovulated ewes using frozen-thawed semen indicates that uterine insemination does considerably increase fertilization rate (10 *versus* 55%) but not to the level obtained using fresh semen (72%) (Armstrong and Evans, 1983).

(2) *Economical semen usage*

Intrauterine insemination not only increases fertilization rates in synchronized and/or superovulated ewes but could potentially allow a more widespread dissemination of valuable genotypes because of economies in semen usage. For example, in the UK, pedigree flock sizes are small (25-100 ewes). This often results in rams of superior genetic merit being used to naturally mate <50 ewes per breeding season. An average ejaculate diluted appropriately and used in a intrauterine insemination programme is more than adequate for 50 ewes. Thus, if 5 ejaculates per day were collected over a six week period during the height of the breeding season, one ram could be used to intrauterine inseminate 10,000 ewes per year (McKelvey and Robinson, 1986). However, for such a scenario to benefit the sheep industry would require a major change of attitudes and co-operation on a national scale. In the UK fibre market, the high cost of quality does together with the relative scarcity of genetically superior males has led to wider use of frozen semen and intrauterine insemination on just such a co-operative basis.

(3) *Out-of-season breeding*

At temperate latitudes the strictly seasonal nature of sheep reproduction results in the bulk of lambs being born in the spring and marketed in the autumn. Supermarkets now aim to sell home produced lambs throughout the year and hence the financial rewards for producing lambs out of season should theoretically be high.

In large scale experiments involving 1900 ewes, cervical insemination during deep seasonal anoestrous (June-July) resulted in a low lambing percentage of 31% to the induced oestrus (Gordon *et al.*, 1969). This low conception rate was primarily due to fertilization failure (Gordon, 1969) suggesting that the uterine environment of the anoestrous ewe is not conducive to sperm transport. Furthermore, semen quantity and quality are reduced during mid-seasonal anoestrus (Colas, 1983). These problems can be overcome by the use of laparoscopic intrauterine insemination and fresh-diluted semen.

For example, 58 Border Leicester x Scottish Blackface ewes (a strictly seasonal commercial breed) which lambed in late March were weaned from their lambs, synchronized to ovulate and inseminated in mid June at 57°N. Three ewes aborted before day 60 of pregnancy when 43 of the remaining 55 ewes were diagnosed pregnant by ultrasonic scanning. Average litter size was 1.9 lambs, suggesting that laparoscopic intrauterine insemination may eliminate the low conception rates that accompany out-of-season breeding.

(4) *Early post partum ewes as embryo donors*

The use of laparoscopic intrauterine insemination to deposit semen into the tip of the uterine lumen, thus by-passing the hostile environment of the involuting uterus, has resulted in fertilization as early as 22 days *post partum* in early weaned and lactating ewes (Wallace *et al.*, 1989). While the resulting embryos do not normally survive when returned to their *post partum* uterine environment, they do survive when transferred to the normal environment of a recipient ewe. This could potentially facilitate the use of early lambing ewes to donate embryos before the end of their breeding season. The resulting embryos could be transferred directly to the recipient ewes or cryopreserved to await transfer at the start of the breeding season the following year.

Timing of intrauterine insemination

While intrauterine insemination of ewes in natural oestrus using frozen-thawed semen results in acceptable conception rates (42-52%) (Azzarini and Valledor, 1988), the procedure is normally used in synchronized females which may or may not have received exogenous gonadotrophin. Females are normally inseminated at a fixed time relative to the synchronization method and not in relation to when ovulation has occurred. The latter is probably more important as it is influenced by a variety of factors including breed, season, synchronization method, gonadotrophin type and dose, whether the male is present and the age and nutritional status of the animal (Evans, 1988). Add to these the variable retention and lifespan of fresh-diluted versus frozen-thawed semen and the difficulty of determining an optimum insemination time is apparent. Nevertheless, in non-superovulated ewes, the general trend is to inseminate at or just prior to the predicted time of ovulation when using fresh-diluted semen (Davis *et al.*, 1984; McKelvey *et al.*, 1985) and after ovulation has occurred when using frozen-thawed semen (Maxwell, 1986a; Haresign, 1990). In superovulated ewes, high fertilization rates have been consistently achieved following intrauterine insemination around the predicted time of ovulation, i.e. 48 to 52 h after progesterone priming removal (Robinson *et al.*, 1989). Unfortunately, a significant reduction in embryo recovery (>40%) was reported, suggesting that insemination at this time interfered with the normal collection and transport of ova into the oviducts. Delaying the insemination time to 60 h significantly improved embryo recovery without any reduction in fertilization rates. However, in a subsequent study when embryos were recovered and assessed at day 5 after insemination, the proportion of transferrable quality embryos was reduced from 100% for the 48 h insemination group to 35% for the 60 h group. Similarly, embryo survival rates following laparoscopic transfer of these embryos were 44 and 18.5% for the 48 and

60 h insemination groups respectively, suggesting that delaying insemination until well beyond ovulation results in the fertilization of aged oocytes of reduced developmental competence (Scudamore *et al.*, 1991a). Attempts to minimize interference with egg collection by administration of a pre-insemination sedative, acepromazine maleate, marginally increased ovum recovery in ewes inseminated at 48 h (53 to 63%) but its use was correlated with reduced embryo survival in recipient ewes. Further work is required to evaluate other sedatives and/or refine the animal handling procedures used in association with the laparoscopic technique.

Dose of semen and site of insemination

When using valuable genotypes, it is obviously desirable to use the minimum effective dose of semen required to ensure maximum fertility. The critical dose of semen is influence by a variety of factors including the quality and type of semen used, its lifespan in various diluents and the site and time of insemination. Of these factors, the variable ability of semen from different individual animals to withstand the freezing and thawing process is one of the main outstanding problems still to be resolved. It is not surprising that semen doses varying from 5 to $>600 \times 10^6$ motile spermatozoa have been used to achieve acceptable pregnancy rates following intrauterine insemination (see review by Evans, 1988).

In non-superovulated ewes, lambing percentages (but not conception rates) are reported to be higher following insemination into the middle rather than the tip or bottom of the uterine horn (84.9 versus 52.7 and 64.7% respectively; Maxwell, 1986b). This probably reflects the relative ease with which the middle of the horn is located and inseminated rather than a difference in semen transport *per se*. Indeed, in superovulated ewes, the deposition of semen in a single uterine horn resulted in normal fertilization rates for ova shed into the contralateral horn (Evans *et al.*, 1984; Robinson *et al.*, 1989). In the latter study, where mean ovulation rates exceeded 10 ovulations per ovary, 100% fertilization rates were recorded in both horns following insemination in the right horn only, implying a high degree of semen migration between horns.

Multiple Ovulation and Embryo Transfer

During the last decade, exciting advances have been made with respect to embryo recovery and transfer techniques. The following sections will review procedures currently practised and emphasize the potential benefits of the new innovations.

Oestrous synchronization and superovulation

The variability in response to current superovulatory regimes is one of the major problems limiting the success of multiple ovulation and embryo transfer (MOET) schemes. To double the rate of genetic improvement for important traits requires that all females contribute 10 progeny to the breeding programme per unit of time (Smith, 1986). Much research effort has therefore been directed towards improving superovulation regimes (see reviews: Moore, 1982; Armstrong and Evans, 1983). To date, the gonadotrophin preparations most frequently used include pregnant mares serum gonadotrophin (PMSG), porcine follicle stimulating hormone (pFSH), horse anterior pituitary extract (HAP) and human menopausal gonadotrophin (hMG). These gonadotrophins are administered as single or multiple injections to females whose oestrous cycles are regulated by progesterone, synthetic progestagens or prostaglandin treatments. The precise timing of exogenous gonadotrophin administration varies relative to the oestrous synchronization method used and normally centres around the last 24 to 72 h of the regulated cycle. Table 1.1 summarizes results for 3 different superovulatory regimes applied to groups of ewes studied over a 3 year period at the Rowett Research Institute. All ewes were synchronized to ovulate using intravaginal pessaries containing synthetic progestagens and were subjected to the same management procedures. The data emphasize the variable nature of the ovulatory response within and between the three treatments studied and the high percentage of ewes (32%) which fail to respond (ovulation rates ≤ 4 corpora lutea). Indeed, 193 of 281 (68.7%) ewes treated had ovulation rates ≤ 10 (Figure 1.1).

At present, the adoption of one method of superovulation *versus* another is often based on cost. PMSG is relatively inexpensive and, unlike the pituitary gonadotrophins, can be administered as a single injection. One of the inherent problems of using this treatment has been the high number of unovulated or cystic follicles. The use of exogenous gonadotrophin releasing hormone (GnRH), to induce an earlier than normal preovulatory LH surge, is effective at increasing the proportion of developing follicles that ovulate (Jabbour and Evans, 1991). GnRH is normally administered at oestrus or just before the time of the endogenous preovulatory LH surge. Repeated laparoscopic observations (6 hour intervals) have shown that GnRH administration improved the synchrony of ovulation by reducing the period during which superovulation begins and the interval from the first to last ovulation (sheep: Walker *et al.*, 1986; Walker *et al.*, 1987; and goats: Cameron *et al.*, 1988). Its use should also remove some of the guess work currently associated with the timing of insemination relative to ovulation.

For the future, the use of homologous gonadotrophin preparations perhaps offers the best hope for a significant advance in superovulation techniques. For example, in sheep the promotion of follicle growth by

Table 1.1 Distribution of ovulation rate following various superovulatory hormone regimes.

Superovulation Treatment	No of ewes treated	No. of ewes with ovulation rates		Mean ovulation rate ± sem		Source
		4 or <	5 or >	all ewes	ewes with 5 or > CL	
hMG	23	3	20	11.6 ± 1.31	13.1 ± 1.21	Robinson *et al.*, unpublished
hMG	10	5	5	7.7 ± 2.30	13.2 ± 2.94	Scudamore *et al.* (1991b)
PMSG + GnRH	73	24	49	8.4 ± 0.75	11.0 ± 0.90	Robinson *et al.*, unpublished
PMSG + GnRH	48	26	22	5.2 ± 0.64	7.8 ± 1.17	Scudamore *et al.* (1991b)
PMSG + GnRH	38	11	27	8.9 ± 1.11	11.5 ± 1.27	Robinson *et al.* (1989)
porcine FSH	43	4	39	14.2 ± 1.24	15.3 ± 1.24	Robinson *et al.* (1989)
porcine FSH	46	17	29	7.6 ± 0.82	10.5 ± 0.95	Scudamore *et al.* (1991b)

Jacqueline M. Wallace

Figure 1.1 Distribution of ovulation rate in a known population of ewes expressed as a percentage of ewes treated with superovulatory hormones.

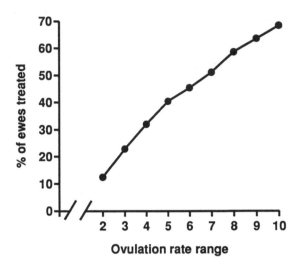

Figure 1.2 The predicted relationship between the ovulation rate of the donor and the number of times she would require to be flushed (----) to yield up to 10 progeny (————) unit time using surgical (–•–) versus laparoscopic (–○–) embryo recovery procedures.

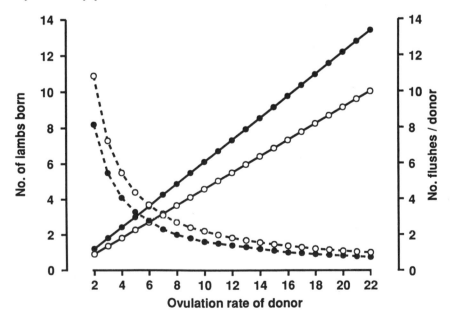

infusion of native FSH produces a controlled increase in ovulation rate (Baird *et al.*, 1984). More recently, ovine FSH was administered in 8 equal doses at 12 hourly intervals, starting 48 h before progestagen withdrawal, to Pedigree Suffolk ewes on two occasions corresponding to early and mid-breeding season. This resulted in mean ovulation rates of 13.1 and 11.3 respectively, with over 69.2% of ewes having in excess of 10 ovulations (Dingwall *et al.*, 1992).

Embryo recovery

The majority of embryo recoveries in MOET schemes are carried out at mid-ventral laparotomy under general anaesthesia. The method used generally depends on the stage of the cycle at which embryos are being recovered. Embryos are collected from the oviduct and tip of the uterine horn, on days 3-4 of the cycle, by retrograde flushing of ovum culture media from the uterine lumen out through the oviduct via a glass cannula. In contrast, embryos are harvested from day 5 of the cycle onwards by flushing the uterine horn from tip to base, the base of the horn being occluded at the external bifurcation by the use of a Foley catheter. Both methods result in acceptable recovery rates (>80%). However, uterine rather than oviduct flushing generally causes fewer post-operative adhesions because less handling of the reproductive tract is required.

The limitations of current superovulation treatments require that females on genetic improvement programmes are flushed repeatedly. The formation of adhesions following surgical embryo recoveries severely limits the number of times an animal can be flushed and often compromises its ability to breed naturally. It was for this reason that non-surgical laparoscopic procedures to repeatedly recover embryos were developed at the Rowett Research Institute (McKelvey *et al.*, 1986). The procedure is essentially equivalent to a surgical uterine flush except that all manipulations of the reproductive tract are carried out with the aid of grasping forceps and fine trochars (5 mm diameter) are used to introduce the flushing catheter and media system into the abdominal cavity. This technique enabled ova to be recovered from superovulated ewes on three occasions over a 2-month period (35%, 76% and 66% ovum recovery on the first, second and third occasions respectively). The lower than normal recovery rates observed using the laparoscopic technique are more than compensated for by its repeatability due to the absence of post-operative adhesions.

Figure 1.2 summarizes the predicted relationship between the ovulation rate of the donor and the number of flushes required to produce up to 10 progeny per unit time using surgical versus laparoscopic recovery procedures. The model assumes a recovery rate of 80 and 60% for the surgical and laparoscopic techniques respectively, a standard fertilization rate of 95% and embryo survival following transfer of 80%. Based on the distribution of ovulation rates presented in Table 1.1 and Figure 1.1, 69.8% of animals with ovulation rates ≤ 10 would in theory need to be flushed from 1.6 to 8.2 times surgically and 2.2 to 10.9 times by laparoscope.

The only drawback of the full laparoscopic technique is the high degree of manual dexterity required. This has led to the development of laparoscope aided techniques. In sheep, the embryo collection catheter is placed in the lumen at the base of the uterine horn by laparoscopy but a small incision is made in the body wall to allow the operator to exteriorize the tip of the uterine horn using forceps. This facilitates easier introduction of the culture media into the uterine lumen, immediately anterior to the utero-tubal junction. This procedure carried out three times at monthly intervals resulted in ovum recovery rates of 63, 71 and 85% for successive flushes, with fewer adhesions to the ovary and fimbria than recorded with full surgical techniques (Tervit, 1989). Indeed, this New Zealand-based research team have recently reported that ewes on an intensive MOET programme have been successfully flushed up to six times per year using the technique (Tervit *et al.*, 1991). Conversely in both sheep and goats, laparoscopy has been used to aid the introduction of culture media into the uterine horn while the embryo collection catheter has been introduced into the uterus via the cervix (Mylne *et al.*, 1992; McKelvey, 1992).

Methods of non-surgical embryo collection via the cervix have been documented in sheep and goats. Using procedures similar to those described for transcervical insemination, Coonrod *et al.* (1986) reported that a catheter system inserted through the retracted cervix allowed embryos to be collected in 11 of 26 ewes studied. The method was not suitable for all ewes, probably because of the tortuous nature of the cervical canal. In contrast, in a large scale commercial trial involving over 300 Angora and Kashmir goats, the cervix was traversed using the same technique in over 90% of attempts and yielded 9.5 ova per donor (Bessoudo *et al.*, 1988). Recently, successful cervical passing was reported in 10 out of 10 ewes whose cervix's had been ripened with a combined prostaglandin E_2 and oestradiol therapy administered 24 h prior to embryo collection. This resulted in an ovum recovery rate of 65% from a mixture of maiden and adult ewes (Barry *et al.*, 1990). Furthermore, the resulting embryos were not influenced by the hormonal regime in that 100% conception rates were recorded following transfer of 8 embryos. These non-invasive techniques offer considerable advantages but as yet no information is available about their repeatability in either species.

Embryo transfer

The evolution of embryo transfer procedures closely follows that of recovery techniques. Currently the majority of commercial MOET companies transfer embryos surgically into the reproductive tract exteriorized at mid-ventral laparotomy. The site of transfer depends on embryo stage. Embryos collected on or before day 4 are transferred to the oviduct in a minimum volume of culture media using a fine glass pipette inserted into the ampulla at the utero-tubal junction. Uterine transfer (day 5 onwards) is achieved by passing a pipette into the uterine lumen approximately 2 cm from the utero-tubal junction and discharging it in the direction of the external bifurcation. When recipient ewes have only one functional corpus luteum, the embryo is transferred to the ipsilateral uterine horn and until recently when two embryos were transferred they were inserted singly into both horns. However, it has been established that survival rates are not influenced by the transfer of two embryos to the ipsilateral versus both uterine horns (Torres and Sevellec, 1987). This is not surprising in that in hemi-ovariectomized multiple ovulating ewes intrauterine migration of embryos to maximize embryo spacing is observed in 100% of ewes and is complete by day 15 of the oestrous cycle (Nephew *et al.*, 1989).

Successful embryo transfer by laparoscopy has been reported by several groups (Mutiga and Baker, 1984; McKelvey and Robinson, 1984; Walker *et al.*, 1985). The simplest procedure involves depositing the embryos via a long finely drawn glass pipette into the uterine lumen while holding the tip of the uterine horn with grasping forceps. When two embryos were transferred to each of 21 ewes using this technique 16 established pregnancies and produced 24 lambs (McKelvey and Robinson, 1984). Laparoscope-aided transfers involve using endoscopy to locate the appropriate uterine horn and exteriorize it via a small incision in the body wall as described for embryo recovery. As the commercial value of recipient ewes is not limiting this technique is likely to become increasingly popular (Tervit, 1989).

Cervical embryo transfer has been reported in goats (Agrawal and Bhattacharyya, 1982) but has the disadvantage that the function or indeed presence of the corpus luteum cannot be confirmed by visual appraisal.

Factors Influencing the Success of MOET

Breeding history

One of the most frustrating components of MOET is the failure of some breeders to provide the commercial company with an adequate breeding history of the donor animal. Several problem areas can easily be diagnosed and avoided. For example, some animals consistently exhibit short oestrous cycles. These are due to premature luteal regression which often occurs in spite of a normal superovulatory response. While the cause of this premature luteal regression can rarely be immediately diagnosed, the practitioner can avoid the problem by recovering the embryos at an earlier stage of development from the oviduct or by administering exogenous progesterone to supplement the ewe's endogenous progesterone signal and recover the embryos at a later stage from the uterus. Conversely, within most flocks, a number of animals exhibit persistent luteal function (Yenikoye *et al.*, 1982; Zarco *et al.*, 1984). This occurs because of a defect in the pulsatile release of prostaglandin $F_{2\alpha}$ which normally signals luteolysis at the end of the cycle and is detected by a failure to return to oestrus at the expected time. Unfortunately, these corpora lutea can persist for over 100 days and the animal is often misdiagnosed as being pregnant or acyclic. Once the occurrence of these corpora lutea has been confirmed by laparoscopic examination of the ovaries or measurement of peripheral progesterone concentrations, luteolysis can be induced by exogenous administration of prostaglandin $F_{2\alpha}$. Other causes of infertility in donor animals can be diagnosed by laparoscopy and these include the occurrence of cystic ovaries, retained placenta, hyperplasia of the oviduct(s) and endometriosis. These conditions cannot all be immediately treated but breeders can avoid wasting money on expensive superovulation hormone therapies.

Synchronous transfer

Close synchrony between the onset of oestrus in donor and recipient animals is probably the most important factor influencing the successful outcome of embryo transfer. Estimates of the degree of asynchrony tolerated by the embryo vary with effects being demonstrated with 8 (McKelvey *et al.*, 1985b), 12 (Moore and Shelton, 1964) and 48 h (Rowson and Moore, 1966). However, to demonstrate an effect of 8 h asynchrony requires oestrus onset to be checked at 4 h intervals. It is unlikely that this would be adopted in commercial practice where 12 hourly checks are more feasible. When embryos recovered on day 4 and 9 of the cycle were transferred to day 7 and 6 recipients respectively (72 h asynchrony), they failed to survive even when a synchronous embryo was developing within the same uterus (Wilmut and Sales, 1981). Similarly, the transfer of two day 4, 6 or 8 embryos into a day 6 uterus showed that advanced embryos were better able to survive and became larger fetuses than retarded embryos (Wilmut *et al.*, 1988). This suggests that if asynchrony is unavoidable it is preferable to select a recipient who showed oestrous behaviour after the donor.

Embryo quality

The decision whether or not to transfer an embryo of inferior quality or retarded development depends on the objectives of transfer. In commercial programmes, the objective is usually to obtain as many progeny as possible from a particular donor. Thus, provided that recipient animals are not a limited resource, embryos at a retarded stage of development would be transferred in the hope that a proportion would "catch up" and survive in the recipient uterus. It is more economically important however to minimize the incidence of retarded development. Recent research suggests that embryo quality is linked to progesterone priming at the time of super-ovulatory hormone administration. Ewes were primed and synchronized to ovulate with a intravaginal pessary containing 30 or 40 mg of synthetic progestagen (fluorogestone acetate). Superovulation was achieved using PMSG and GnRH and ewes were inseminated at 48 h after pessary withdrawal. Ovulation rate was not influenced by the level of progestagen priming. Embryo development assessed at day 3 was similar for both levels of priming with 96% of embryos having ≥ 16 cells. However, at day 6 only, 44% of embryos recovered from the 30 mg progestagen group were at the expected stage of development (morulae and blastocysts) compared with 88% of embryos recovered from the 40 mg group. Moreover, the proportion of embryos surviving following transfer to synchronous recipients on days 3 and 6 of the cycle was reduced by 13 and 19% respectively for embryos arising from donors primed with 30 versus 40 mg progestagen (Scudamore *et al.*, 1992). Similarly, both ovulation rate and the number of embryos of transferrable quality (assessed on day 6) increased when ewes were primed for 12 days with two consecutive versus a single controlled internal drug release device (CIDR, containing 9% progesterone) (Thompson *et al.*, 1990). These studies both suggest that the level of progesterone/progestagen priming prior to superovulation may be influencing follicle growth and development and/or the oviduct-uterine environment by mechanisms as yet unknown.

Embryo survival

The many factors shown to influence embryo survival in mated ewes (Wallace and Ashworth, 1990) undoubtedly have similar effects in donor and recipient animals involved in MOET programmes. Of these, age, nutrition, stress, season and disease are perhaps most likely to influence the outcome of transfer programmes but no documented data exist on the influence of any of these factors on embryo survival from superovulated animals of valuable genotypes. This is ironic in that embryo transfer technology has long been used to study embryo survival *per se*.

For example, two independent embryo transfer studies have demonstrated that ewe lambs are not suitable as embryo donors. Survival of ewe lamb embryos following transfer into ewe lamb (McMillan and McDonald, 1985) or adult ewe (Quirke and Hanrahan, 1977) uterine environments was 25 and 33% respectively. Ewe lambs are however suitable as recipients in that 52 and 73% of adult embryos survived when transferred to the ewe lamb and adult uterus respectively.

Similarly, embryo transfer studies have been used to separate the effects of nutritional status pre and post mating on embryo growth and survival (McKelvey *et al.*, 1988). In the first study, embryos were recovered from donor ewes in thin (condition score < 1½), medium (2.5 - 3), or fat (>4) body condition and transferred to recipients in medium condition. Embryos from thin donors were of inferior quality to those recovered from medium or fat donors but their survival following transfer was not influenced by donor body condition. However, fetuses derived from donors in medium condition were heavier at day 60 of pregnancy emphasizing that the nutritional status of the donor can influence early conceptus growth. Secondly, embryos from donors in moderate body condition were transferred to recipient ewes in thin, medium or fat condition. Following transfer, recipients were fed either 0.5 or 1.5 x maintenance rations. Pregnancy and embryo survival rates were independent of the body condition of recipients at transfer but dependent on post-transfer nutrition being lower in high-plane ewes (pregnancy and embryo survival rates of 87 and 75% versus 97 and 87% in high and low plane ewes respectively).

In these and other studies, high plane feeding post-mating has been associated with a reduction in peripheral plasma progesterone concentrations. Indeed when progesterone levels were restored post-mating using exogenous progesterone, pregnancy rates returned to normal levels (Parr *et al.*, 1987). Careful attention to both donor and recipient nutrition is therefore required to maximize the yield of good quality embryos and their survival following transfer.

A seasonal trend in embryo survival has been observed within a single breeding season in prolific Damline ewes (Ashworth *et al.*, 1989). Ovulation rate and embryo survival were monitored in each ewe on 2 or 3 occasions during a single breeding season. This revealed that embryo loss was highest at the end of the breeding season in March than after mating in September, November or January but not all variation in embryo loss could be explained by seasonal changes in ovulation rate. Conversely, a seasonal difference in ovulation and embryo recovery rate but not embryo survival following transfer has been documented for ewes superovulated in the breeding versus non-breeding seasons (Torres *et al.*, 1987). However, two different batches of recipient ewes were utilized in the latter study. The use of repeatable laparoscopic embryo recovery procedures together with *in vitro* culture to assess embryo viability/development should facilitate multiple within animal comparisons of ovulatory response, embryo recovery and predicted survival throughout the seasonal cycle.

The advances in procedures for artificial insemination and embryo transfer developed during the 1980s offer a bright future for real genetic improvement within the sheep and goat sectors during the next decade. The challenge to breeders, geneticists and veterinarians alike is to develop co-operative improvement schemes to reap the benefits from these new techniques. The laparoscopic procedures detailed in this chapter, as applied to livestock production, are in their infancy and their full potential has not yet been realized. In ensuring that they achieve this potential, we should perhaps remind ourselves of how the technique of laparoscopy has transformed and simplified the whole field of human medicine. Raoul Palmer, a gynaecologist, developed the laparoscope from the principle of culdoscopy in the late 1950s. Initially used as a diagnostic tool in gynaecology and urology, it was later routinely used to carry out operative procedures such as sterilization. The pioneering work of Patrick Steptoe allowed the aspiration of oocytes from ovarian follicles by laparoscopy for the first time. This in turn allowed *in vitro* fertilization techniques to be developed and resulted ultimately in the birth of Louise Brown, the first "test-tube baby", in 1978.

References

Agrawal, K.P. and Bhattacharyya, N.K. (1982) Nonsurgical transplantation of embryos in goats. *Proceedings 3rd International Congress on Goat Production and Disease.* Dairy Goat Journal Publishing Company p. 340.

Armstrong, D.T. and Evans, G. (1983) Factors influencing success of embryo transfer in sheep and goats. *Theriogenology* 19, 31-42.

Ashworth, C.J., Sales, D.I. and Wilmut, I. (1989) Evidence of an association between the survival of embryos and the periovulatory plasma progesterone concentration in the ewe. *Journal of Reproduction and Fertility* 87, 23-32.

Azzarini, M. and Valledor, F. (1988) Intra-uterine or intra-cervical insemination of ewes in natural oestrus with frozen or fresh semen. *Produccío Ovina* 1, 1-8.

Baird, D.T., McNeilly, A.S., Wallace, J.M. and Webb, R. (1984) Infusion of FSH increases ovulation rate in Welsh Mountain ewes. *Proceedings Vth Reinier de Graaf Symposium.* Abstract 13.

Barry, D.M., Niekek, van, C.H., Rust, J. and Walt, ven der T. (1990) Cervical embryo collection in sheep after ripening of the cervix with prostaglandin E_2 and estradial. *Theriogenology* 33, 190.

Bessoudo, E., Davies, L., Coonrod, S., Gamex, J. and Kraemer, D.C. (1988) Non-surgical collection of caprine embryos under commercial quarantine conditions. *Theriogenology* 29, 221.

Boland, M.P. and Gordon, I. (1978) Recovery and fertilization of eggs following natural service and uterine insemination in the Galway ewe. *Irish Veterinary Journal* 32, 123-125.

Cameron, A.W.N., Battye, K.M. and Trounson, A.O. (1988) Time of ovulation in goats (*Capra hircus*) induced to superovulate with PMSG. *Journal of Reproduction and Fertility* 83, 747-752.

Colas, G. (1983) Factors affecting the quality of ram semen. In: Haresign, W. (ed.) *Sheep Production.* Butterworths, London, pp. 453-465.

Coonrod, S.A., Coren, B.R., McBride, B.L., Bowen, M.J. and Kraemer, D.C. (1986) Successful non-surgical collection of ovine embryos. *Theriogenology* 25, 149.

Davis, I.F., Kerton, D.J., McPhee, S.R., White, M.B., Banfield, J.C. and Cahill, L.P. (1984) Uterine artificial insemination in ewes. In: Lindsay, D.R. and Pearce, D.T. (eds.) *Reproduction in Sheep.* Cambridge University Press, pp. 304-305.

Dingwall, W.S., Fernie, K., FitzSimons, J. and McKelvey, W.A.C. (1992) MOET in Suffolks: Increasing the superovulation rate. *Animal Production* (In Press).

Dzuik, P.J., Lewis, J.M., Graham, E.F. and Mayer, R.H. (1972) Comparison between natural service and artificial insemination with fresh or frozen semen at an appointed time in the ewe. *Journal of Animal Science* 35, 572-575.

Eppleston, J., Maxwell, W.M.C., Battye, K.M. and Roberts, E.M. (1986) Effect of thawed motility and intra-uterine dose of motile sperm on fertility in ewes. *Proceedings 18th Annual Conference of the Australian Society of Reproductive Biology.* p. 19.

Evans, G. (1988) Current topics in artificial insemination of sheep. *Australian Journal of Biological Science* 41, 103-116.

Evans, G. and Armstrong, D.T. (1984) Reduction of sperm transport in ewes by superovulation treatments. *Journal of Reproduction and Fertility* 70, 47-53.

Evans, G. and Maxwell, W.M.C. (eds.) (1987) *Salamon's artificial insemination of sheep and goats.* Butterworths, Sydney.

Evans, G., Holland, M.K., Nottle, H.B., Sharpe, P.H. and Armstrong, D.T. (1984) Production of embryos in sheep using FSH preparations and laparoscopic intra-uterine insemination. In: Lindsay, D.R. and Pearce, D.T. (eds.) *Reproduction in Sheep.* Cambridge University Press, pp. 313-315.

Foote, R.H. (1974) Artificial insemination. In: Hafez, E.E.S. (ed.) *Reproduction in Farm Animals.* Lea and Febiger, Philadelphia, pp. 212-246.

Gordon, I. (1969) Factors affecting response of anoestrous sheep to progestagen treatment. *Journal of the Department of Agriculture and Fisheries, Republic of Ireland* 66, 232-267.

Gordon, I., Caffrey, W. and Morrin, P. (1969) Induction of early breeding in sheep following treatment with progestagen-impregnated pessaries and PMSG. *Journal of the Department of Agriculture and Fisheries, Republic of Ireland* 66, 232-267.

Grigoryan, S. and Nazaryan, V. (1981) Effect of antioxidants during deep freezing of ram spermatozoa. *Zhivotnovodstvo* 9, 49-50.

Halbert, G.W., Dobson, H., Walton, J.S. and Buckrell, B.C. (1990a) A technique for transcervical intra-uterine insemination of ewes. *Theriogenology* 33, 993-1010.

Halbert, G.W., Dobson, H., Walton, J.S., Sharpe, P. and Buckrell, B.C. (1990b) Field evaluation of a technique for transcervical intra-uterine insemination of ewes. *Theriogenology* 33, 1231-1243.

Haresign, W. (1990) Controlling reproduction in sheep. In: Slade, C.F.R. and Lawrence, T.L.J. (eds.) *New Development in Sheep Production.* British Society of Animal Production, Occasional Publication No. 14, pp. 23-37.

Hawk, H.W., Cooper, B.S. and Conley, H.H. (1987) Inhibition of sperm transport and fertilization in superovulating ewes. *Theriogenology* 28, 139-153.

Hernandez, J.J.P., Rodriquez, R.O. and Gonzalez, P.E. (1976) Evaluation of four methods of collecting semen from Tobasco and Pelibuey rams. *Tecnica Pecuaria en Mexico* 30, 45-51.

Jabbour, H.N. and Evans, G. (1991) Ovarian and endocrine responses of merino ewes following treatment with PMSG and GnRH or PMSG antiserum. *Animal Reproduction Science* 24, 259-270.

Kerton, D.J., McPhee, S.R., Davis, I.F., White, M.B., Banfield, J.C. and Cahill, L.P. (1984) A comparison of insemination techniques in corriedale ewes. *Proceedings of the Australian Society of Animal Production* 15, 701.

Killeen, I.D. and Caffrey, G.J. (1982) Uterine insemination of ewes with the aid of a laparoscope. *Australian Veterinary Journal* 59, 95.

Killeen, I.D. and Moore, N.W. (1970) Fertilization and survival of fertilized eggs in the ewe following surgical insemination at various times after the

onset of oestrus. *Australian Journal of Biological Science* 23, 1279-1287.
Lightfoot, R.J. and Salamon, S. (1970) Fertility of ram spermatozoa frozen by the pellet method. 1. Transport and viability of spermatozoa within the genital tract of the ewe. *Journal of Reproduction and Fertility* 22, 385-398.
McKelvey, W.A.C. (1992) Reproductive techniques in goats. *The Goat Veterinary Journal* (In Press).
McKelvey, W.A.C. and Robinson, J.J. (1984) Normal lambs born following transfer of embryos by laparoscopy. *Veterinary Record* 115, 230.
McKelvey, W.A.C. and Robinson, J.J. (1986) Embryo survival and growth in the ewe - recent studies on the effects of nutrition and on novel techniques for the recovery and transfer of embryos. *The Rowett Research Institute, Annual Report* 1986, pp. 9-25.
McKelvey, W.A.C., Robinson, J.J. and Aitken, R.P. (1985a) A simplified technique for the transfer of ovine embryos by laparoscopy. *Veterinary Record* 117, 492-494.
McKelvey, W.A.C., Robinson, J.J. and Aitken, R.P. (1988) The use of reciprocal embryo transfer to separate the effects of pre- and post-mating nutrition on embryo survival and growth of the ovine conceptus. *11th International Congress on Animal Reproduction and Artificial Insemination*, Dublin, Abstract 176.
McKelvey, W.A.C., Robinson, J.J., Aitken, R.P. and Henderson, G. (1985b) The evaluation of a laparoscopic insemination technique in ewes. *Theriogenology* 24, 519-535.
McKelvey, W.A.C., Robinson, J.J., Aitken, R.P. and Robertson, I.S. (1986) Repeated recoveries of embryos from ewes by laparoscopy. *Theriogenology* 25, 855-865.
McMillan, W.H. and McDonald, M.F. (1985) Survival of fertilized ova from ewe lambs and adult ewes in the uteri of ewe lambs. *Animal Reproduction Science* 8, 235-240.
Mattner, P.E., Entwistle, K.W. and Martin, I.C.A. (1969) Passage, survival and fertility of deep-frozen ram semen in the genital tract of the ewe. *Australian Journal of Biological Science* 22, 181-187.
Maxwell, W.M.C. (1986a) Artificial insemination of ewes with frozen-thawed semen at a synchronised oestrus 1. Effect of time of onset of oestrus, ovulation and insemination on fertility. *Animal Reproduction Science* 10, 301-308.
Maxwell, W.M.C. (1986b) Artificial insemination of ewes with frozen-thawed semen at a synchronised oestrus 2. Effects of dose of spermatozoa and site of intra-uterine insemination on fertility. *Animal Reproduction Science* 10, 309-316.
Maxwell, W.M.C. and Hewitt, L.J. (1986) A comparison of vaginal, cervical and intra- uterine insemination of sheep. *Journal of agricultural Science*, Cambridge 106, 191-193.
Maxwell, W.M.C., Curnock, R.M., Logue, D.N. and Reed, H.C.B. (1980) Fertility of ewes following artificial insemination with frozen semen in pellets or straws: a preliminary report. *Theriogenology* 14, 83-89.
Memon, M.A. and Ott, R.S. (1981) Methods of semen preservation and artificial insemination in sheep and goats. *World Review of Animal Production* 107, 19-25.
Milovanov, V.K. and Sokolovskaya, I.I. (1980) Long-term storage of ram semen and new possibilities of large scale selection in sheep breeding. *Vestnik Selskok. Nauki* 12, 122-132.
Milovanov, V.K., Kaljczova, E., Shajdullin, I. and Varnavskaya, A. (1981) Chemical characteristics of antioxidants and their effect in freezing ram semen. *Zhivotnovodstvo* 9, 45-46.
Moore, N.W. (1982) Egg transfer in the sheep and goat. In: Adams, C.E. (ed.)

Mammalian Egg Transfer. C.R.C. Press Inc., Florida.

Moore, N.W. and Shelton, J.N. (1964) Effect of degree of synchronisation between donor and recipient, age of egg and site of transfer on the survival of transferred eggs. *Journal of Reproduction and Fertility* 7, 145-152.

Moore, R.W., Dow, B.W. and Staples, L.D. (1989) Artificial insemination of farmed feral goats with frozen-thawed semen. *Proceedings of the New Zealand Society of Animal Production* 49, 171-173.

Moore, R.W., Miller, C.M., Hall, D.R.H. and Dow, B.W. (1988) Cervical versus laparoscopic AI of goats after PMSG injection at or 48 hours before CIDR removal. *Proceedings of the New Zealand Society of Animal Production* 48, 69-70.

Mutiga, E.R. and Baker, A.A. (1984) Transfer of sheep embryos through a laparoscope. *Veterinary Record* 114, 401-402.

Mylne, M.J.A., McKelvey, W.A.C., Fernie, K. and Matthews, K. (1992) A report on the use of a transcervical technique for embryo recovery in sheep. *Veterinary Record* 130, 59.

Nephew, K.P., McLure, K.E. and Pope, W.F. (1989) Embryonic migration relative to maternal recognitiion of pregnancy in sheep. *Journal of Animal Science* 67, 999-1005.

O'Shea, T. and Wales, R.G. (1966) Effect of casein, lecithin, glycerol, and storage at 5°C on diluted ram and bull semen. *Australian Journal of Biological Sciences* 19, 871-882.

Parr, R.A., Davis, I.F., Fairclough, R.J. and Milers, M.A. (1987) Overfeeding during early pregnancy reduced peripheral progesterone concentration and pregnancy rate in sheep. *Journal of Reproduction and Fertility* 80, 317-320.

Pavlok, A. and Fléchon, J.E. (1985) Some factors influencing the interaction of ram spermatozoa with zone-free hamster eggs. *Journal of Reproduction and Fertility* 74, 597-604.

Pavlok, A., Peteliková, J. and Fléchon, J.E. (1983) Interaction of zone-free hamster eggs with fresh and frozen ram spermatozoa in vitro. In: André, J. (ed.) *The Sperm Cell*. Martinus Nijhoff, The Hague, pp. 51-54.

Quinn, P.J., Salamon, S. and White, I.G. (1968) The effect of cold shock and deep-freezing on ram spermatozoa collected by electrical ejaculation and by an artificial vagina. *Australian Journal of Agricultural Science* 19, 119.

Quirke, J.F. and Hanrahan, J.P. (1977) Comparison of the survival in the uteri of adult ewes of cleared ova from adult ewes and ewe lambs. *Journal of Reproduction and Fertility* 51, 487-489,

Ritar, A.J. and Salamon, S. (1982) Effects of seminal plasma and of its removal and of egg yolk in the diluent on the survival of fresh and frozen-thawed spermatozoa of the Angora goat. *Australian Journal of Biological Sciences* 35, 305-312.

Ritar, A.J. and Salamon, S. (1983) Fertility of fresh and frozen-thawed semen of the Angora goat. *Australian Journal of Biological Science* 36, 49-59.

Ritar, A.J., Ball, P.D. and O'May, P.J. (1990a) Artificial insemination of cashmere goats: Effects on fertility and fecundity of intravaginal treatment, method and time of insemination, semen freezing process, number of motile spermotozoa and age of females. *Reproduction, Fertility and Development* 2, 377-384.

Ritar, A.J., Ball, P.D. and O'May, P.J. (1990b) Examination of methods for the deep freezing of goat semen. *Reproduction, Fertility and Development* 2, 27-34.

Rival, M.D., Chenoweth, P.J. and McMicking, L.I. (1984) Semen deposition and fertility in ovine artificial breeding programmes. In: Lindsay, D.R. and Pearce, D.T. (eds.) *Reproduction in Sheep*. Cambridge University Press, pp. 301-303.

Robinson, J.J., Wallace, J.M. and Aitken, R.P. (1989) Fertilization and ovum recovery rates in superovulated ewes following cervical insemination or laparoscopic intra-uterine insemination at different times after progestagen withdrawal and in one or both uterine horns. *Journal of Reproduction and Fertility* 87, 771-782.

Rowson, L.E.A. and Moore, R.M. (1966) Embryo transfer in the sheep: the significance of synchronising oestrus in the donor and recipient animal. *Journal of Reproduction and Fertility* 11, 207-212.

Roy, A. (1957) Egg yolk coagulating enzyme in the semen and cowper's gland of the goat. *Nature* 179, 318.

Scudamore', C.L., Robinson, J.J. and Aitken, R.P. (1991a) The effect of timing of laparoscopic insemination in superovulated ewes, with or without sedation, on the recovery of embryos, their stage of development and subsequent viability. *Theriogenology* 35, 907-914.

Scudamore, C.L., Robinson, J.J., Aitken, R.P., Kennedy, D.J., Ireland, S. and Robertson, I.S. (1991b) Laparoscopy for intrauterine insemination and embryo recovery in superovulated ewes at a commercial embryo transfer unit. *Theriogenology* 35, 330-337.

Scudamore, C.L., Robinson, J.J., Aitken, R.P. and Robertson, I.S. (1992) The effect of different levels of progestagen priming on the quality of embryos recovered from superovulated ewes. *Theriogenology* (submitted).

Smith, C. (1986) Use of embryo transfer in genetic improvement of sheep. *Animal Production* 42, 81-87.

Tervit, H.R. (1989) Embryo transfer and sperm sexing. *2nd International Congress for Sheep Veterinarians*. Proceedings of the Society's 19th Seminar, pp. 144-157.

Tervit, H.R., Godd, P.G., James, R.W. and Frazer, M.D. (1984) The insemination of sheep with fresh or frozen semen. *Proceedings of the New Zealand Society of Animal Production* 44, 11-13.

Tervit, H.R., Thompson, J.G., McMillan, W.H. and Amyes, N.C. (1991) Repeated surgical embryo recovery from Texel donor ewes. *Theriogenology* 35, 282.

Thompson, J.G.E., Simpson, A.C., James, R.W. and Tervit, H.R. (1990) The application of progesterone-containing CiDR™ devices to superovulated ewes. *Theriogenology* 33, 1297-1304.

Torres, S. and Sevellec, C. (1987) Repeated superovulation and surgical recovery of embryos in the ewe. *Reproduction Nutrition and Development* 27, 941-944.

Torres, S., Cognie, Y. and Colas, G. (1987) Transfer of superovulated sheep embryos. *Theriogenology* 22, 407-419.

Trounson, A.O. and Moore, N.W. (1974) Fertilization in the ewe following multiple ovulation and uterine insemination. *Australian Journal of Biological Science* 27, 301-304.

Walker, S.K., Smith, D.H. and Seamark, R.F. (1986) Timing of multiple ovulations in the ewe after treatment with FSH or PMSG with and without GnRH. *Journal of Reproduction and Fertility* 77, 135-142.

Walker, S.K., Smith, D.H., Seamark, R.F. and Godfrey, B. (1987) Variation in the timing of multiple ovulations following gonadotrophin releasing hormone treatment and its relevance to collecting pronuclear embryos of sheep. *Theriogenology* 28, 129-137.

Walker, S.K., Warnes, G.M., Quinn, P., Seamark, R.F. and Smith, D.H. (1985) Laparoscopic technique for the transfer of embryos in sheep. *Australian Veterinary Journal* 62, 105-106.

Wallace, J.M. and Ashworth, C.J. (1990) Embryo loss. *Proceedings of the Sheep Veterinary Society* 14, 134-136.

Wallace, J.M., Robinson, J.J. and Aitken, R.P. (1989) Successful pregnancies after transfer of embryos recovered from ewes induced to ovulate 24-29 days *post partum. Journal of Reproduction and Fertility* 86, 627-635.

Wilmut, I. and Sales, D.I. (1981) Effect of asynchronous environment on embryonic development in sheep. *Journal of Reproduction and Fertility* 61, 179-184.

Wilmut, I., Ashworth, C.J., Springbett, A.J. and Sales, D.I. (1988) Effect of variation in embryo stage on the establishment of pregnancy, and embryo survival and growth in ewes with two embryos. *Journal of Reproduction and Fertility* 83, 233-237.

Yanagimachi, R. (1984) Zona-free hamster eggs: their use in assessing fertilizing capacity and examining chromosomes of human spermotozoa. *Gamete Research* 10, 187-232.

Yenikoye, A., Pelletier, J., Andre, D. and Mariana, J.C. (1982) Anomalies in ovarian function of Peulh ewes. *Theriogenology* 17, 355-364.

Zarco, L., Stabenfeldt, G.H., Kindahl, H., Quirke, J.F. and Granström, E. (1984) Persistance of luteal activity in the non-pregnant ewe. *Animal Reproduction Science* 7, 245-267.

Chapter 2

Nutrition: its Effects on Reproductive Performance and its Hormonal Control in Female Sheep and Goats

S.M. Rhind

The Macaulay Land Use Research Institute, Pentlandfield, Roslin, Midlothian, EH26 6RF, UK

Introduction

Sheep and goats have become adapted to exploit pastoral resources which vary widely according to the nature of the climate, terrain and vegetation. The availability and quality of vegetation in their habitats is often highly variable and, in many breeds, physiological mechanisms have evolved which ensure that the animal's nutrient requirements for maintenance, pregnancy and lactation are matched to variations in the available nutritional resources.

Effects of nutritional state on reproductive performance have undoubtedly been recognized by shepherds and goat-herds for centuries and management practices have evolved which take account of that knowledge. Since the late 19th Century, the arts of animal husbandry have become a science and, beginning with the work of Heape (1899) and Marshall (1904, 1908), the effects of nutritional inputs on the animal's physiology have been described and quantified in ever-increasing detail.

Although the nature of nutritional resources, and therefore the nature of the management problems, differs with climate, soil and vegetation, the physiological processes which govern the animal's reproductive performance are the same; the female must reach puberty, show behavioural oestrus and shed one or more ova which have then to be fertilized. Thereafter, the resultant embryos have to be maintained throughout a 5-month pregnancy. Nutritional inadequacies can potentially act on any of these stages of the reproductive process. Much of this review concerns the effects of nutrition on ovulation rate and embryo mortality; this reflects the fact that these components of the reproductive process are affected most and therefore offer the greatest scope for manipulation.

While there is a large literature on this subject with respect to the ewe, relatively little is known about the goat. Although, like sheep, goats have been economically important in many parts of the world for centuries, they are only now attracting significant scientific interest and research funding because of their potential value as producers of fine fibre such as cashmere and mohair. Consequently, for this species there is only a very small body of information concerning the effects of nutrition on reproductive physiology and all of the descriptions of endocrine mechanisms are based on work done only with sheep.

For the purposes of this review, effects of nutrition on each component of the reproductive process (oestrus, ovulation, etc.) are considered separately. However, it should be remembered that these components are interdependent. Thus, for example, the effects of nutritional state on embryo survival may depend on the ovulation rate.

Onset of Puberty

Prerequisites for the onset of puberty in the female sheep and goat include the achievement of a critical body size and, in the higher latitudes where breeding activity is seasonal, a short-day photoperiod. In both sheep (Foster *et al.*, 1985; Dyrmundsson, 1987) and goats (Shelton, 1961; Riera, 1982; McCall *et al.*, 1989; Wolde-Michael *et al.*, 1989) restricted nutrition during early life and failure to achieve the necessary live weight before the end of the first breeding season after birth results in a delay in the onset of puberty until at least the following breeding season when the animal has usually reached the critical live weight.

Hormonal control

Before puberty, secretion of luteinizing hormone (LH) is pulsatile but the interval between pulses is relatively long (2-3 hours) (Foster, 1988). The onset of puberty is dependent on the initiation of a relatively high frequency of LH pulses which in turn stimulate follicle development, a sustained increase in oestradiol concentrations, a preovulatory LH surge and ovulation; FSH secretion does not seem to be deficient (Foster *et al.*, 1985; Foster, 1988). In lambs which are below the threshold live weight, initiation of frequent LH pulse secretion is inhibited, irrespective of photoperiod.

Incidence of Oestrus

The incidence of oestrus during the year is dependent primarily on latitude and therefore photoperiodic stimulus, animal age, genotype and lactation (Hafez, 1952) but nutritional state can also affect it.

It has been shown that, under conditions of severe undernutrition, oestrus can be inhibited in the ewe (Hafez, 1952) or the breeding season can be prematurely terminated (McKenzie and Terrill, 1937; Knight *et al.*, 1983). There is also limited evidence that oestrous activity can be inhibited in ewes in excessive body condition (Rhind *et al.*, 1984). There appear to be no reports of such effects in goats but since they are frequently reared in areas where feed resources are limited, suppression of oestrus, as a result of low levels of nutrition, may be an important determinant of reproductive performance in this species.

The underlying endocrine mechanisms responsible for these effects have not been characterized. However, under most management conditions, during the breeding season, the nutritional state of the animal is unlikely to influence the incidence of behavioural oestrus and it is therefore unlikely to be a limiting factor on reproductive performance.

Ovulation Rate

Ovulation rate is one of the most important determinants of reproductive performance in female sheep and goats. In ewes, it depends on age (Knight *et al.*, 1975), genotype (Wheeler and Land, 1977) and stage of the breeding season (Hulet *et al.*, 1974; Davis *et al.*, 1976) as well as nutrition (Gunn, 1983). Less is known about goats but as in the sheep there is evidence of effects of age (Adu *et al.*, 1979), genotype (Sands and McDowell, 1978), and stage of the breeding season (Shelton, 1961) on reproductive performance.

Unlike oestrous activity, ovulation is never completely suppressed by undernutrition even when this is severe (Hafez, 1952; Allen and Lamming, 1961) but selective breeding, combined with the powerful pressures of natural selection, have resulted in a range of ewe genotypes showing a variety of patterns of ovulatory responses to nutrition. Some breeds, such as the Finnish Landrace and its crosses, have been selected for prolificacy and are generally reared under favourable conditions under which they are not subject to natural selection pressures; these ewes have a high lambing rate irrespective of nutritional state (Doney and Gunn, 1973). On the other hand, ewes of breeds such as the Scottish Blackface, which are often maintained under conditions of minimal management in a climatically harsh environment, are subject to intense selection pressures. Consequently, the ovulation and lambing rates of these ewes are generally lower and have become closely geared to nutritional resources including available pasture and ewes' body fat reserves, which affect the capacity of the ewe to sustain pregnancy and lactation independently of extraneous management factors including nutritional supplementation (Gunn, 1983).

A wide range of litter sizes, and by inference, ovulation rates, has also been reported in goats of different breeds in a wide range of environments

(Devendra and Burns, 1983). While a substantial component of this variation is attributable to effects of plane of nutrition/body condition on reproductive performance (Sachdeva et al., 1973), specific effects of nutrition on ovulation rate and embryo mortality have not been quantified.

Timescale of action

Nutritional effects on reproductive performance operate over periods of years, months and days before ovulation (Gunn, 1983). Virtually all of the detailed studies have been performed with sheep.

Long-term effects

Experiments designed to measure the effects of nutrition in early life on reproductive performance several years later are necessarily difficult and often costly to conduct. Nevertheless, there is experimental evidence that severe undernutrition in the early life of the ewe can reduce lambing rates in adult life (Gunn, 1977; Allden, 1979; Williams, 1984).

Long-term effects of undernutrition are not confined to very young animals. Fletcher (1974) reported a significant reduction in ovulation rate in ewes which had undergone a period of severe nutritional restriction 6 to 12 months earlier compared with ewes which had been adequately fed at that time. This difference in ovulation rate occurred despite the fact that the underfed ewes had returned to a liveweight similar to that of the adequately fed ewes by 1 month before the ovulation rate was measured.

Effects of body condition and level of feed intake

There is a very large literature on this topic with respect to the ewe, much of which appears at first to be confusing and contradictory (Coop, 1966; Killeen, 1967; Cockrem, 1979). In general, ovulation rate increases with increasing ewe liveweight and body condition (medium-term effect); it is also increased in response to an increased level of premating feed intake (flushing; short-term effect). However, the reported responses are not consistent across all experiments, probably because the effects of the two components, body condition and intake, are almost always confounded. Gunn (1983) put forward a unifying theory which may account for most of the anomalies. He suggested that ovulation rate is dependent on the short-term effect of flushing only in ewes which are within a certain intermediate range of body condition and that above and below the critical range, energy or feed intake has no effect. The critical range of body condition, like the effect of body condition on ovulation rate, is genotype dependent.

The ovulation rate and kidding rate of goats can be increased by improved pre-mating nutrition (Henniawati and Fletcher, 1986; Wentzel, 1987; Chaniago, 1988). There is also a positive effect of liveweight on reproductive performance which is particularly important in young does during their first breeding season (Wentzel, 1987).

Effect of specific nutrients

While energy is regarded by many as the most significant dietary component with respect to the determination of ovulation rate, there have been numerous reports of increased ovulation rates associated with feeding of protein-rich supplements, particularly lupin grain (Davis *et al.*, 1981; Fletcher, 1981; Kenney and Smith , 1985; Nottle *et al.*,1988; Smith, 1988). Like effects of increased energy intake, the effects of increasing protein levels in the diet have also been inconsistent (Smith, 1984, 1988) with levels of energy and protein in the ration, type of protein in the diet, ewe genotype, ewe liveweight and body condition and duration of the nutritional treatment all being implicated as possible reasons.

Like effects of body condition and level of feed intake, the effects of dietary protein and energy are difficult to separate and are confounded in many experiments. Nevertheless, as a result of a series of experiments specifically designed to separate the effects of the protein and energy components of the lupin supplement, one group of workers have concluded that the effect of lupin supplementation on ovulation rate can be explained primarily by the increased supply of energy-yielding substrates (Teleni *et al.*, 1984; Teleni and Rowe, 1986), i.e. they concluded that the effect of protein *per se* on ovulation rate, if any, was minimal.

More recently, Waghorn *et al.* (1990) reported evidence of a quite different mechanism. They showed that, in wethers which had a high protein diet, nitrogen retention was increased and plasma concentrations of some essential amino acids were increased. These observations were linked to results obtained from ewes in a separate trial designed to examine the ovulatory response to different protein and energy intakes. It was shown that ovulation rate was highly correlated with plasma concentrations of essential amino acids but in particular to concentrations of branched chain amino acids (BCAA). Furthermore, it was found that it was not absolute concentration of BCAA which was related to ovulation rate but change in concentration; following an increase in protein intake, ewes that had an increase in ovulation rate from one to two had increases in plasma BCAA which were twice as large as those in ewes which showed no change in ovulation rate.

Results of many earlier experiments concerning the effects of dietary protein on ovulation rate cannot readily be explained in terms of the recent hypotheses because the experiments were not designed to answer the questions which are now raised. There is still a need to conduct experiments in which plasma or serum concentrations of specific nutrients, such as BCAA, and other energy-providing substrates are monitored in animals fed a range of dietary energy and protein levels.

Ovarian follicle populations

The ovulation rate of ewes depends on the number of ovarian follicles that grow and mature. The developmental process and therefore the number of ovarian follicles which ovulate can potentially be altered at many different stages.

Long-term effects

The pool of primordial follicles from which all follicles are recruited throughout the animals' lifetime is established by the time of birth or shortly after in most species including the sheep and goat (Greenwald and Terranova, 1988); no further mitotic activity occurs in the germ cells thereafter. The effect of nutrition *in utero*, if any, on the population of primordial follicles present in early life is not known, but it is possible that severe undernutrition in the first few weeks of life, which is known to affect subsequent reproductive performance (Gunn, 1977), may act through changes in the physiology of the primordial follicle population or in the hormonal milieu to which they are subject.

In the ewe, the process of follicle development from initiation of growth of the primordial follicle to ovulation is thought to take about 6 months (Cahill and Mauleon, 1980). The latter stages of growth are, however, very rapid with follicles growing from 0.5 mm diameter to 4 or 5 mm diameter in just 8 or 9 days (Turnbull *et al.*, 1977). The effects of nutrition on follicle development up to 0.5 mm diameter have not been reported but the observations of Fletcher (1974) of a reduction in ovulation rate, months after a period of severe undernutrition may be explicable in terms of a reduction in the size of the follicle cohorts recruited from the primordial pool or a reduction in their rate of development.

Effects of body condition and level of feed intake

Follicle recruitment and development is a continuous process. Thus, irrespective of the stage of the oestrous cycle, one or more potentially ovulatory follicles is always present but the final stages of maturation, ovulation and formation of a functional corpus luteum are dependent on exposure of these follicles, at an appropriate time, to an appropriate hormonal profile which normally occurs only following luteal regression (McNeilly *et al.*, 1991). The follicles which are destined to ovulate within a few days of luteal regression are drawn from a pool of follicles which can be of any size >2 mm diameter (Driancourt and Cahill, 1984). The effects of nutrition on ovulation rate can potentially be mediated either through an effect on the number of follicles which are in the appropriate size class and physiological state at the time of luteolysis or through differences in the proportion of these follicles which is induced to undergo the final stages of maturation and ovulation.

Allison (1977) reported greater numbers of follicles >2 mm diameter and larger mean follicle diameters in ewes of high live weight compared with ewes of low live weight. In studies of Scottish Blackface ewes, larger numbers of follicles >4 mm diameter (Rhind and McNeilly, 1986) or >2.5 mm diameter (Rhind *et al.*, 1989a) have been reported in ewes in high body condition compared with those in low condition. It has also been shown that a higher proportion of the large follicles in the high-condition animals compared with the low-condition animals was oestrogenic and potentially ovulatory (McNeilly *et al.*, 1987). These differences in follicle populations paralleled the differences in ovulation rate associated with differences in body condition in these ewes (Rhind and McNeilly, 1986). It was concluded that differences in ovulation rate associated with differences in body condition are primarily due to the difference in the number of potentially ovulatory follicles present at luteal regression when the final selection of the ovulatory follicle begins. These findings were subsequently confirmed by histological studies of ovarian follicle populations (Xu *et al.*, 1989) which also showed that a lower proportion of the follicles in ewes in high condition became atretic late in the follicular phase, compared with those of low-condition ewes.

In a study of ovarian follicle populations during different stages of oestrus, Haresign (1981) found no difference with level of feed intake in the numbers of large (>3 mm diameter) follicles present except at 48 h after the onset of oestrus, just a few hours before ovulation. The ewes which had the higher level of feed intake also had a significantly higher ovulation rate (2.6 *v* 1.8, P<0.05). It is clear that, unlike in ewes in different condition score classes, this difference was not due to differences between treatments in the numbers of potentially ovulatory follicles present at luteal regression but was due to differences in the proportion of these follicles which was induced to mature and ovulate.

These observations can begin to explain the results of many previous studies. Gunn (1983) concluded that the effects of flushing were confined to ewes in an intermediate range of body condition. The absence of a response to flushing in thin ewes, particularly in breeds such as the Scottish Blackface which generally has only one ovulation when in low body condition, may be due to the fact that only a single, large, oestrogenic and potentially ovulatory follicle is present at any stage of the cycle. Thus, irrespective of level of feed intake and associated metabolic and endocrine stimuli in the days or weeks before mating, the ovulation rate is likely to be limited to one in such ewes by the fact that there is only one potentially ovulatory follicle present.

At intermediate levels of body condition, some additional potentially ovulatory follicles are likely to be present. Increasing the level of feed intake for say 2 or 3 weeks before mating is unlikely to affect the number of potentially ovulatory follicles but can enhance the growth and/or reduce atresia of those present, resulting in the final maturation of a greater proportion of the follicles. At high levels of body condition, the observed responses are less readily explicable but the absence of a flushing response may be due to the fact that the effect of high body condition on the metabolic and endocrine status of the ewes overrides the effects of contemporary nutrition, resulting in a relatively high ovulation rate irrespective of contemporary levels of feed intake.

The effects of nutrition on the pattern of both ovarian follicle growth and development and ovulation rate are not the same in ewes of all genotypes. The magnitude of the increase in ovulation rate with increasing body condition is much greater in ewes of conventional breeds (Gunn and Doney, 1975, 1979) than in more fecund breeds such as the Finnish Landrace and its crosses (Rhind and Schanbacher, 1991). Numerous large, oestrogenic follicles develop in ewes of the more fecund breed, largely irrespective of the ewes' body condition, and a high proportion of these follicles is induced to ovulate irrespective of level of feed intake (Rhind and Schanbacher, 1991).

Effect of specific nutrients

The effect of dietary protein content on ovulation rate appears to operate over a short timescale. While many studies in which an ovulatory response to protein supplementation has been demonstrated have been conducted over one or more oestrous cycles (Radford *et al.*, 1980; Davis *et al.*, 1981; Fletcher, 1981) responses have also been reported following lupin feeding for only 6 days (Lindsay, 1976; Oldham and Lindsay, 1984); responses to increased energy intake over such a short timescale have not been recorded.

This observation is consistent with the conclusion of Nottle *et al.* (1985) that the increase in ovulation rate associated with increased dietary protein is attributable to a reduction in the rate of follicle atresia following luteolysis, i.e. like the effects of increased feed and/or energy intake, the effect of dietary protein on ovulation rate seems to be mediated at a late stage of the oestrous cycle and follicle development.

Hormonal control

Effects of nutrition on follicle development and ovulation rate could potentially be mediated through changes in circulating blood metabolites (glucose, non-esterified fatty acids, amino acids, etc.), metabolic hormones (insulin, growth hormone, etc.), gonadotrophins (LH and FSH) or a combination of several of these factors. With increasing knowledge of the effects of nutrition on these factors and on associated patterns of ovarian activity, it is becoming apparent that many factors are involved in determining ovulation rate and at present understanding of them is at best simplistic.

Long-term effects

Data pertaining to the effects of ewe nutrition on ovulation rate months or years later are scarce and none of the reports of such effects include measures of the associated hormonal changes, either at the time of nutritional restriction or at the time of ovulation rate measurements.

Medium and short-term effects

Due to the known roles of LH and FSH in the control of ovarian function, much of the research effort concerning the nutritional control of ovulation rate has centred on these hormones. While FSH secretion is fairly constant throughout most of the oestrous cycle, LH secretion is pulsatile with the incidence of pulses being approximately one every 3 hours during the luteal phase and increasing to approximately one per hour during the follicular phase (Baird and McNeilly, 1981). The increasing LH pulse frequency at this time culminates in a massive release of LH in a preovulatory surge which is associated with a surge in FSH secretion.

Effects of ewe body condition and level of feed intake

In an early study of the role of gonadotrophins in the induction of the different ovulation rates of ewes of different genotypes the timing of the preovulatory LH surge was implicated (Thimonier and Pelletier, 1971). It was subsequently shown that differences in ovulation rate with level of feed intake were not associated with differences in either timing or size of the surge (Lishman *et al.*, 1974; Haresign, 1981).

More detailed studies of LH secretion (Yuthasastrakosol *et al.*, 1975, 1977) showed that this hormone was normally secreted episodically and that the frequency of pulses differed according to the animals' reproductive state. In subsequent studies, the separate effects of body condition and level of feed intake on the pattern of LH pulse secretion and of FSH secretion have been studied and shown to be different.

The effects of ewe body condition on gonadotrophin profiles were reported by Rhind and McNeilly (1986) and Rhind et al. (1989a). In these studies, two groups of ewes were fed to achieve levels of body condition approximately equivalent to the upper and lower extremes normally recorded in practice in the UK. The ewes were then fed to maintain these levels of body condition so that they had similar levels of feed intake at the time of study and the effects of intake and body condition were not confounded. In both experiments, there was no effect of ewe body condition on mean LH concentrations or mean LH pulse frequency during either the luteal or follicular phases of the cycle. However, in one of the experiments (Rhind and McNeilly, 1986) there was a slightly lower mean LH pulse amplitude in the low-condition ewes during the follicular phase. In that study, mean FSH concentrations were consistently lower in the low-condition ewes, particularly during the luteal phase. These differences were absent in the later study and in other studies in which ewes of different live weights and body condition were compared (Findlay and Cumming, 1976; Xu et al., 1989).

In both experiments the differences in body condition were associated with differences in populations of large follicles. While the results of the first study appeared to suggest that body condition affected follicle populations through changes in circulating FSH concentrations, the results of the second suggest that additional factors must be involved in this process. It seems likely that FSH has an important role in the control of follicle development and ovulation rate but it acts in concert with other hormones.

In a study of the effect of level of premating feed intake on reproductive performance and gonadotrophin profiles, Rhind et al. (1985) fed two groups of Cheviot ewes so that at the time of study they were in similar levels of body condition but their intake differed by a factor of three with the lower intake group being fed a live weight maintenance ration. Ewes of the high-intake group had a higher mean ovulation rate (1.95 v 1.40, P < 0.01). A frequent blood sample collection regime revealed no differences between the groups in the pattern of FSH or LH secretion during the luteal phase of the cycle preceding the determination of ovulation rate. Mean FSH and LH concentrations were also unaffected during the follicular phase of the cycle which followed luteal regression. However, between 72 and 25 h before the preovulatory LH surge, the incidence of LH pulses was approximately twice as high in the high intake group compared with ewes fed a maintenance ration. These findings are supported by the work of Rhind et al. (1986). While in the latter experiment the intake effects were confounded with a difference in ewe body condition, the previously reported absence of an effect of body condition on gonadotrophins would indicate that the effect on LH pulse frequency was attributable only to the effects of level of feed intake.

It was postulated that the enhanced LH pulse frequency of ewes with a high intake, compared with ewes fed on live weight maintenance ration, stimulates the final stages of maturation of a greater proportion of the pool of potentially ovulatory follicles and so increases the ovulation rate.

Effect of specific nutrients

The endocrine mechanisms through which the rate of follicle atresia can be affected by dietary protein remain uncertain. Brien *et al.* (1976) reported an effect of lupin supplementation on FSH concentrations; higher values were recorded at days 12 to 14 of the cycle in supplemented ewes and there was a greater fall in concentrations over the following 2 days in supplemented animals. The work of Knight *et al.* (1981) confirmed these observations. However, others have reported effects of lupin supplementation on ovulation rate without any associated change in FSH profiles (Radford *et al.*, 1980; Scaramuzzi and Radford, 1983; Downing and Scaramuzzi, 1991). It remains unclear whether FSH profiles have a physiological role in the determination of ovulation rate or they are merely consequences of changes in the pattern of follicle development and oestrogen production. Dietary protein level does not affect LH secretion, including LH pulse frequency (Radford *et al.*, 1980).

Recent experiments designed to study the effects of specific branched-chain amino acids (Downing and Scaramuzzi, 1991) showed an increase in ovulation rate in response to infusion of a mixture of amino acids but no change in FSH or LH concentrations. It therefore seems unlikely that the effects of dietary protein on ovulation rate are mediated through changes in gonadotrophin profiles alone.

Fertilization rate

After the ovum is shed, one of the first determinants of the rate of reproductive success is the fertilization rate. Under normal circumstances, 80 to 95% of ova are likely to be fertilized in both sheep and goats (Hancock, 1962). There are very few studies of the effect of nutrition on the rate of fertilization but it does not appear to be affected significantly by nutritional state (Casida, 1963).

Prenatal Mortality

In most studies of the effects of nutrition on ova wastage, losses due to fertilization failure have not been separated from those due to embryo mortality. However, since the incidence of failure of fertilization is often relatively low and largely independent of nutrition, it may be assumed that most reported effects on ova wastage are due primarily to effects on the rate of embryo mortality.

Normal ova wastage rates in the ewe are reported to be between approximately 5% and 45% (Edey, 1969) of which a proportion is attributable to fertilization failure. The rates of wastage have not been assessed in goats but it seems likely that they fall in a broadly similar range. While the majority of ova wastage/embryo death occurs during the first month after mating (Edey, 1969), significant numbers of embryos may be lost at later stages in highly fecund ewes (Rhind *et al.*, 1980), perhaps because uterine capacity becomes a limiting factor, and in ewes of conventional breeds when they are severely undernourished (Kelly *et al.*, 1989).

While nutrition is an important determinant of embryo mortality rate in the ewe, there are many additional factors which interact with nutrition and have considerable influence on it (see reviews by Hanly, 1961; Edey, 1969, 1976b; Wilmut *et al.*, 1986). Rates of mortality are dependent on breed (Cumming *et al.*, 1975). Losses are higher in young animals (Quirke and Hanrahan, 1977), in lactating ewes (Cognie *et al.*, 1975) and in ewes subject to climatic stresses including heat (Alliston and Ulberg, 1961) and high rainfall and cold (Griffiths *et al.*, 1970). They are also higher in ewes with high ovulation rates (Edey, 1969).

The effects of ewe nutrition on prenatal development, and ova and embryo wastage have been reviewed previously (Edey, 1976b; Robinson, 1986; Wilmut *et al.*, 1986).

Timescale of action

Long-term effects

It is not known whether or not embryo mortality rates can be affected by severe nutritional excesses or deficiencies several years or months before mating and particularly in early life. However, it would be unwise to discount the possibility that hypothalamic/pituitary function could be affected for a long period, or indeed permanently, so that patterns of secretion of the luteotrophic hormone LH could also be affected long after the nutritional event. This in turn could affect progesterone secretion and ova wastage rates.

Effects of body condition and feed intake

Following an extensive review of the literature, Edey (1976b) concluded that very low live weights, and therefore low levels of body condition, were associated with increased embryo mortality and an increased incidence of barrenness. However, it was concluded that there was no relationship between rates of wastage and live weight *per se* at higher live weights and condition scores. Detrimental effects of very high levels of body condition on ova wastage rates (Rhind *et al.*, 1984) and on the incidence of barrenness (Gunn *et al.*, 1983) have also been reported.

The literature concerning the effects of level of feed intake on embryo mortality is confused. Undernutrition has been found to increase it in some experiments (Edey 1966; Rhind *et al.*, 1989c) but not in others (Gunn *et al.*, 1972; Parr *et al.*, 1982). Edey (1976b) concluded that some of these anomalies may be due to the fact that experimental groups were not sufficiently large to demonstrate statistically significant differences in the rate of induced wastage which often amounted to < 15%, to insufficiently severe nutritional treatments and to breed and strain differences. Some of the variation in response may also be due to the fact that premating nutrition can influence subsequent embryonic growth and survival post mating. For example, Gunn *et al.* (1979) showed in Cheviot ewes which had a low level of feed intake before mating that the loss of multiple-shed ova was greater than that of single-shed ova while in ewes fed at high level before mating there was no difference in wastage rates.

Like excessively high levels of body condition, high levels of feed intake in early pregnancy may also have a detrimental effect on ova wastage rates (Cumming *et al.*, 1975; Brien *et al.*, 1977).

Effects of specific nutrients

Although the beneficial effects of increased dietary protein on ovulation rate have been repeatedly demonstrated, there have been few studies specifically designed to determine the effect of dietary protein level on embryo survival rates. Van der Westhuysen (1971) fed high protein supplement in early pregnancy but failed to improve rates of survival. Edey (1976b) concluded that the only other specific nutrient known to have a direct role in embryo wastage was selenium; in certain areas, fertility problems, thought to be due to embryonic mortality, were cured by selenium administration (Hartley, 1963; Mudd and Mackie, 1973).

The specific effects on the reproductive performance of goats of dietary energy and protein have not yet been researched but, as in other species, deficiencies of some other specific nutrients can have a detrimental effect on reproductive performance. Severe iodine deficiency results in hypothyroidism and anoestrus (Reddi and Rajan, 1986), and selenium deficiency can seriously depress conception rates and kid survival (Anke *et al.*, 1989).

Hormonal control

The mechanisms involved in the control of ova and embryo wastage, and in particular the mechanisms by which they are influenced by nutrition, remain unclear. Potential causes of wastage include abnormalities of the ovum or embryo, failure of the systems of maternal recognition of pregnancy and suppression of the maternal immune system, luteal inadequacy and failure of the supply of progesterone to the uterus and the conceptuses; all of these parameters and many more are inextricably linked. Most aspects of the process of pregnancy establishment are ultimately dependent on the supply of progesterone to the uterus without which the pregnancy fails. Consequently much of the research into the effects of nutrition on ova and embryo wastage in the ewe has centred on progesterone profiles. The pivotal role of progesterone in the maintenance of pregnancy and in the determination of ova wastage rate has been demonstrated by the studies of Parr *et al.* (1982) involving progesterone replacement therapy and ovariectomized ewes. It was found that as the dose of progesterone injected to replace the endogenous steroid was decreased from 25 to 5 mg/day, embryo survival rates at day 11 decreased from 96 to 74%. However, there was no critical threshold concentration; the rate of embryonic death increased steadily as the dose of progesterone decreased. This may indicate that each embryo may require a different minimum progesterone concentration for its survival or may reflect differences in the pattern of progesterone delivery to different parts of the uterus.

While both overnutrition and undernutrition have been found to increase ova and embryo wastage, undernutrition is associated with elevated concentrations of progesterone in jugular venous plasma (Parr *et al.*, 1987; Rhind *et al.*, 1989d) and a high plane of nutrition is associated with reduced progesterone concentrations (Brien *et al.*, 1981). These differences in circulating concentrations can be explained by differences in the rate of clearance of progesterone from the circulation and not necessarily by differences in secretion rate (Parr *et al.*, 1982).

The anomalous relationships between level of feed intake, circulating progesterone concentrations and rates of ova wastage, together with the inconsistent effects of progesterone supplementation on embryo survival rates (Parr *et al.*, 1987; Disken and Niswender,1989) clearly show that the effect of nutrition on embryo survival cannot be explained solely in terms of mean progesterone concentrations in the peripheral circulation. Mean progesterone concentrations may be unimportant while short-term changes in the pattern of delivery to the uterus could be critical. (Short-term changes would not be apparent with the infrequent sample collection regimes used in most studies.)

Experiments involving frequent sample collection have shown that there are frequent, large, short-term increases in progesterone concentrations. Some, but not all, of these can be explained by the occurrence immediately beforehand of transient increases (pulses) in the luteotrophic hormone, LH; all LH pulses are followed by transient increase in progesterone although some increases in progesterone also occur independently (McNeilly and Fraser, 1987; Rhind *et al.*, 1989d, 1991a).

Rhind *et al.* (1989b) showed that, in underfed ewes with an increased rate of ova wastage, the mean frequency of LH pulses was reduced. While this reduction was not translated into a significant reduction in the frequency of progesterone pulses in the peripheral circulation, probably because the occurrence of some progesterone pulses is independent of LH, it was postulated that the change in LH profiles could have mediated the effect of undernutrition on reproductive performance. Specifically, they suggested that since progesterone concentrations in the ovarian vein are about 100 times higher in the ovarian vein than in the peripheral circulation (Baird *et al.*, 1981) and progesterone can pass from the ovarian vein to the ovarian artery via a countercurrent exchange mechanism (Einer-Jensen and McCracken, 1981), the progesterone concentration in the ovarian artery is likely to be higher than in the peripheral circulation. Since a branch of the ovarian artery supplies the oviduct and upper uterine horn (Hunter, 1987), the supply of progesterone to these tissues is also likely to be greater than concentrations in the peripheral circulation would indicate. It follows that increases in progesterone concentrations stimulated by LH pulses would also be larger in the ovarian vein and artery than in the peripheral circulation. Consequently, the size of the reduction in progesterone delivery to the uterus and oviduct, associated with the reduction in LH pulse frequency in underfed ewes would also be exacerbated and could be sufficient to affect embryo growth and survival. The nature of the putative mechanism is such that it suggests that measurements of progesterone concentrations in the peripheral circulation are relatively meaningless.

A reduction in LH pulse frequency has also been reported on ewes in excessively high body condition and implicated in the occurrence of a higher rate of ova wastage in such ewes (Rhind *et al.*, 1984). Very fat ewes, unlike ewes in moderate body condition, showed a nocturnal reduction in LH pulse frequency and it was suggested that this reduction may have a critical influence on the pattern of progesterone delivery to the uterus, oviduct and conceptus (Rhind *et al.*, 1991a).

The precise mechanisms through which changes in progesterone profiles could influence early embryo development and survival remain unclear but Ashworth and Bazer (1989) showed that administration of progesterone to ewes increased levels of ten specific endometrial proteins. It has not yet been demonstrated that the pattern of synthesis of these proteins can determine embryo survival rates but this represents a possible mechanism through which nutritionally induced changes in progesterone profiles could affect reproductive performance.

Effects of Nutrition on Hypothalamic/Pituitary Function

Nutritional factors could influence the hypothalamus/pituitary, and so gonadotrophin profiles, directly (through effects of nutrients or metabolic hormones such as insulin acting on the target organ), or through changes in the sensitivity of these organs to oestradiol or other hormonal feedback.

Measurement of GnRH secretion by the hypothalamus is technically difficult but there is a very close temporal relationship between LH and GnRH secretion (Clarke and Cummins, 1982) and so LH pulse frequency can be used as an index of GnRH pulse frequency and hypothalamic activity. Since pituitary activity and gonadotrophin output is highly dependent on hypothalamic activity and GnRH output, nutritionally induced changes in pituitary activity are not readily separated from changes in hypothalamic activity. However, pituitary responsiveness to GnRH, an index of pituitary activity, can be measured by injecting a physiological dose of GnRH and measuring the subsequent changes in gonadotrophin concentrations.

Direct (oestrogen-independent) effects

Studies of ovariectomized ewes and lambs have shown that a reduction in ewe live weight and body condition is associated with a reduction in LH pulse frequency (Foster *et al.*, 1989; Rhind *et al.*, 1989a; Thomas *et al.*, 1990). There was no effect of low body condition on pituitary concentrations of mRNA for LH or FSH (Thomas *et al.*, 1990), no difference with body condition in the pituitary response to a GnRH challenge (Rhind *et al.*, 1989a) and no difference in numbers of GnRH receptors in the pituitary or in pituitary weight (Tatman *et al.*, 1990) and so it appears that the effect is mediated through changes in hypothalamic, but not pituitary, activity.

Short-term increases in feed intake, of insufficient duration to affect body condition, resulted in rapid increases in LH pulse frequency (indicating an effect on hypothalamic function) and in mean circulating LH concentration in ovariectomized, growth-restricted female lambs (Foster *et al.*, 1989). There was also a rapid increase in pituitary concentrations of mRNA coding for gonadotrophin subunits (Landerfeld *et al.*, 1989). However, in a study of ovariectomized adult ewes (Rhind *et al.*, 1989b), there was no effect of level of feed intake on the hypothalamic activity, as indicated by LH pulse frequency, or on pituitary response to a GnRH challenge. While the results of work on the growth-restricted lamb apparently indicate that there is a direct effect of level of feed intake on hypothalamic activity and perhaps on pituitary activity, the results obtained with adult ewes do not support this. The reason for the discrepancy is unclear at present.

Indirect (oestrogen-dependent) effects

While the ovariectomized lamb and ewe models have provided much information concerning the direct action of nutritional factors at the hypothalamus/pituitary, they specifically exclude oestradiol, the ovarian steroid which may have the greatest inhibitory influence on the hypothalamic/pituitary system. Oestradiol has a pivotal role in the control of the onset of puberty. The hypothalamic/pituitary system of ewe lambs is capable of generating a high frequency of LH pulses well before puberty and the ovary is capable of responding to such a stimulus. However, before the onset of puberty hypothalamic/pituitary activity remains suppressed owing to hypersensitivity to oestradiol (Foster, 1988). At puberty, the sensitivity to oestradiol of the hypothalamic/pituitary system decreases and so results in an increased LH pulse frequency.

The effect of nutritional state on the onset of puberty can also be explained in terms of changes in hypothalamic sensitivity; Foster *et al.* (1985) found that in growth-restricted, ovariectomized, female lambs with oestradiol implants, an increase in feed intake and live weight resulted in an increased LH pulse frequency while the oestradiol stimulus was unchanged.

Similar effects of nutritional status on oestradiol feedback have been reported in adult ewes. In a recent study involving ovariectomized ewes in different levels of body condition or with different levels of feed intake, oestradiol was replaced using subcutaneous implants. Both hypothalamic and pituitary activity were found to be much lower in ewes in low body condition indicating that both of these tissues were more sensitive to oestradiol feedback in the low condition ewes compared with those in high condition (Rhind *et al.*, 1991b). In contrast to the effect of body condition, differences in level of intake resulted in differences only in hypothalamic activity. The increased pituitary sensitivity associated with low levels of body condition probably reflects an increase in the concentration of oestradiol receptors; such an inverse relationship has been demonstrated previously in ovariectomized ewes (Adams and Ritar, 1986).

Thus, there is evidence that both low levels of feed intake and low body condition are associated with enhanced hypothalamic sensitivity to oestradiol. The effects of steroids on hypothalamic activity and LH secretion are mediated through endorphins and enkephalins, opioid peptides produced in the brain, and through neurotransmitters such as dopamine, adrenaline and serotonin (Haynes *et al.*, 1989) but the importance of the different mechanisms may differ with season, age of the animal and stage of the cycle.

Haynes *et al.* (1989) concluded that the inhibitory effects of oestradiol were probably not mediated through opioid peptides but dopamine inhibition was probably involved. The recent reports of Ebling *et al.* (1990) and Recabarren *et al.* (1990) are consistent with this suggestion; they showed that treatment of prepubertal lambs (in which the hypothalamic/pituitary system is highly sensitive to oestradiol) with opioid antagonists did not affect LH secretion, indicating that opioid inhibition was not the primary cause of hypogonadotrophism. However, the relationships between nutritional state and opioid inhibition have not been reported in the intact, adult ewe and the interactions between opioid and oestradiol feedback mechanisms in ewes in different nutritional states remain to be elucidated.

In summary, the mechanisms through which nutrition affects hypothalamic/pituitary activity are complex and are not completely understood. It appears that the effects operate primarily, but not solely, through changes in hypothalamic activity and that changes in sensitivity of this organ to oestradiol are involved.

What are the Roles of Blood Metabolites and Metabolic Hormones in the Control of Reproduction?

Increases in level of feed intake are associated with depressed concentrations of growth hormone (GH) and increased insulin concentrations but not necessarily with altered glucose concentrations (Wynn *et al.*, 1988; Polkowska, 1989; Thomas *et al.*, 1990). Concentrations of thyroxine (T_4) and triiodothyronine (T_3) and the T_4/T_3 ratio are also affected (Wynn *et al.*, 1988).

The effects of differences in dietary energy and protein intake and changes in body condition on hypothalamic and pituitary activity could be mediated either through changes in circulating blood metabolites or through associated changes in metabolic hormone profiles. While there are clear associations between nutritional state, gonadotrophin secretion and metabolic hormone/blood metabolite profiles, causal relationships have not yet been identified.

In studies involving nutrient-restricted prepubertal lambs, Foster *et al.* (1988) showed that reductions in LH pulse frequency associated with dietary restriction could be overcome by infusion of glucose and amino acids. However, associated changes in metabolic hormone profiles were not recorded and so it remains unclear whether the effect was directly attributable to these nutrients or to associated changes in metabolic hormone profiles.

Short-term infusion of free fatty acids into ovariectomized ewes and lambs, resulting in circulating levels similar to those found in starved sheep, suppressed GH secretion but did not alter LH secretion (Estienne *et al.*, 1989, 1990). These results indicate that acute changes in circulating FFA or GH concentrations do not directly drive changes in hypothalamic activity, but they do not show the effect of long-term infusions or elucidate the mechanisms by which nutrients influence hypothalamic activity.

In another study (Downing and Scaramuzzi, 1991) involving intact ewes, glucose infusions over a 5 day period increased the ovulation rate but did not affect LH secretion and reduced circulating FSH concentrations. It was concluded that the effect on ovulation rate could not be explained in terms of elevation of gonadotrophin concentrations and that changes in insulin, GH and prolactin concentrations associated with glucose infusion may have been the cause of the observed ovulatory response.

Ebling *et al.* (1990) postulated that amino acid neurotransmitters, aspartate and glutamate, may directly stimulate GnRH neurons and LH release and demonstrated stimulatory effects of pharmacological doses of N-methyl-D-L-aspartate on LH secretion. However, Downing and Scaramuzzi (1991) reported no significant changes in LH or FSH profiles following intravenous infusion of tyrosine, tyrosine and phenylalanine, aspartic acid or tryptophan indicating that such a direct mechanism is unlikely to operate.

In summary, while there is evidence that nutritional state can alter both hypothalamic activity and ovulation rate, the factors which affect the former remain unclear. Furthermore, it is becoming increasingly clear that although hypothalamic activity and gonadotrophin secretion are central to the control of follicle development, ovulation and pregnancy maintenance, the effects of nutrition on ovulation rate are not mediated solely through these factors. The effect of gonadotrophins on the ovaries of sheep, and in particular on the ovulation rate, is apparently modified by circulating concentrations of metabolic hormones and/or specific nutrients, as in non-ruminant species (Veldhuis *et al.*, 1986; Tonetta and di Zerega, 1989).

The Future

While some progress has been made in understanding the effects of nutrition on hypothalamic/pituitary function, endocrine status and follicle populations in the ewe, this information is lacking for the goat. The available data indicate that the effects of nutrition are broadly similar in the two species. The effects of nutrition on ovarian follicle populations, hypothalamic/pituitary function and endocrine status await investigation in the goat but in view of the demonstrable similarities in ovarian activity between these species, it is suggested that many aspects of the underlying physiology will prove to be similar. It is already clear that the effects of nutrition on ovulation rate cannot be explained simply in terms of gonadotrophin profiles and there is a need for studies of the effects of nutritional status on intra-ovarian regulatory substances, some of which have only recently been identified or purified (Tsafriri, 1988). In particular, there is a need to determine the effects of nutritionally induced changes in metabolic hormones, if any, on intrafollicular peptides and steroids and to characterize the effect of such changes on the ovarian response to gonado-trophins.

With regard to effects of nutrition on rates of embryo mortality, there is a need to determine the relationships between nutritional state, patterns of progesterone delivery to the uterus (which may differ from patterns in the peripheral circulation) and patterns of production of uterine proteins including those involved in the maternal recognition of pregnancy.

Responses to nutritional, and indeed pharmacological, treatments have become more predictable owing to knowledge of the effects of nutrition on ovarian follicle populations and endocrine control mechanisms. However, the discovery of many new regulatory substances, particularly within the ovary, means that much work remains to be done in order to properly understand the endocrine mechanisms through which nutrition affects reproductive performance.

References

Adams, N.R. and Ritar, A.J. (1986) Measurement of estrogen receptors in the ovariectomized ewe as affected by body condition and secondary binding sites. *Biology of Reproduction* 35, 828-832.

Adu, I.F., Buvanendran, V. and Lakpini, C.A.M. (1979) The reproductive performance of Red Sokoto goats in Nigeria. *Journal of agricultural Science,* Cambridge 93, 563-566.

Allden, W.G. (1979) Undernutrition of the Merino sheep and its sequelae. V. The influence of severe growth restriction during early post-natal life on reproduction and growth in later life. *Australian Journal of Agricultural Research* 30, 939-948.

Allen, D.M. and Lamming, G.E. (1961) Nutrition and reproduction in the ewe. *Journal of agricultural Science*, Cambridge 56, 69-79.

Allison, A.J. (1977) Effect of nutritionally induced liveweight differences on the ovulation rates and the population of ovarian follicles in ewes. *Theriogenology* 8, 19-24.

Alliston, C.W. and Ulberg, L.C. (1961) Early pregnancy loss in sheep at ambient temperatures of 70° and 90°F as determined by embryo transfer. *Journal of Animal Science* 20, 608-613.

Anke, M., Angelow, L., Groppel, B., Arnhold, W. and Gruhn, K. (1989) The effect of selenium deficiency on reproduction and milk performance of goats. *Archives of Animal Nutrition* 39, 483-490.

Ashworth, C.J. and Bazer, F.W. (1989) Changes in ovine conceptus and endometrial function following asynchronous embryo transfer or administration of progesterone. *Biology of Reproduction* 40, 425-433.

Baird, D.T. and McNeilly, A.S. (1981) Gonadotrophic control of follicular development and function during the oestrous cycle of the ewe. *Journal of Reproduction and Fertility*, Supplement 30, 119-133.

Baird, D.T., Swanston, I.A. and McNeilly, A.S. (1981) Relationship between LH, FSH, and prolactin concentration and the secretion of androgens and estrogens by the preovulatory follicle in the ewe. *Biology of Reproduction* 24, 1013-1025.

Brien, F.D., Baxter, R.W., Findlay, J.K. and Cumming, I.A. (1976) Effect of lupin grain supplementation on ovulation rate and plasma follicle stimulating hormone (FSH) in maiden and mature ewes. *Proceedings of the Australian Society of Animal Production* 11, 237-240.

Brien, F.D., Cumming, I.A. and Baxter, R.W. (1977) Effect of feeding a lupin grain supplement on reproductive performance of maiden and mature ewes.

Journal of agricultural Science, Cambridge 89, 437-443.

Brien, F.D., Cumming, I.A., Clarke, I.J. and Cocks, C.S. (1981) Role of plasma progesterone concentrations in early pregnancy of the ewe. *Australian Journal of Experimental Agriculture and Animal Husbandry* 21, 562-565.

Cahill, L.P. and Mauleon, P. (1980) Influences of season, cycle and breed on the follicular growth rates in sheep. *Journal of Reproduction and Fertility* 58, 321-328.

Casida, L.E. (1963) The level of fertility in the female as influenced by feed level and energy intake. *Proceedings of the Sixth International Congress on Nutrition*, Edinburgh, pp. 366-375.

Chaniago, T.D. (1988) Comparison of the reproductive performance of Etawah cross goats and Kacang goats maintained on low quality feed with and without supplementation during mating and peri-kidding period. *Proceedings of the VI World Conference on Animal Production*, Helsinki, Finland, p. 588.

Clarke, I.J. and Cummins, J.T. (1982) The temporal relationship between gonadotrophin releasing hormone (GnRH) and luteinizing hormone (LH) secretion in ovariectomized ewes. *Endocrinology* 111, 1737-1739.

Cockrem, F.R.M. (1979) A review of the influence of liveweight and flushing on fertility made in the context of efficient sheep production. *Proceedings of the New Zealand Society of Animal Production* 39, 23-42.

Cognie, Y., Hernandez-Barreto, M. and Saumande, J. (1975) Low fertility in nursing ewes during the non-breeding season. *Annales de Biologie Animale Biochimie Biophysique* 15, 329-343.

Coop, I.E. (1966) Effect of flushing on reproductive performance of ewes. *Journal of agricultural Science*, Cambridge 67, 305-323.

Cumming, I.A., Blockey, M.A de B., Winfield, C.G., Parr, R.A. and Williams, A.H. (1975) A study of relationships of breed, time of mating, level of nutrition, live weight, body condition and face cover to embryo survival in ewes. *Journal of agricultural Science*, Cambridge 84, 559-565.

Davis, I.F., Kenney, P.A. and Cumming, I.A. (1976) Effect of time of joining and rate of stocking on the production of Corriedale ewes in southern Victoria. 5. Ovulation rate and embryonic survival. *Australian Journal of Experimental Agriculture and Animal Husbandry* 16, 13-18.

Davis, I.F., Brien, F.D., Findlay, J.K. and Cumming, I.A. (1981) Interactions between dietary protein, ovulation rate and follicle stimulating hormone level in the ewe. *Animal Reproduction Science* 4, 19-28.

Devendra, C. and Burns, M. (1983) *Goat Production in the Tropics*. Commonwealth Agricultural Bureau, Slough, UK.

Diskin, M. and Niswender, G.D. (1989) Effect of progesterone supplementation on pregnancy and embryo survival in ewes. *Journal of Animal Science* 67, 1559-1563.

Doney, J.M. and Gunn, R.G. (1973) Progress in studies on the reproductive performance of hill sheep. *Hill Farming Research Organisation Sixth Annual Report*, pp. 69-73.

Downing, J.A. and Scaramuzzi, R.J. (1991) Nutrient effects on ovulation rate, ovarian function and the secretion of gonadotrophic and metabolic hormones in sheep. *Journal of Reproduction and Fertility*, Supplement 43, 209-227.

Driancourt, M.A. and Cahill, L.P. (1984) Preovulatory follicular events in sheep. *Journal of Reproduction and Fertility* 71, 205-211.

Dyrmundsson, O.R. (1987) Advancement of puberty in male and female sheep. In: Marai, I.F.M. and Owen, J.B. (eds.) *New Techniques in Sheep Production*. Butterworths, London. pp. 65-76.

Ebling, F.J.P., Wood, R.I., Karsch, F.J., Vannerson, L.A., Suttie, J.M., Bucholtz, D.C., Schall, R.E. and Foster, D.L. (1990) Metabolic interfaces between growth and reproduction. III Central mechanisms controlling pulsatile luteinizing hormone secretion in the nutritionally growth-limited female lamb. *Endocrinology* 126, 2719-2727.

Edey, T.N. (1966) Nutritional stress and preimplantation embryonic mortality in Merino sheep. *Journal of agricultural Science*, Cambridge 67, 287-302.

Edey, T.N. (1969) Prenatal mortality in sheep: a review. *Animal Breeding Abstracts* 37, 173-190.

Edey, T.N. (1976a) Embryo mortality. In: Tomes, G.J., Robertson, D.E. and Lightfoot, R.J. (eds.) *Sheep Breeding* (Proceedings of International Sheep Breeders Conference, Muresk, Western Australia), pp.400-410.

Edey, T.N. (1976b) Nutrition and embryo survival in the ewe. *Proceedings of the New Zealand Society of Animal Production* 36, 231-239.

Einer-Jensen, N. and McCracken, J.A. (1981) The transfer of progesterone in the ovarian vascular pedicle of the sheep. *Endocrinology* 109, 685-690.

Estienne, M.J., Schillo, K.K., Green, M.A. and Boling, J.A. (1989) Free fatty acids suppress growth hormone but not luteinizing hormone secretion in sheep. *Endocrinology* 125, 85-91.

Estienne, M.J., Schillo, K.K., Hileman, S.M., Green, M.A., Hayes, S.H. and Boling, J.A. (1990) Effects of free fatty acids on luteinizing hormone and growth hormone secretion in ovariectomized lambs. *Endocrinology* 126, 1934-1940.

Findlay, J.K. and Cumming, I.A. (1976) FSH in the ewe: effect of season, liveweight and plane of nutrition in plasma FSH and ovulation rate. *Biology of Reproduction* 15, 335-342.

Fletcher, I.C. (1974) An effect of previous nutritional treatment on the ovulation rate of Merino ewes. *Proceedings of the Australian Society of Animal Production* 10, 261-264.

Fletcher, I.C. (1981) Effects of energy and protein intake on ovulation rate associated with feeding of lupin grain to Merino ewes. *Australian Journal of Agricultural Research* 32, 79-87.

Foster, D.L. (1988) Puberty in the female sheep. In: Knobil, E. and Neill, J. (eds.), *The Physiology of Reproduction*, Vol. 2. Raven Press Ltd, New York, pp. 1739-1762.

Foster, D.L., Ebling, F.J.P., Micka, A.F., Vannerson, L.A., Bucholtz, D.C., Wood, R.I., Suttie, J.M. and Fenner, D.E. (1989) Metabolic interfaces between growth and reproduction. I. Nutritional modulation of gonadotrophin, prolactin, and growth hormone secretion in the growth-limited female lamb. *Endocrinology* 125, 342-350.

Foster, D.L., Ebling, F.J.P., Vannerson, L.A., Bucholtz, D.C., Wood, R.I., Micka, A.F., Suttie, J.M. and Vennvliet, B.A. (1988) Modulation of gonadotrophin secretion during development by nutrition and growth. *Proceedings of 11th International Congress on Animal Reproduction and Artificial Insemination*, Vol. 5. Dublin, Ireland, pp. 101-108.

Foster, D.L., Yellon, S.M. and Olster, D.H. (1985) Internal and external determinants of the timing of puberty in the female. *Journal of Reproduction and Fertility* 75, 327-344.

Greenwald, G.S. and Terranova, P.F. (1988) Follicular selection and its control. In: Knobil, E. and Neill, J. (eds.), *The Physiology of Reproduction*, Vol. 2, Raven Press, New York, pp. 387- 445.

Griffiths, J.G., Gunn, R.G. and Doney, J. (1970) Fertility in Scottish Blackface ewes as influenced by climatic stress. *Journal of agricultural Science*, Cambridge 75, 485-488.

Gunn, R.G. (1977) The effects of two nutritional environments from 6 weeks prepartum to 12 months of age on lifetime performance and reproductive potential of Scottish Blackface ewes in two adult environments. *Animal Production* 25, 155-164.

Gunn, R.G. (1983) In: Haresign, W. (ed.), *Sheep Production* (Proceedings of the 35th Easter School in Agricultural Science), University of Nottingham,

London, Butterworths, pp. 99-110.

Gunn, R.G. and Doney, J.M. (1975) The interaction of nutrition and body condition at mating on ovulation rate and early embryo mortality in Scottish Blackface ewes. *Journal of agricultural Science*, Cambridge 85, 465-470.

Gunn, R.G. and Doney, J.M. (1979) Fertility in Cheviot ewes. 1. The effect of body condition at mating on ovulation rate and early embryo mortality in North and South Country Cheviot ewes. *Animal Production* 29, 11-16.

Gunn, R.G., Doney, J.M. and Russel, A.J.F (1972) Embryo mortality in Scottish Blackface ewes as influenced by body condition at mating and by post mating nutrition. *Journal of agricultural Science*, Cambridge 79, 19-25.

Gunn, R.G., Doney, J.M. and Smith, W.F. (1979) Fertility in Cheviot ewes. 2. The effect of level of pre-mating nutrition on ovulation rate and early embryo mortality in North and South Country Cheviot ewes in moderately-good condition at mating. *Animal Production* 29, 17-23.

Gunn, R.G., Smith, W.F., Senior, A.J., Barthram, E. and Sim, D.A. (1983) Premating pasture intake and reproductive responses in North Country Cheviot ewes in different body conditions. *Animal Production* 36, 509 (Abstract).

Hafez, E.S.E. (1952) Studies on the breeding season and reproduction of the ewe. *Journal of agricultural Science*, Cambridge 42, 189-265.

Hancock, J.L. (1962) Fertilization in farm animals. *Animal Breeding Abstracts* 30, 285-310.

Hanly, S. (1961) Prenatal mortality in farm animals. *Journal of Reproduction and Fertility* 2, 182-194.

Haresign, W. (1981) The influence of nutrition on reproduction in the ewe. 1. Effects on ovulation rate, follicle development and luteinizing hormone release. *Animal Production* 32, 197-202.

Hartley, W.J. (1963) Selenium and ewe fertility. *Proceedings of the New Zealand Society of Animal Production* 23, 20-27.

Haynes, N.B., Lamming, G.E., Yang, K-P., Brooks, A.N. and Finnie, A.D. (1989) Endogenous opioid peptides and farm animal reproduction. In; Milligan, S.R. (ed.) *Oxford Reviews of Reproductive Biology* 11. Oxford University Press, pp. 111-145.

Heape, W. (1899) Abortion, barrenness and fertility in sheep. *Journal of the Royal Agricultural Society* (3rd series) 10, 217- 248.

Henniawati and Fletcher, I.C. (1986) Reproduction in Indonesian sheep and goats at two levels of nutrition. *Animal Reproduction Science* 12, 77-84.

Hulet, C.V., Price, D.A. and Foote, W.C. (1974) Effect of month of breeding and feed level on ovulation and lambing rates of Panama ewes. *Journal of Animal Science* 39, 73-78.

Hunter, R.H.F. (1987) *The Fallopian Tubes. Their Role in Fertility and Infertility*. Springer Verlag, London, p.47.

Kelly, R.W., Wilkins, J.F. and Newnham, J.P. (1989) Fetal mortality from day 30 of pregnancy in Merino ewes offered different levels of nutrition. *Australian Journal of Experimental Agriculture* 29, 339-342.

Kenney, P.A. and Smith, R.S. (1985) Effects of including lupins with cereal grain rations on the production of lambing ewes during drought. *Australian Journal of Experimental Agriculture* 25, 529-535.

Killeen, I.D. (1967) The effects of body weight and level of nutrition before, during and after joining on ewe fertility. *Australian Journal of Experimental Agriculture and Animal Husbandry* 7, 126-136.

Knight, T.W., Hall, D.R.H. and Wilson, L.D. (1983) Effects of teasing and nutrition on the duration of the breeding season in Romney ewes. *Proceedings of the New Zealand Society of Animal Production* 43, 17-19.

Knight, T.W., Oldham, C.M., Smith, J.F. and Lindsay, D.R. (1975) Studies in ovine infertility in agricultural regions in Western Australia: analysis of reproductive wastage. *Australian Journal of Experimental Agriculture and Animal Husbandry* 15, 183-188.

Knight, T.W., Payne, E. and Peterson, A.J. (1981) Effect of diet and liveweight on FSH and oestradiol concentrations in Romney ewes. *Proceedings of the Australian Society of Reproductive Biology* 13, 19.

Landerfeld, T.D., Ebling, F.J.P., Suttie, J.M., Vannerson, L.A., Padmanabhan, V., Beitins, I.Z. and Foster, D.L. (1989) Metabolic interfaces between growth and reproduction. 2. Characterization of changes in messenger ribonucleic acid concentrations of gonadotrophin subunits, growth hormone, and prolactin in nutritionally growth-limited lambs and the differential effects of increased nutrition. *Endocrinology* 125, 351-356.

Lindsay, D.R. (1976) The usefulness to the animal producer of research findings in nutrition on reproduction. *Proceedings of the Australian Society of Animal Production* 11, 217-224.

Lishman, A.W., Stielau, W.J. Dreosti, I.E., Botha, W.A., Stewart, A.M. and Swart, C.E. (1974) The release of luteinizing hormone at oestrus in ewes on two planes of nutrition during lactation. *Journal of Reproduction and Fertility* 41, 227-230.

Marshall, F.H.A. (1904) Fertility in sheep. *Transactions of the Highland Agricultural Society* 16, 34-43.

Marshall, F.H.A. (1908) Fertility in Scottish sheep. *Transactions of the Highland Agricultural Society* 20, 139-151.

McCall, D.G., Clayton, J.B. and Dow, B.W. (1989) Nutrition effects on live weight and reproduction of Cashmere doe hoggets. *Proceedings of the New Zealand Society of Animal Production* 49, 157-161.

McKenzie, F. and Terrill, C.E. (1937) Estrus, ovulation and related phenomena in the ewe. *Missouri Agricultural Experiment Station Research Bulletin* No 264.

McNeilly, A.S. and Fraser, H.M. (1987) Effect of GnRH agonist-induced suppression of LH and FSH on follicle growth and corpus luteum function in the ewe. *Journal of Endocrinology* 115, 273-282.

McNeilly, A.S., Jonassen, J.A. and Rhind, S.M. (1987) Reduced ovarian follicicular development as a consequence of low body condition in ewes. *Acta Endrocrinologica* 115, 75-83.

McNeilly, A.S., Picton, H.M., Campbell, B.K. and Baird, D.T. (1991) Gonadotrophic control of follicle growth in the ewe. *Journal of Reproduction and Fertility*, Supplement.

Mudd, A.J. and Mackie, I.L. (1973) The influence of vitamin E and selenium on ewe prolificacy. *Veterinary Record* 93, 197-199.

Nottle, M.B., Armstrong, D.T., Setchell, B.P. and Seamark, R.F. (1985) Lupin feeding and folliculogenesis in the Merino ewe. *Proceedings of the Nutrition Society of Australia* 10, 145.

Nottle, M.B., Hynd, P.I., Seamark, R.F. and Setchell, B.P. (1988) Increases in ovulation rate in lupin-fed ewes are initiated by increases in protein digested post-ruminally. *Journal of Reproduction and Fertility* 84, 563-566.

Oldham, C.M. and Lindsay, D.R. (1984) The minimum period of intake of lupin grain required by ewes to increase their ovulation rate when grazing dry summer pasture. In: Lindsay, D.R. and Pearce, D.T. (eds.) *Reproduction in Sheep*. Cambridge University Press, Cambridge, pp.274-276.

Parr, R.A., Cumming, I.A. and Clarke, I.J. (1982) Effects of maternal nutrition and plasma progesterone concentrations on survival and growth of the sheep embryo in early gestation. *Journal of agricultural Science*, Cambridge 98, 39-46.

Parr, R.A., Davis, I.F., Fairclough, R.J. and Miles, M.A. (1987) Overfeeding during early pregnancy reduces peripheral progesterone concentration and pregnancy rate in sheep. *Journal of Reproduction and Fertility* 80, 317-320.

Polkowska, J. (1989) Effect of protein deficiency on some hypothalamic and pituitary hormones in growing male lambs. An immunohistochemical study. *Reproduction, Nutrition, Development* 29, 347-356.

Quirke, J.F. and Hanrahan, J.P. (1977) Comparison of the survival in the uteri of adult ewes of cleaved ova from adult ewes and ewe lambs. *Journal of Reproduction and Fertility* 51, 487- 489.

Radford, H.M., Donegan, S. and Scaramuzzi, R.J. (1980). The effect of supplementation with lupin grain on ovulation rate and plasma gonadotrophin levels in adult Merino ewes. *Proceedings of the Australian Society of Animal Production* 13, 457.

Recabarren, S.E., Zapata, P. and Parilo, J. (1990) Disappearance of opioidergic tone on LH secretion in underfed prepubertal sheep. *Hormone and Metabolic Research* 22, 225-228.

Reddi, N.M. and Rajan, A. (1986) Reproductive behaviour and semen characteristics in experimental hypothyroidism in goats. *Thereogenology* 25, 263-274.

Rhind, S.M., Gunn, R.G., Doney, J.M. and Leslie, I.D. (1984) A note on the reproductive performance of Greyface ewes in moderately fat and very fat condition at mating. *Animal Production* 38, 305-307.

Rhind, S.M., Leslie, I.D., Gunn, R.G. and Doney, J.M. (1985) Plasma FSH, LH, prolactin and progesterone profiles of Cheviot ewes with different levels of intake before and after mating, and associated effects on reproductive performance. *Animal Reproduction Science* 8, 301-313.

Rhind, S.M., Leslie, I.D., Gunn, R.G. and Doney, J.M. (1986) Effects of high levels of body condition and food intake on plasma follicle stimulating hormone, luteinizing hormone, prolactin and progesterone profiles around mating in Greyface ewes. *Animal Production* 43, 101-107.

Rhind, S.M., McKelvey, W.A.C., McMillen, S.R., Gunn, R.G. and Elston, D.A. (1989c) Effect of restricted food intake, before and/or after mating, on the reproductive performance of Greyface ewes. *Animal Production* 48, 149-155.

Rhind, S.M., McMillen, S. and McKelvey, W.A.C. (1991b) Effects of levels of food intake and body condition on the sensitivity of the hypothalamus and pituitary to ovarian steroid feedback in ovariectomised ewes. *Animal Production* 52, 115-125.

Rhind, S.M., McMillen, S.R., McKelvey, W.A.C., Rodriguez-Herrejon, F.F. and McNeilly, A.S. (1989a) Effect of the body condition of ewes on the secretion of LH and FSH and the pituitary response to gonadotrophin-releasing hormone. *Journal of Endocrinology* 120, 497-502.

Rhind, S.M., McMillen, S.R., Wetherill, G.Z., McKelvey, W.A.C. and Gunn, R.G. (1989d) Effects of low levels of food intake before and/or after mating on gonadotrophin and progesterone profiles in Greyface ewes. *Animal Production* 49, 267-273.

Rhind, S.M. and McNeilly, A.S. (1986) Follicle populations, ovulation rates and plasma profiles of LH, FSH and prolactin in Scottish Blackface ewes in high and low levels of body condition. *Animal Reproduction Science* 10, 105-115.

Rhind, S.M., Martin, G.B., McMillen, S., Tsonis, C.G. and McNeilly, A.S. (1989b) Effect of level of food intake of ewes on the secretion of LH and FSH and on the pituitary response to gonadotrophin-releasing hormone in ovariectomized ewes. *Journal of Endocrinology* 121, 325-330.

Rhind, S.M., Robinson, J.J., Fraser, C. and McHattie, I. (1980) Ovulation and embryo survival rates and plasma progesterone concentrations of prolific ewes treated with PMSG. *Journal of Reproduction and Fertility* 58, 139-144.

Rhind, S.M. and Schanbacher, B.D (1991) Ovarian follicle populations and ovulation rates of Finnish Landrace cross ewes in different nutritional states and associated profiles of gonadotrophins, inhibin, growth hormone (GH) and insulin-like growth factor-1. *Domestic Animal Endocrinology* 8, 281-291.

Rhind, S.M., Wetherill, G.Z. and Gunn, R.G. (1991a) Diurnal profiles of LH, prolactin and progesterone and their inter-relationships in ewes in high or moderate levels of body condition. *Animal Reproduction Science* 24, 119-126.

Riera, S. (1982) Reproductive efficiency and management in goats. *Proceedings of the 3rd International Conference on Goat Production and Disease*, Tucson, Arizona, USA pp. 162-174.

Robinson, J.J. (1986) Nutrition and embryo loss in farm animals. In: Sreenan, J.M. and Diskin, M.G. (eds.) *Embryonic Mortality in Farm Animals*. Martinus Nuijhoff, The Hague, pp. 235-248.

Sachdeva, K.K., Sengar, O.P.S., Singh, S.N. and Lindahl, I.L. (1973) Studies on goats. I. Effect of plane of nutrition on the reproductive performance of does. *Journal of agricultural Science*, Cambridge 80, 375-379.

Sands, M. and McDowell, R.E. (1978) The potential of the goat for milk production in the tropics. *Cornell International Agriculture, Mimeograph* 60, Department of Science, Ithaca, New York.

Scaramuzzi, R.J. and Radford, H.M. (1983) Factors regulating ovulation rate in the ewe. *Journal of Reproduction and Fertility* 69, 353-367.

Shelton, M. (1961) Kidding behaviour of Angora goats. *Texas Agricultural Experimental Station Progress Report* 2189, pp. 1-4.

Smith, J.F. (1984) Protein, energy and ovulation rate. In: Land, R.B. and Robinson, D.W. (eds.) *Genetics of Reproduction in Sheep*. Butterworths, London.

Smith, J.F. (1988) Influence of nutrition on ovulation rate in the ewe. *Australian Journal of Biological Sciences* 41, 27-36.

Tatman, W.R., Judkins, M.B., Dunn, T.G. and Moss, G.E. (1990). Luteinizing hormone in nutrient-restricted ovariectomized ewes. *Journal of Animal Science* 68, 1097-1102.

Teleni, E. and Rowe, J.B. (1986) Ovulation rate of ewes: role of energy and protein. *Journal of Agriculture, Western Australia* 27, 36-38.

Teleni, E., Rowe, J.B. and Croker, K.P. (1984) Ovulation rates in ewes: the role of energy-yielding substrates. In: Lindsay, D.R. and Pearce, D.T. (eds.) *Reproduction in Sheep*. Cambridge University Press, Cambridge.

Thimonier, J. and Pelletier, J. (1971) A genetic difference in the preovulatory release of LH in ewes of the Ile-de-France breed: its relation to number of ovulations. *Annales de Biologie Animale Biochemie Biophysique* 15, 329-343.

Thomas, G.B., Mercer, J.E., Karalis, T., Rao, A., Cummins, J.T. and Clarke, I.J. (1990) Effect of restricted feeding on the concentrations of growth hormone (GH) gonadotrophins, and prolactin (PRL) in plasma and on the amounts of messenger ribonucleic acid for GH, gonadotropin subunits, and PRL in the pituitary glands of adult ovariectomized ewes. *Endocrinology* 126, 1361-1367.

Tonetta, S.A. and di Zerega, G.S. (1989) Intragonadal regulation of follicular maturation. *Endocrine Reviews* 10, 205-229.

Tsafriri, A. (1988) Local nonsteroidal regulators of ovarian function. In: Knobil, E. and Neill, J. (eds.) *The Physiology of Reproduction, Vol 1*. Raven Press Ltd., New York, pp. 527-566.

Turnbull, K.E., Braden, A.W.H. and Mattner, P.E. (1977) The pattern of follicular growth and atresia in the ovine ovary. *Australian Journal of Biological Sciences* 30, 229-241.

Van der Westhuysen, J.M. (1971) Effect of dietary energy and protein levels on early embronic mortality in young and mature Merino ewes. *Agroanimalia* 3, 91-94.

Veldhuis, J.D., Nestler, J.E., Strauss, J.F. and Gwynne, J.T. (1986) Insulin regulates low density lipoprotein by swine granulosa cells. *Endocrinology* 118, 2242-2253.

Waghorn, G.C., Smith, J.F. and Ulyatt, M.J. (1990) Effect of protein and energy intake on digestion and nitrogen metabolism in wethers and on ovulation in ewes. *Animal Production* 51, 291-300.

Wentzel, D. (1987) Effects of nutrition on reproduction in the Angora goat. *Proceedings of the IV International Conference on Goats*, Vol 1, Brazilia, Brazil, pp. 571-575.

Wheeler, A.G. and Land, R.B. (1977). Seasonal variation in oestrus and ovarian activity of Finnish Landrace, Tasmanian Merino and Scottish Blackface ewes. *Animal Production* 24, 363-376.

Williams, A.H. (1984). Long-term effects of nutrition of ewe lambs on the neonatal period. In: Lindsay, D.R. and Pearce, D.T. (eds.) *Reproduction in Sheep*. Cambridge University Press, Cambridge.

Wilmut, I., Sales, D.I. and Ashworth, C.J. (1986) Maternal and embryonic factors associated with prenatal loss in mammals. *Journal of Reproduction and Fertility* 76, 851-864.

Wolde-Michael, T., Miller, H.M., Holmes, J.H.G., McGregor, B.A. and Galloway, D.B. (1989) Effect of supplementary feeding and zeranol on puberty in feral Cashmere goats. *Australian Veterinary Journal* 66, 124-126.

Wynn, P.C., Reis, P.J., Fleck, E., Ward, W., Tunks, D.A. and Munro, S.G. (1988) The influence of protein and energy supply on ovine metabolic hormone status. *Proceedings of the Nutrition Society of Australia* 13, 125.

Xu, Z., McDonald, M.F. and McCutcheon, S.N. (1989) The effects of nutritionally-induced liveweight differences on follicular development, ovulation rate, oestrous activity and plasma follicle-stimulating hormone levels in the ewe. *Animal Reproduction Science* 19, 67-78.

Yuthasastrakosol, P., Palmer, W.M. and Howland, B.E. (1975) Luteinizing hormone, oestrogen and progesterone levels in peripheral serum of anoestrous and cyclic ewes as determined by radioimmunoassay. *Journal of Reproduction and Fertility* 43, 57-65.

Yuthasastrakosol, P., Palmer, W.M. and Howland, B.E. (1977) Release of LH in anoestrous and cyclic ewes. *Journal of Reproduction and Fertility* 50, 319-321.

Chapter 3

Photoperiodic and Nutritional Influences on Maternal and Fetal Endocrine Mechanisms Regulating Fetal Development During the Second Half of Gestation

J.M. Bassett

University of Oxford, Growth and Development Unit, University Field Laboratory, Wytham, Oxford OX2 8QJ, UK

Introduction

In seasonally breeding species such as sheep and goats, pregnancy and fetal development frequently take place against a background of seasonal cycles in body weight and limited nutrient availability, which has an enormous influence on the successful outcome of the pregnancy and subsequent perinatal survival.

During the past twenty five years, great advances in our understanding of the requirements of the fetal lamb for successful prenatal development and subsequent post-natal survival have been made possible by the adoption of the chronically-cannulated fetal lamb as an experimental preparation for the investigation of a wide range of physiological systems regulating prenatal development. This experimental preparation has proved particularly suitable for the long-term study in conscious and healthy fetuses of the regulation and functional development of endocrine systems influencing the growth of lambs throughout the last month of gestation and has permitted quantitative studies of the nutritional transactions between the ewe and the developing fetuses.

To an extent, however, the success of these approaches has tended to divert attention away from the important role played by maternal adaptation in supporting the pregnancy. This review will try, therefore, to underline the importance of the maternal role by examining recent advances in our understanding of communication of seasonal information to the lamb before birth and by considering how both maternal and fetal endocrine systems may be involved in nutritional regulation of fetal lamb growth. Although the systems considered may be only distantly related, maternal transfer of information to the fetus is vital to the successful functioning of fetal systems regulating normal development in both.

Photoperiodic Regulation of Maternal and Fetal Endocrine Systems

Endocrine rhythms and their entrainment by photoperiod

While the annual cycle of reproductive activity and its endocrine regulatory mechanisms (Karsch *et al.,* 1984; Foster *et al.,* 1986) in sheep and goats is very clearly entrained by the annual cycle of changes in photoperiod, many other seasonal changes, such as those in wool or hair growth, moulting, food intake and fattening, are also strongly influenced by the seasonal changes in photoperiod. Furthermore, investigations in recent years have shown that each of these changes are associated with large alterations in the secretion of pituitary hormones. Circannual rhythms in the plasma concentrations of ACTH and cortisol (Ssewannyana *et al.,* 1990), beta-endorphin (Ebling and Lincoln, 1987), prolactin (Ravault and Ortavant, 1977), growth hormone (GH) (Barenton *et al.,* 1987), thyrotropin (TSH) and thyroxine (Wallace, 1979; Lincoln *et al.,* 1980; Webster *et al.,* 1991) and the reproductive hormones follicle stimulating hormone (FSH) and lutenizing hormone (LH) (Lincoln, 1991) have been described in sheep, while similar seasonal changes in prolactin have also been observed in goats (Buttle, 1974).

Although the circannual rhythms of reproductive activity, plasma LH and plasma prolactin persist in the absence of daylength change (Karsch *et al.,* 1989; Jackson and Jansen, 1991), the annual rhythms in pituitary hormone secretion are normally entrained to the changing photoperiod. However, despite this entrainment, the most striking feature is the extent to which the phase relationships of these rhythms differ among the various hormones (Lincoln, 1991), considering that the effects of photoperiod on these rhythms all seem to be transduced through the diurnal rhythm in pineal melatonin secretion (Karsch *et al.,* 1984; Arendt, 1986; Lincoln, 1991). The mechanisms by which daylength controls this daily rhythm in nocturnal melatonin release by the pineal have been reviewed in detail by Karsch *et al.* (1984) and Arendt (1986).

Transfer of photoperiodic information to the fetus

Observations in small mammals have shown that circadian rhythms in the fetus and the circannual rhythm in reproductive activity after birth are entrained by the daily photoperiod experienced by the mother during pregnancy (Reppert *et al.*, 1985; Stetson *et al.*, 1986). Subsequent investigations showed that photic information is transferred to the fetus through the placental transfer of melatonin (Weaver and Reppert, 1986) but some recent studies (Stetson *et al.*, 1991) suggest additional mechanisms dependent on melatonin actions in the mother may also be involved. In a larger species such as the sheep with its long period of gestation, far more detailed investigations of fetal circadian rhythms and responses to manipulation of the photoperiod to which the mother is exposed are possible through the use of chronically-cannulated and instrumented preparations. Circadian rhythms in fetal breathing and rapid low voltage electrocortical activity (Boddy *et al.*, 1973), fetal heart rate and heart rate variability (Dalton *et al.*, 1977) have been observed, as well as circadian rhythms in fetal cerebrospinal fluid (CSF) vasopressin concentration (Stark and Daniel, 1989) and in the plasma concentrations of prolactin (McMillen *et al.*, 1987; Bassett *et al.*, 1989b; Vergara *et al.*, 1989), ACTH and cortisol (Jones, 1979; McMillen *et al.*, 1987). Recent investigations (Yellon and Longo, 1987; Zemdegs *et al.*, 1988) have shown that the diurnal rhythm of melatonin seen in the ewe is faithfully reflected in the fetus, though with some attenuation of amplitude, despite the immaturity of pineal innervation and pathways for melatonin synthesis in the fetus. Subsequent investigations showed that the melatonin rhythm in the fetus is abolished by maternal pinealectomy (Yellon and Longo, 1988; McMillen and Nowak, 1989), while inversion of the photoperiod, so that ewes received light from 22.00 to 08.00 h from 100 days gestation, resulted in inversion of both maternal and fetal melatonin rhythms (Figure 3.1). Rapid equilibration of melatonin across the ovine placenta has also been clearly demonstrated (Yellon and Longo, 1988; Zemdegs *et al.*, 1988).

The evidence therefore indicates that a diurnal rhythm of melatonin, capable of synchronizing circadian and circannual clocks in the fetus with those in the mother and with the photoperiod, is imposed on the fetal lamb throughout gestation. The absence, after maternal pinealectomy, of any diurnal rhythm in fetal melatonin, even close to term, together with the very low amplitude and great irregularity of diurnal melatonin rhythms in lambs during the first two weeks after birth (Claypool *et al.*, 1989; Nowak *et al.*, 1990), reinforce the significance of the maternal rhythm in the entrainment of seasonal rhythms in the infant lamb. However, it is evident from changes in plasma prolactin in lambs subjected to altered photoperiod from birth or one week of age (Ebling *et al.*, 1988, 1989) that the central neuroendocrine mechanisms regulating prolactin release can respond to photoperiodic signals during the perinatal period despite the irregularity of diurnal melatonin rhythms.

Figure 3.1 Pattern of melatonin in the circulation of the ewe and fetus during the last month of pregnancy when subjected to a reversed photoperiod regime. Top panel: plasma melatonin (mean + S.E. of nine 48 h periods) from 6 pineal-intact ewes (open circles) and their fetuses (closed triangles). Bottom panel: mean melatonin concentrations (+ S.E. of twelve 48 h periods) in 6 pinealec-tomized ewes (open squares) and their fetuses (closed trangles). The dark bar indicates 14 h night.

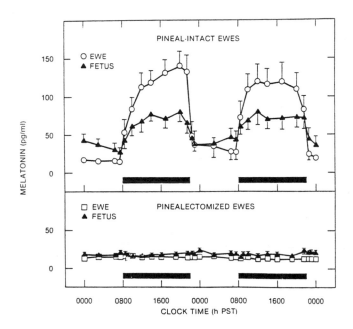

Reproduced from Yellon and Longo (1988) with permission of the author and *Biology of Reproduction*.

Entrainment of fetal endocrine rhythms by photoperiod

Recent evidence (McMillen and Walker, 1991) shows that reversal of the melatonin rhythm by inversion of the daily photoperiod results in reversal of the circadian rhythm of fetal breathing, confirming that the melatonin rhythm does entrain circadian pacemakers in the fetal lamb. In contrast, these studies failed to demonstrate similar inversion in the circadian rhythm of prolactin in the fetus, even though it did so in the mother. However, other observations show that regulation of prolactin release by the diurnal melatonin rhythm in the late-gestation ovine fetus is more complex and is already highly responsive to changes in the duration of the daily photoperiod and the nocturnal increase in plasma melatonin.

Observations in our laboratory (Bassett *et al.*, 1988) showed that prolactin concentrations of individual chronically-cannulated fetuses in ewes bred by out-of-season breeding methods varied over a very wide range during the last month of gestation and, like prolactin concentrations in the ewes, depended almost totally on the time of year when the studies were carried out, with over 80% of the variation between fetuses being associated with variation in the natural daily photoperiod to which the ewes were exposed during late pregnancy (Figure 3.2). Environmental temperature in the animal rooms was not controlled, so seasonal variation in temperature could have influenced the maternal plasma prolactin concentration. However, Ravault and Ortavant (1977) reported that abolition of the seasonal temperature rhythm had no significant effect on the circannual rhythm of plasma prolactin concentration in rams and it seems likely that there would be little direct influence of seasonal temperature variation on the fetus, although this cannot be ruled out.

Figure 3.2 Mean daily plasma prolactin concentrations during the last month of gestation in chronically-cannulated fetal lambs and their mothers plotted relative to the time when they reached 130 days gestation. (Each point is the mean of 5-20 daily observations.) The relationship between time and plasma prolactin (continuous line) was calculated by least squares analysis. Ambient photoperiod at Oxford is shown by the dotted line. Note the logarithmic scale for prolactin concentrations. Modified from Bassett *et al.* (1988) with permission of *The Quarterly Journal of Experimental Physiology.*

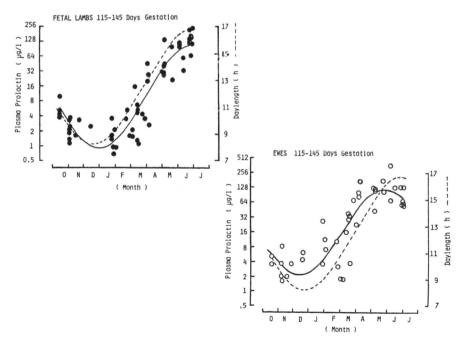

Supplemental lighting to provide a summer long-day photoperiod of 16 h light [16L:8D] for 3-6 weeks before fetal cannulation at 118 days gestation during midwinter increased both maternal and fetal plasma prolactin concentrations to mid-summer values, despite the low winter environmental temperatures (Bassett *et al.*, 1989a), while supplemental lighting from the time of cannulation resulted in increasing concentrations of prolactin throughout the period of observation (Figure 3.3). By contrast, intravenous infusion of melatonin for 16 h overnight into similar ewes with cannulated fetuses during summer, decreased plasma prolactin concentrations in both the ewe and her fetuses to low values within 2-3 days. Seron-Ferre *et al.* (1989) similarly observed that melatonin implants in the ewe decreased fetal plasma prolactin, but without significantly altering the diurnal rhythm.

Figure 3.3 Effect on mean daily plasma prolactin concentrations in (a) pregnant ewes and (b) their fetuses of changing daily photoperiod from a natural winter short day (<9 h light; open symbols; seven ewes, nine fetuses) to an artificial long day (16 h light) either 3-6 weeks before (solid symbols and solid lines, seven ewes, 12 fetuses) or after (solid symbols and broken lines; three ewes, six fetuses) vascular cannulation at 118 days of gestation. Values are means ±SE of log-transformed prolactin concentrations.

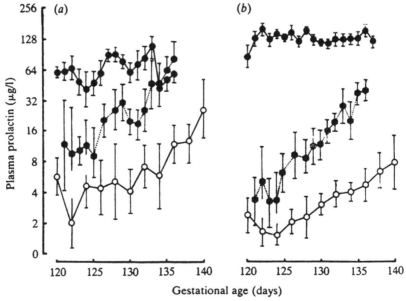

Reproduced from Bassett *et al.* (1989a) with permission of *The Journal of Endocrinology*.

These observations show clearly that while a daily melatonin rhythm may be imposed on the fetal lamb through the response of the maternal pineal to the light:dark cycle and despite the immaturity of mechanisms regulating fetal pineal function, neuroendocrine mechanisms in the fetal brain regulating circadian and circannual responses to seasonal changes in the duration of the nocturnal melatonin signal are fully functional and are the principal regulators of prolactin secretion from the fetal pituitary from 100 days of gestation and possibly earlier. Indeed, in the absence of detailed information about the pre-experimental photoperiodic history of the sheep used, it is probably the functioning of this mechanism, rather than some ontogenetic developmental programme which explains the increase in the size of secretion granules within pituitary lactotrophs from newborn lambs and near-term fetuses (Parry *et al.*, 1979), as well as the increase in plasma prolactin during late gestation in the fetal lambs studied by Mueller *et al.* (1979).

It is also the functioning of this neuroendocrine regulatory system which confounds interpretation of the observations reported by McMillen and Walker (1991), as both the fetuses and ewes in their study clearly interpreted the initial decrease from a natural summer photoperiod to a [12L:12D] regime as a short-day signal, while the subsequent inversion to [12D:12L] was interpreted as a change to a long day. Since the alterations in prolactin secretion following an increase in photoperiod take far longer to develop and clearly involve a different mechanism from the rapid switch-off in secretion consequent on imposition of short days, it is far from surprising that no clear change was observed.

Effects of seasonal changes in the daily photoperiod on the fetal plasma concentrations of other hormones influenced by photoperiod have not been reported. Plasma concentrations of ß-endorphin and α-melanocyte-stimulating hormone in the fetal lamb are high and, like prolactin secretion in the fetus, are under inhibitory dopaminergic control (Newman *et al.*, 1987) but, although seasonal photoperiodic effects on ß-endorphin in adult sheep are well established (Ssewannyana *et al.*, 1990) and are thought to be involved in regulation of seasonal changes in appetite and fat deposition (Ebling and Lincoln, 1987), photoperiodic effects on fetal plasma levels have not been determined.

In a rather different investigation, Stark and Daniel (1989) reported that exposure of ewes to continuous light virtually abolished the diurnal rhythm in the CSF concentration of vasopressin in fetal lambs. This is of particular interest since a diurnal rhythm of CSF vasopressin has also been recognized in the adult goat (Seckl and Lightman, 1987) and it is known that the regulation of CSF vasopressin concentration involves the suprachiasmatic nucleus (Schwartz and Reppert, 1985). An extensive network of sex-steroid-dependent vasopressinergic nerve fibres, probably the source of the CSF vasopressin, originates from cell bodies in the hypothalamic paraventricular and supraoptic nuclei, as well as the suprachiasmatic nucleus and varies with season in ways which seem closely related to photoperiodically-controlled events (Pevet, 1991). While the functional significance of this system during fetal life remains unknown, the influence of testosterone and oestrogens on vasopressinergic innervation suggests the possibility that it may be involved in the neural regulation of reproduction and sexual behaviour (Clarke, 1977), since these are susceptible to manipulation by inappropriate androgenization during fetal life. Recent investigations (Wood *et al.*, 1991a) show that these androgen-determined systems include that regulating the response to environmental cues timing the onset of reproduction in post-natal life.

Possible functions of prenatal photoperiodic entrainment of endocrine rhythms

Alterations in prolactin secretion by the fetal pituitary attest to the functionality of neural and neuroendocrine mechanisms responsible for entrainment of hypothalamic mechanisms controlling pituitary hormone release from 100 days of gestation (Bassett *et al.*, 1989a). But, while study of hypothalamic-hypophyseal vascular connections suggests that hypothalamic releasing factors could control pituitary function from as early as 45 days gestation (Levidiotis *et al.*, 1989), other investigations indicate that significant synthesis of prolactin by the pituitary does not begin until after 50 days gestation (Leisti *et al.*, 1982). However, the relevance for fetal development of the very large difference between winter and summer pregnancies in prolactin secretion by the pituitary is not known.

Ortavant *et al.* (1988), citing observations of Bocquier (1985), reported that maintenance of ewes under a long photoperiod from day 100 of gestation significantly increased the birth weight of their twin lambs, even though food intake was identical. This result appears consistent with the suggestion of Forbes *et al.* (1979) that high plasma prolactin concentrations might explain the higher growth rate of weaned lambs kept under long daylength. However, two farm-scale experiments at Wytham failed to find any effect of additional lighting during the last 6-8 weeks of pregnancy on the birth weight of lambs born in early January or late March, despite evidence that plasma prolactin concentrations in the lambs at birth were significantly increased (Bassett, 1992). Ebling *et al.* (1989) also failed to find any effect of long photoperiod during late pregnancy on either birth weight or subsequent post-natal growth.

However, in our experiments there was some evidence that the long photoperiod may have increased the rate of udder development in the ewe but the magnitude of the effect on udder size was less than that due to increased fetal number (Bassett, 1992) and seems unlikely to have led to increases in milk production comparable to those reported by Ortavant *et al.* (1988).

Studies in the goat (Forsyth *et al.*, 1985) showed that suppression of prolactin release throughout much of pregnancy led to some delay in mammary development in does with a single fetus but it appeared that the production of placental lactogen by the fetal placenta compensated for the loss of prolactin and ultimate mammary size and milk yields were not reduced, so it is not clear that large changes in plasma prolactin, consequent on alteration in photoperiod during pregnancy, would necessarily alter udder development and subsequent milk production. On the other hand, these investigations, like earlier studies of Robinson *et al.* (1978), illustrate the very great quantitative importance of the increased amounts of progesterone, placental lactogen and oestrogen secreted into the maternal circulation by the fetal placenta during multiple pregnancies, in determining the increased growth and later the accumulation of colostrum by the mammary gland before parturition.

Lincoln (1990) has reported that the seasonal changes in plasma prolactin are closely associated with the seasonal cycle of wool, horn and hair growth in sheep, especially in the more primitive breeds. Lynch and Russel (1990) have reported that high plasma prolactin concentrations are necessary for reactivation of hair follicles in Kashmir goats and have demonstrated that administration of prolactin can advance the onset of moulting, while blockade of its secretion with bromocriptine delays it. It is perhaps of significance that the second half of gestation is the period of prenatal development when the secondary wool follicles are initiated in the fetus but, although it is known that the growth of secondary fibres is totally dependent on the presence of the thyroid (Thorburn and Hopkins, 1973) and that the number of secondary follicles produced can be greatly influenced by the level of maternal nutrition at this time (Alexander, 1974), it is not known whether prolactin affects their development. Other studies (Barlet, 1985) showed that prolactin can alter intestinal and placental calcium transport during late pregnancy in sheep but the relevance of these findings to the changes seen with altered photoperiod, or out-of-season pregnancies, has not been determined.

The most important role played by the imposition of maternal melatonin rhythms on the fetus probably lies in the prenatal entrainment of circannual rhythms which will govern the response of the growing lamb or kid to its environment. In sheep and goats, one of the most important of these is undoubtedly the influence of photoperiod on attainment of sexual maturity. To date, most discussions of photoperiodic effects on the attainment of sexual maturity in the sheep have ignored the prenatal photoperiodic history. However, with the recognition that the fetal brain is not only entrained by photoperiod but actively responds to photoperiodic cues, it is necessary to consider whether prenatal photoperiodic experience influences reproductive maturation. Recent limited observations on goats (Deveson *et al.*, 1991) suggest that prenatal photoperiodic history is capable of influencing the timing of puberty in female kids but there would appear to be less dependence of reproductive maturity on physical size in the goat than in the sheep.

It is evident that there is a marked difference between male and female sheep in the influence of photoperiod on the change in neuroendocrine sensitivity to inhibitory steroid feedback and the onset of the increase in pulsatile LH secretion which brings about maturation of gonadal function and that the establishment of this sex difference is dependent on androgen levels between 60 and 120 days of gestation (Wood *et al.*, 1991a). However, although post-natal attainment of sexual maturity in the male is influenced little by alterations in photoperiod during post-natal life (Wood *et al.*, 1991b) and attainment of sexual maturity in the female is known to depend on the experience of a short-day photoperiod following exposure to a period of long days (Yellon and Foster, 1985), the relevance of the prenatal photoperiodic experience has not been considered.

The recognition that information about the prevailing photoperiod is transferred to the developing fetal lamb well before full term, and not only entrains circadian and circannual rhythms in the fetus but actively regulates pituitary secretory activity *in utero*, is clearly of very great importance for our understanding of factors regulating development. However it is also evident that a great deal has still to be learned about the role of these processes in determining neuroendocrine and physical development in sheep and goats.

Maternal and Fetal Endocrine Mechanisms Modulating the Influence of Nutrition on Fetal Development

The need to provide nutrients for the growth of the conceptus imposes substantial additional demands on metabolic homeostasis in the mother throughout pregnancy. But while the effects of pregnancy on the nutritional requirements of ewes have been discussed elsewhere in this volume, it should be appreciated that pregnancy in sheep and goats also results in large and progressive alterations in the distribution of nutrient use within maternal tissues to accomodate the increasing requirements of the developing conceptus. This is partly because ewes appear unable to increase their food intake during late pregnancy to meet the increasing nutrient requirements of the conceptus and indeed food intake may actually decline, especially near term, but neither the restriction of space within the abdominal cavity nor the endocrine changes associated with pregnancy provide a complete explanation for the changes observed (Forbes, 1986; Weston, 1988).

Promotion of maternal adaptation to pregnancy by fetoplacental hormones

From the earliest stages of embryonic development, the conceptus influences the course of its development through secretion of hormones and growth factors into the lumen of the uterus (Bazer *et al.*, 1989). After placental attachment, this manipulation of the maternal environment is continued through hormone secretion directly into the maternal circulation. The mechanisms by which alterations in nutrient partitioning within the ewe are manipulated by these hormones to support fetal development, include major changes in blood volume, cardiac output and in the redistribution of blood flow to provide increased flow to the uteroplacenta (Rosenfeld, 1984). These alterations in the endocrine balance of the ewe are achieved through secretion into the maternal circulation of steroids, like progesterone and oestradiol, and polypeptides, including chorionic somatomammotrophin (placental lactogen), by the fetal placenta. However, while the significance of these alterations for nutrition of the developing fetus have been discussed in earlier reviews (Baumann and Currie, 1980; Bassett, 1986, 1991) and progesterone is considered to be one of the important determinants of the metabolic alterations during human pregnancy (Kalkhoff *et al.*, 1970), the consequences for maternal metabolic regulation of the major difference between sheep and goats in placental progesterone production has not been examined. Whereas plasma progesterone concentrations in ewes increase steadily from mid-pregnancy and are proportional to fetal number (Bassett *et al.*, 1969), concentrations do not change in goats and are totally dependent on production by ovarian corpora lutea (Thorburn and Schneider, 1972).

The maintenance of maternal glucose homeostasis

Because of the central importance of glucose for fetal metabolism and the dependence of uteroplacental glucose transport on the maternal plasma glucose concentration (Hay *et al.*, 1983), one focus of attention has been the maternal endocrine and metabolic changes directed towards conserving glucose for placental transfer. This is particularly important during late pregnancy when fetal demands are highest and when there is the added load imposed by accelerating mammary development. The consequences for fetal growth of failure to maintain normoglycaemia in the ewe have been vividly illustrated by the dramatic decline, or even cessation, in the rate of linear growth of fetal lambs, as measured by changes in their crown-rump length measured *in utero* (Mellor and Matheson, 1979; Mellor and Murray, 1981, 1982a and b). In human pregnancy the principal consequence of the endocrine adaptations is hyperglycaemia, insulin antagonism and the need for maintained hypersecretion of insulin to overcome the antagonism

(Gillmer and Persson, 1979). In contrast, the extent to which antagonism of insulin action on peripheral tissue glucose utilization could lead to hyperglycaemia in sheep or goats must be far more restricted, because of their inability to increase food intake to meet the increased nutrient demands of the pregnant uterus and developing mammary gland (Weston, 1988). The nature of ruminant digestion and the consequence that glucose is derived largely via gluconeogenic pathways must also be significant.

There is strong evidence for a steady reduction in the capacity for lipid synthesis and in the ability of insulin to stimulate lipogenesis during the last third of pregnancy in sheep which is accompanied by an increase in the maximal rate of lipolysis stimulated by ß-adrenergic agonists (Vernon *et al.*, 1981; Guesnet *et al.*, 1991). Despite these observations, it is not evident that changes in glucose availability lead to significant alterations in the utilization of glucose in hind limb muscle of pregnant ewes relative to that in non-pregnant animals (Oddy *et al.*, 1985), nor in the distribution of glucose utilization between maternal tissues and the fetus (Hay *et al.*, 1983; Leury *et al.*, 1990). Alterations in the plasma growth hormone:insulin concentration ratio have been implicated in the development of the changes in lipid metabolism (Vernon *et al.*, 1981; Mellor *et al.*, 1987). However, the increase in plasma growth hormone often appears to follow, rather than precede, the changes in lipid mobilization during late pregnancy (see below). It has also been proposed that placental lactogen may explain many of the metabolic adaptations (Bauman and Currie, 1980; Thordarson *et al.*, 1987). However, although concentrations of placental lactogen are higher in multiple pregnancies and increase during undernutrition in goats and sheep (Hayden *et al.*, 1980; Brinsmead *et al.*, 1981), it is also evident that hypoglycaemia is also more severe during undernutrition of animals with multiple pregnancies. Intravenous infusion of placental lactogen into non-pregnant sheep for 36 h resulted in increased concentrations and turnover of FFA during the first 8 h of infusion (Thordarson *et al.*, 1987) but alterations in glucose metabolism were not consistent with the proposed role of placental lactogen as an insulin antagonist. Further, when the action of placental lactogen in pregnant sheep was blocked by infusion of a potent antiserum against it, there were no changes consistent with any role of placental lactogen as an antagonist of maternal glucose metabolism (Waters *et al.*, 1985), raising serious doubts about its postulated role as a modulator of homeorhesis during late pregnancy.

Wastney *et al.* (1982) reported evidence for insulin resistance in some twin-pregnant ewes on the basis of responses to intravenous glucose administration but somewhat similar studies in our laboratory (Bassett, 1989) did not provide evidence for significant impairment in in the ability of late-pregnant ewes to remove glucose loads, nor in the ability of injected insulin to decrease their plasma glucose concentration acutely. There was some evidence for a reduction in the ability of insulin to decrease plasma free fatty acids (FFA) in pregnant ewes but more recent investigations using far smaller amounts of insulin have not confirmed this (J. M. Bassett, unpublished observations). However, such approaches which depend on substantially augmenting plasma insulin concentrations may be inappropriate for sheep, as plasma insulin concentrations are usually low and the principal problem for the ewe appears to be the inability to produce sufficient glucose and to conserve it in order to maintain normoglycaemia throughout the day in the face of the continuing needs of fetal tissues, rather than the ability to remove glucose in maternal tissues. Yet, while the counter-regulatory endocrine mechanisms leading to recovery from hypoglycaemia have been a subject of intensive clinical research (Cryer, 1981; Gerich, 1988), the contribution of such mechanisms to the maintenance of normoglycaemia in late-pregnant sheep has received little consideration.

Glucose counter-regulation in the ewe during late pregnancy

Hypoglycaemia has long been recognized as the inevitable correlate of maternal undernutrition in the sheep and great stress has been laid on the need to maintain normoglycaemia in the ewe during late pregnancy if fetal growth is not to be compromised. However, direct studies of the endocrine counter-regulatory mechanisms have not been reported. Sequential changes in plasma metabolite and hormone concentrations in underfed ewes during late pregnancy (Blom *et al.*, 1976; Hove and Blom, 1976; Mellor *et al.*, 1987) showed that increased mobilization of FFA occurred well prior to hypoglycaemia, indicating a shift in the balance of substrate use by maternal tissues and which, as in non-pregnant sheep (Bassett, 1974), is associated with decreased plasma insulin and increased plasma growth hormone concentrations. Measurements in our laboratory of changes in plasma metabolite and hormone concentrations during late pregnancy and lactation in ewes fed sufficient to prevent hypoglycaemia in ewes with a single fetus (Figure 3.4) have confirmed the steadily worsening hypoglycaemia in ewes with two or more fetuses reported by others, but they also emphasize the lack of any clear dependence of the steadily increasing FFA mobilization which occurred in all ewes without significant changes in plasma growth hormone. The increase in plasma GH really only occurred close to parturition and there was no significant relationship between plasma growth hormone concentrations during the last six weeks of pregnancy and the total weight of lamb born to each ewe ($r = 0.12$, $n = 28$). Mean plasma pancreatic glucagon concentrations in the ewes were not significantly related to the number of fetuses or the weight of lamb born but values were highest at the start of observations and decreased steadily till parturition. Mean plasma glucose ($r = -0.69$, $p < 0.001$) and plasma insulin ($r = -0.56$, $p < 0.01$) were negatively correlated with total lamb birth weight, while mean plasma FFA was positively correlated ($r = 0.82$, $p < 0.001$) with the weight of lamb born. These relationships further illustrate the effect of fetal size and number in

Figure 3.4 Mean pre-feeding plasma FFA, glucose, insulin and growth hormone concentrations in pregnant ewes with a single fetus (o), twins (●), or three or more fetuses (□) during the last 6 weeks of gestation. Ewes were shorn two weeks before the first observation and were fed 650 g of a concentrate mixture (16% CP) daily and hay *ad libitum*. Further details of the ewes and their care have been reported elsewhere (Bassett, 1992; J. M. Bassett, unpublished observations).

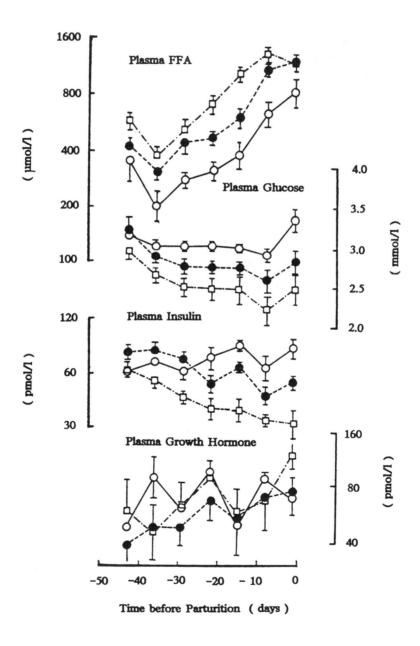

determining the magnitude of the metabolic and endocrine alterations during late pregnancy. In addition, they emphasize the importance of increased lipid mobilization in maintaining fetal growth and in supporting normo-glycaemia, through provision of an alternative oxidizable substrate, even in ewes with a single fetus. It is also significant that, although glucose flux rates in ewes suckling twins may be substantially higher during lactation than before delivery (Bergman and Hogue, 1967), hypoglycaemia was not observed in any of the ewes during lactation even though most suckled twins. Plasma insulin concentration also continued to decrease after delivery, despite the restoration of normoglycaemia and suggests that there may be fundamental differences between pregnancy and lactation in the way glucose homeostasis is regulated.

Decreasing plasma glucose concentrations may be directly responsible for the declining insulin concentrations during late pregnancy. However, because normal plasma glucose concentrations are close to the threshold for stimulation of insulin release, it seems far more likely that central neural mechanisms, like those mediating responses to more acute hypoglycaemic stress (Havel and Taborsky, 1989) may be involved. Increased autonomic nervous activity and the release of catecholamines from the adrenal medulla are undoubtedly the most important short-term mechanisms bringing about the recovery of normoglycaemia, with part of the effect being mediated through stimulation of glucagon secretion (Cryer, 1981; Gerich, 1988), but changes in plasma adrenaline or noradrenaline concentrations in pregnant sheep have apparently not been reported. Changes in plasma cortisol following insulin-induced hypoglycaemia in pregnant ewes (Bassett, 1989) indicate that there is a normal catechol response to hypoglycaemia but plasma growth hormone concentrations, unlike those in man, are not greatly increased by acute hypoglycaemia in pregnant sheep. When adrenaline is infused into fasted ewes during late pregnancy (Figure 3.5) it rapidly increases plasma glucose and maintains high concentrations for many hours as in non-pregnant and lactating ewes (Bassett, 1971; Leenanuruksa and McDowell, 1985), even though its acute strongly inhibitory influence on insulin release is at least partially overcome during prolonged adminis-tration. On the other hand noradrenaline infusion at the same rate has little or no effect on plasma glucose and insulin concentrations even though, like adrenaline, it increased plasma FFA concentrations greatly throughout the infusion (data not shown). Interestingly, however, the addition of cortisol to the noradrenaline infusion resulted in a slow but continuous increase in plasma glucose and insulin concentrations during the latter part of the study (Figure 3.5). While each of the infusions clearly inhibited any marked changes in growth hormone secretion, as evidenced by the rapid increase in plasma growth hormone after cessation of the infusions, it is apparent that increased sympathetic activity and adrenomedullary adrenaline secretion could contribute to the restoration of normoglycaemia in late pregnant sheep in a way consistent with their actions in other species (Cryer, 1981).

Figure 3.5 Mean plasma glucose, growth hormone and insulin in late pregnant ewes during prolonged intravenous infusion with adrenaline at 0.05 μmol/min (■-■ , n=4), noradrenaline at 0.05 μmol/min (●-● , n=4), or noradrenaline (0.05 μmol/min) and cortisol at 0.28 μmol/min (O-O , n=4). The duration of the infusion is indicated by the broken vertical lines. Mean S.E. values are shown by the vertical bars. (J. M. Bassett, unpublished observations.)

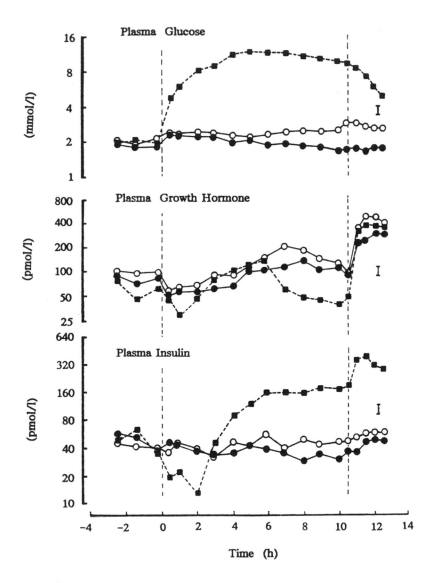

The ready increase in plasma glucose in both well-fed and under-nourished pregnant ewes during exercise (Chandler *et al.*, 1985; Leury *et al.*, 1990), as well as the increased secretion of glucagon and cortisol and decreased insulin and growth hormone during exercise (Bell *et al.*, 1983), together with the similar responses of shorn ewes in late pregnancy to a cold stress (Thompson *et al.*, 1982), confirm that increased sympathetic activity can bring about changes similar to those observed on catecholamine administration. Increases in sympathetic activity during exercise also result in a reduction in uterine blood flow. However, despite this, the increases in maternal plasma glucose concentration resulted in significant augmentation of uterine and umbilical glucose uptake in underfed ewes (Chandler *et al.*, 1985; Leury *et al.*, 1990). In this context, it is significant that winter shearing of pregnant ewes 6-8 weeks before lambing significantly increases lamb birth weight (Thompson *et al.*, 1982). The metabolic and endocrine changes in shorn animals (Symonds *et al.*, 1988a and b) are consistent with increased sympathetic activity and include increases in plasma glucose and FFA concentrations and increased rates of glucose and fat turnover. In more recent investigations (Symonds and Lomax, 1992), it has been shown that the increased growth in fetuses of shorn ewes is also associated with increased accumulation of brown adipose tissue and improved perinatal thermogenesis.

The role of parasympathetic mechanisms in glucose counterregulation during late pregnancy

Investigations in man and other species indicate that the parasympathetic arm of the autonomic nervous system also plays an important role in the central co-ordination of responses to mild hypoglycaemia (Havel and Taborsky, 1989) and although its effects are largely mediated through the vagus via actions at post-synaptic cholinergic or peptidergic nerve terminals within target tissues, changes in its level of activity can be monitored conveniently through changes in the secretion of pancreatic polypeptide (PP), a hormone secreted by the F-cells of the pancreatic islets and totally dependent on the integrity of the vagal innervation of the pancreas. Studies in our laboratory (J. M. Bassett, unpublished observations) indicate that while PP concentrations in adult sheep are often higher than those reported for man and the pig, concentrations throughout the last 6 weeks of pregnancy and during lactation are far higher than those reported for other species (see review by Taylor, 1989). Also, there is no evidence for any significant increase above these high levels following insulin injection into late-pregnant ewes (Figure 3.6), even though a response similar to that reporte in other species is observed following insulin injections into non-

J.M. Bassett

Figure 3.6 Mean plasma concentrations of pancreatic polypeptide and glucose in four non-pregnant dry ewes (broken lines) and in 10 twin-pregnant ewes, two to three weeks before parturition (continuous lines), administered either 1.25 Units (open symbols) or 2.5 Units (closed symbols) of insulin by intravenous injection at the time indicated by the arrow. Glucose and pancreatic polypeptide concentrations were converted to logarithms before calculation of means and are plotted logarithmically. Vertical bars show the average S.E. (J. M. Bassett, unpublished observations).

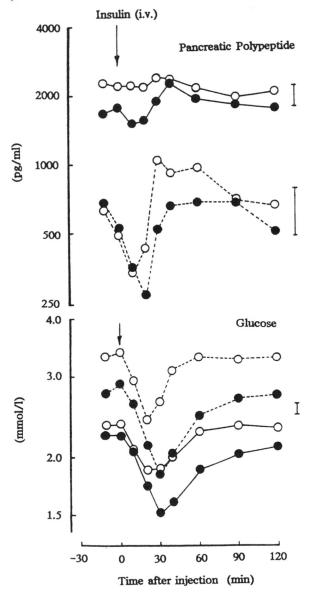

pregnant, dry sheep whose basal plasma PP concentrations are markedly lower. The role of PP in the sheep remains unknown, but atropine has been shown to abolish its secretion (Shulkes and Hardy, 1982; J. M. Bassett, unpublished observations) and one possibility must be that the high concentrations are indicative of maintained central glucopenia. High plasma PP concentrations have been related to impaired glucose handling in man (Hallgren and Lundquist, 1983) and there have been suggestions that PP may have some, as yet undefined, role in regulation of glucose tolerance (Taylor, 1989), but the full significance of these observations remains uncertain. However, they do suggest that the role of the autonomic nervous system in regulation of glucose homeostasis during late pregnancy in sheep should be examined more closely.

The role of the placenta in determining fetal nutrition

The great importance of placental size in determining the growth rate and perinatal survival of fetal lambs has been detailed in a number of reviews (Mellor, 1983; Bassett, 1986, 1991; Hay, 1991), as has the essentiality of glucose for the support of fetal metabolism (Battaglia and Meschia, 1978) and it is not proposed to discuss these topics further. However, it is important to note that while the placenta is the major interface between the developing fetus and the mother, the fetal part of the cotyledonary placenta is a fetal organ perfused by blood from the fetal circulation and, although in intimate contact with the maternal tissues to facilitate exchange of nutrients and waste products with the maternal circulation, it appears to be nourished by glucose from the fetal circulation rather than by glucose in transit from the mother (Bassett *et al.*, 1985; Bassett, 1986; Hay, 1991; Hodgson *et al.*, 1991). This glucose utilization accounts for virtually all of the glucose consumed by the uteroplacental tissues (Hay, 1991). Like glucose transfer across the placenta from mother to fetus, the uptake of glucose from the fetal circulation by placental tissue is regulated by concentration and the number of transporter sites in the cell membrane and is not subject to control by insulin (DiGiacomo and Hay, 1989; Hay, 1991). Unlike glucose, amino acids are accumulated in the fetal blood by active transport across the placenta (reviewed by Hay, 1991) but there is insufficient information about their metabolism within the fetus to indicate whether they play any important part in the limitation or regulation of the response of fetal growth to altered maternal nutrition. Placental metabolism is therefore effectively controlled by glucose metabolism within the fetus itself and it is this interdependence of fetal somatic and placental development which determines the close relationships between fetal and placental size, despite their very different patterns of development throughout fetal life (Bassett, 1986, 1991). However, it must be appreciated that the primary determinant of placental size under most circumstances probably will have been the ability of the developing trophoblast in promoting adequate maternal vascularization of the cotyledonary attachments early in pregnancy (Bassett, 1991).

The role of insulin in modulating the effect of alterations in glucose availability on fetal metabolism and growth

As noted earlier, the principal determinant of the rate of glucose transfer to the conceptus is the maternal plasma glucose concentration, under most conditions, and it is only in late pregnancy when very small placental size, whether consequent on experimental reduction in the number of uterine caruncles before mating (Harding *et al.*, 1985; Owens *et al.*, 1987a and b) or resulting from prolonged heat stress during pregnancy (Bell *et al.*, 1987; Bell *et al.*, 1989), may seriously limit the supply of glucose for the support of fetal oxidative metabolism and growth. However, because the rate of placental glucose transfer is dependent on the maintenance of a concentration gradient between the maternal and fetal sides of the placenta, the rate of glucose utilization within the fetus has a very important influence on glucose transfer. As a consequence, the concentration of insulin within the fetal circulation plays a vital role in regulating fetal tissue glucose utilization and thereby the plasma concentration and rate of transfer of glucose from the maternal circulation. Indeed studies on fetal lambs deprived of insulin by either pancreatectomy or streptozotocin administration have shown that fetal glucose utilization is only 40-60% of that in intact fetuses (Fowden and Hay, 1988; Hay and Meznarich, 1988). More detailed investigations of the relationships between fetal glucose utilization, and the fetal plasma concentrations of glucose and insulin show that the majority of fetal glucose utilization is dependent on the plasma glucose concentration but that this is augmented by increasing concentrations of insulin (Hay *et al.*, 1988). However, while these investigations have provided an important quantitative insight into the role of insulin as a regulator of fetal glucose metabolism and thereby in bringing about the changes in maternal metabolism discussed earlier, they provide little information about the sites of action within the fetus, or the quantitative changes in fetal tissue growth which ultimately depend on the alterations in glucose utilization resulting from insulin action within the fetus.

Given the very low concentration range of glucose in the plasma of fetal lambs (0.4-1.5 mmol/l) perhaps the most surprising aspect of the role of insulin as a modulator of fetal glucose metabolism is the fact that it occurs at all. In virtually all post-natal mammals, including lambs, insulin secretion is only stimulated by glucose concentrations of 3-5 mmol/l or greater. In the fetal lamb, it is evident from numerous studies that increases in plasma glucose concentrations within the physiologic range, whether associated with variations in nutrition (Bassett and Madill, 1974a; Phillips *et al.*, 1978; Schreiner *et al.*, 1980), manipulation of placental size (Robinson *et al.*, 1980) or infusions of exogenous glucose (Bassett and Thorburn, 1971; Bassett and Madill, 1974b; Phillips *et al.*, 1978) are accompanied by increased insulin release and increased utilization of glucose. Incubation or perifusion *in vitro* of pancreas pieces from fetal lambs confirmed the exquisite sensitivity of insulin release from fetal ß-cells to stimulation by glucose and showed that this sensitivity was lost at birth (Bassett *et al.*, 1973). Subsequent observations (Bassett, 1977) showed that fetal pancreatic ß-cells were capable of responding to glucose in this low concentration range from as early as 50 days of gestation, so it is evident that insulin secretion from the fetal pancreas may be involved in modulating fetal metabolism throughout the last two thirds of prenatal development. Recent ultrastructural investigations of the development of the islets of Langerhans in the sheep pancreas (Titlbach *et al.*, 1985) show that the islets from the fetal lamb pancreas, differed from those of post-natal sheep, both in organizational structure and in the nature of the ß-cells present. While, at birth, large islet bodies consisting mainly of large argyrophil cells containing insulin represented more than half of the islet tissue, these structures regressed rapidly within the first 10 days after birth and had virtually disappeared by 2 months of age, being replaced by far smaller more typical islets. These observations suggest the sensitivity to glucose is associated with the fetal generation of pancreatic islet cells and that the change in glucose threshold is associated with the appearance of a new generation of ß-cells with different functional characteristics. However, this possibility remains to be investigated.

As in post-natal life, a major role of insulin is in redirecting substrate utilization among the tissues of the fetal body. This is most dramatically illustrated by the change in the distribution of glucose use between the fetal body and the uteroplacenta after pancreatectomy or streptozotocin administration (Fowden and Hay, 1988; Hay and Meznarich, 1988) but is also shown by the altered distribution of cardiac output during administration of insulin to fetal lambs, with an increased portion being distributed to the tissues of the fetal carcass and being compensated for by a reduced proportion distributed to the placenta (Milley, 1987). More recent studies (Milley and Papacostas, 1989) have shown that increased blood flow to the hind limbs in fetal sheep is associated with increased oxygen and glucose uptake and the authors suggest that increased oxygen use by non-visceral tissues could explain all the increased oxygen use observed in hyper-insulinaemic fetuses. However, the most impressive evidence of the effects of insulin comes from the comparison of growth and development in control and pancreatectomized fetal lambs (Fowden *et al.*, 1986; Fowden *et al.*, 1989). Using the technique for measuring crown-rump length growth *in utero* described by Mellor and Matheson, (1979), they have shown (Figure 3.7) that pancreatectomy causes a very substantial decrease in the rate of linear growth as well as a reduction in weight gain (Fowden *et al.*, 1989)

Figure 3.7 The mean (± S.E.) increment in fetal crown rump length (CRL) with respect to time from the day of the second operation (↑) at which the fetuses were pancreatectomized (●, n = 6), or sham operated (○, n = 5) (*A*), or pancreatectomized and given insulin treatment at either a low (◐, n = 4) or high dose (◑, n = 4) (*B*). The shaded area shows the CRL increment of control and pancreatectomized fetuses from panel *A*.

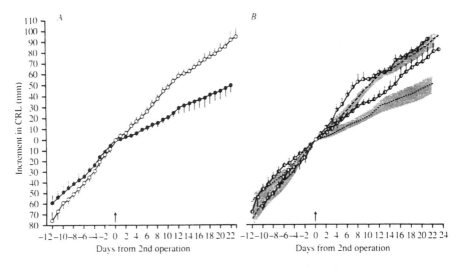

Reproduced from Fowden *et al.* (1989) with permission of the authors and *The Quarterly Journal of Experimental Physiology.*

and that this reduction in growth can be completely prevented by insulin infusions which restore plasma insulin concentrations to the normal range.

These experiments therefore provide compelling evidence for a major role of fetal insulin in modulating the effects of nutrient availability on the rate of fetal growth, brought about by the relationship of insulin secretion to the fetal glucose concentration. Since fetal plasma glucose concentrations are increased after pancreatectomy, it is evident that neither glucose availability alone nor glucose with any of the other hormones can bring about this stimulation of growth. Fowden (1989) has reviewed evidence for the mechanisms by which insulin may regulate fetal growth and it is clear that this may involve direct effects on tissue anabolism as well as effects mediated through the production of growth factors such as insulin-like growth factor-I (IGF-I). Determination of plasma IGF-I concentrations in the pancreatectomized fetuses showed that IGF-I production in the fetus, like growth, was dependent on the plasma concentration of insulin (Gluckman *et al.*, 1987) and is therefore involved in mediation of the growth promoting activity of insulin. Owens (1991) has recently stressed the important role of the IGFs in the regulation of growth in the fetal lamb.

However, it is clear from the studies on the pancreatectomized fetal lamb that the pathway linking fetal nutritional state to fetal growth is dependent primarily on the responsiveness of insulin secretion from the fetal pancreatic beta cell to changes in the fetal plasma glucose concentration. This must be so, whatever the role of growth hormone (Gluckman, 1986; Browne and Thorburn, 1989) or placental lactogen (Chene *et al.*, 1988) as a fetal growth promoter. Indeed, increases in the fetal plasma GH concentration during fasting (Bassett and Madill, 1974a) and evidence that growth hormone functions as an insulin antagonist (Parkes and Bassett, 1985) and an important regulator of fat mobilization in the fetus (Stevens and Alexander, 1986) suggest that a primary role of growth hormone in the fetus, as in post-natal life, may be to act as a glucose counterregulatory hormone. The maintenance of normoglycaemia in the pregnant ewe throughout gestation is therefore vital for the maintenance of growth in the fetal lamb, irrespective of limitations imposed by small placental size, or alterations in uterine blood flow.

Conclusions

While two apparently unrelated aspects of maternal and fetal endocrine function have been considered, both illustrate the dependence of fetal endocrine and developmental responses on information provided by the mother about her environment. Transfer of photoperiodic information entrains fetal neuroendocrine mechanisms regulating endocrine rhythms responsible for the post-natal adaptation of the lamb to the seasonal changes in its environment, while the transfer of glucose provides signal information for regulation of fetal growth as well as substrate. Both are vital for successful development and for adaptation to post-natal life. It is apparent, however, that there are many aspects of both regulatory systems which require further investigation if their function and the limits to successful fetal development are to be understood.

Acknowledgements

Experimental work in the author's laboratory described has been supported variously by CSIRO and by project grants from the MRC, AFRC, British Diabetic Association and the University of Oxford Medical Research Fund. Their support and the assistance of those who participated in the experiments is gratefully acknowledged.

References

Alexander, G. (1974) Birth weight of lambs: influences and consequences. *Ciba Foundation Symposium* 27, 215-239.

Arendt, J. (1986) Role of the pineal gland and melatonin in seasonal reproductive function in mammals. *Oxford Reviews of Reproductive Biology* 8, 266-320.

Barenton, B., Chabanet, C. and Pelletier, J. (1987) Influence of photoperiod and protein diet on growth hormone secretion in rams. *Proceedings of the Society for Experimental Biology and Medicine* 185, 312-317.

Barlet, J.P. (1985) Prolactin and calcium metabolism in pregnant ewes. *Journal of Endocrinology* 107, 171-175.

Bassett, J.M. (1971) Metabolic effects of catecholamines in sheep. *Australian Journal of Biological Sciences* 23, 903-914.

Bassett, J.M. (1974) Diurnal patterns of plasma insulin, growth hormone, corticosteroid and metabolite concentrations in fed and fasted sheep. *Australian Journal of Biological Sciences* 27, 167-181.

Bassett, J.M. (1977) Glucagon, insulin and glucose homeostasis in the fetal lamb. *Annales De Recherches Veterinaires* 8, 362-373.

Bassett, J.M. (1986) Nutrition of the conceptus. *Proceedings of the Nutrition Society* 45, 1-10.

Bassett, J.M. (1989) Metabolic and endocrine responses of pregnant and lactating ewes to intravenous glucose or insulin. *Journal of Agricultural Science (Cambridge)* 113, 173-182.

Bassett, J.M. (1991) Current perspectives on placental development and its integration with fetal growth. *Proceedings of the Nutrition Society* 50, 311-319.

Bassett, J.M. (1992) Effects of additional lighting to provide a summer photoperiod for winter-housed pregnant ewes on plasma prolactin, udder development and lamb birthweight. *Journal of Agricultural Science (Cambridge)* (In press).

Bassett, J.M., Bomford, J. and Mott, J.C. (1988) Photoperiod: An important regulator of plasma prolactin concentration in fetal lambs during late gestation. *Quarterly Journal of Experimental Physiology* 73, 241-244.

Bassett, J.M., Burks, A.H. and Pinches, R.A. (1985) Glucose metabolism in the ovine conceptus. In: Jones C.T. (ed.) *The Physiological Development of the Fetus and Newborn* Academic Press, London. pp 71-75.

Bassett, J.M., Curtis, N., Hanson, C. and Weeding, C.M. (1989a) Effects of altered photoperiod or maternal melatonin administration on plasma prolactin concentrations in fetal lambs. *Journal of Endocrinology* 122, 633-643.

Bassett, J.M. and Madill, D. (1974a) The influence of maternal nutrition on plasma hormone and metabolite concentrations of foetal lambs. *Journal of Endocrinology* 61, 465-477.

Bassett, J.M. and Madill, D. (1974b) Influence of prolonged glucose infusions on plasma insulin and growth hormone concentrations of foetal lambs. *Journal of Endocrinology* 62, 299-309.

Bassett, J.M., Madill, D., Nicol, D.H. and Thorburn, G.D. (1973) Further studies on the regulation of insulin release in foetal and post-natal lambs: a role of glucose as a physiological regulator of insulin release *in utero*. In: Comline, K.S., Cross, K.W., Dawes, G.S. and Nathanielsz, P.W. (eds.) *Foetal and neonatal physiology, Barcroft Centenary Symposium*. Cambridge University Press, Cambridge, pp. 351-359.

Bassett, J.M., Oxborrow, T.J., Smith, I.D. and Thorburn, G.D. (1969) The concentration of progesterone in the peripheral plasma of the pregnant ewe. *Journal of Endocrinology* 45, 449-457.

Bassett, J.M. and Thorburn, G.D. (1971) The regulation of insulin secretion by the ovine foetus *in utero. Journal of Endocrinology* 50, 59-74.

Bassett, N.S., Bennett, L., Ball, K.T. and Gluckman, P.D. (1989b) Presence of a diurnal rhythm in fetal prolactin secretion and influence of maternal nutrition. *Biology of the Neonate* 55, 164-170.

Battaglia, F.C. and Meschia, G. (1978) Principal substrates of fetal metabolism. *Physiological Reviews* 58, 499-527.

Baumann, D.E. and Currie, W.B. (1980) Partitioning of nutrients during pregnancy and lactation: A review of mechanisms involving homeostasis and homeorhesis. *Journal of Dairy Science* 63, 1514-1529.

Bazer, F.W., Vallet, J.L., Harney, J.P., Gross, T.S. and Thatcher, W.W. (1989) Comparative aspects of maternal recognition of pregnancy between sheep and pigs. *Journal of Reproduction and Fertility* 37, Suppl., 85-89.

Bell, A.W., Bassett, J.M., Chandler, K.D. and Boston, R.C. (1983) Fetal and maternal endocrine responses to exercise in the pregnant ewe. *Journal of Developmental Physiology* 5, 129-141.

Bell, A.W., McBride, B.W., Slepetis, R., Early, R.J. and Currie, W.B. (1989) Chronic heat stress and prenatal development in sheep: 1. Conceptus growth and maternal plasma hormones and metabolites. *Journal of Animal Science* 67, 3289-3299.

Bell, A.W., Wilkening, R.B. and Meschia, G. (1987) Some aspects of placental function in chronically heat-stressed ewes. *Journal of Developmental Physiology* 9, 17-29.

Bergman, E.N. and Hogue, D.E. (1967) Glucose turnover and oxidation rates in lactating sheep. *American Journal of Physiology* 213, 1378-1384.

Blom, A.K., Hove, K. and Nedkvitne, J.J. (1976) Plasma insulin and growth hormone concentrations in pregnant sheep II: Post-absorptive levels in mid- and late pregnancy. *Acta Endocrinologica* 82, 553-560.

Bocquier, F. (1985) Influence de la photoperiode et de la temperature sur certains equilibres hormonaux et sur les performances zootechniques de la brebis en gestation et lactation. *Thése Doctoral Ingresse Institut National Agricole Paris-Grignon (INAPG), 78850 Thiveral-Grignon, France.*

Boddy, K., Dawes, G.S. and Robinson, J.S. (1973) A 24 hour rhythm in the foetus. In: Comline, K.S., Cross, K.W., Dawes, G.S. and Nathanielsz, P.W. (eds.), *Foetal and Neonatal Physiology, Proceedings of the Sir Joseph Barcroft Centenary Symposium,* Cambridge University Press, Cambridge, pp. 63-66.

Brinsmead, M.W., Bancroft, B.J., Thorburn, G.D. and Waters, M.J. (1981) Fetal and maternal ovine placental lactogen during hyperglycaemia, hypoglycaemia and fasting. *Journal of Endocrinology* 90, 337-343.

Browne, C.A. and Thorburn, G.D. (1989) Endocrine control of fetal growth. *Biology of the Neonate* 55, 331-346.

Buttle, H.L. (1974) Seasonal variations of prolactin in plasma of male goats. *Journal of Reproduction and Fertility* 37, 95-99.

Chandler, K.D., Leury, B.J., Bird, A.R. and Bell, A.W. (1985) Effects of undernutrition and exercise during late pregnancy on uterine, fetal and uteroplacental metabolism in the ewe. *British Journal of Nutrition* 53, 625-635.

Chene, N., Martal, J. and Charrier, J. (1988) Ovine chorionic somatomammotropin and foetal growth. *Reproduction, Nutrition, Development* 28, 1707-1730.

Clarke, I.J. (1977) The sexual behaviour of prenatally androgenized ewes observed in the field. *Journal of Reproduction and Fertility* 49, 311-315.

Claypool, L.L., Wood, R.I., Yellon, S.M. and Foster, D.L. (1989) The ontogeny of melatonin secretion in the lamb. *Endocrinology* 124, 2135-2143.

Cryer, P.E. (1981) Glucose counterregulation in man. *Diabetes* 30, 261-264.

Dalton, K.J., Dawes, G.S. and Patrick, J.E. (1977) Diurnal, respiratory, and other rhythms of fetal heart rate in lambs. *American Journal of Obstetrics and Gynecology* 127, 414-424.

Deveson, S., Arendt, J. Forsyth, I.A. (1991) Prepartum light treatment delays puberty in goats born in autumn. *Advances in Pineal Research* 5, 233-235

DiGiacomo, J.E. and Hay, W.W. Jr. (1989) Regulation of placental glucose transfer and consumption by fetal glucose production. *Pediatric Research* 25, 429-434.

Ebling, F.J.P., Claypool, L.E. and Foster, D.L. (1988) Neuroendocrine responsiveness to light during the neonatal period in the sheep. *Journal of Endocrinology* 119, 211-218.

Ebling, F.J.P. and Lincoln, G.A. (1987) B-Endorphin secretion in rams related to season and photoperiod. *Endocrinology* 120, 809-818.

Ebling, F.J.P., Wood, R.I., Suttie, J.M., Adel, T.E. and Foster, D.L. (1989) Prenatal photoperiod influences neonatal prolactin secretion in the sheep. *Endocrinology* 125, 384-391.

Forbes, J.M. (1986) The effects of sex hormones, pregnancy, and lactation on digestion, metabolism, and voluntary food intake. In: Milligan, L.P., Grovum, W.L. and Dobson, A. (eds.), *Control of Digestion and Metabolism in Ruminants*, Reston, Englewood Cliffs, USA. pp. 420-435.

Forbes, J.M., Driver, P.M., Brown, W.B. and Scanes, C.G. (1979) The effect of daylength on the growth of lambs. 2. Blood concentrations of growth hormone, prolactin, insulin and thyroxine, and the effect of feeding. *Animal Production* 29, 43-51.

Forsyth, I.A., Byatt, J.C. and Iley, S. (1985) Hormone concentrations, mammary development and milk yield in goats given long-term bromocriptine treatment in pregnancy. *Journal of Endocrinology* 104, 77-85.

Foster, D.L., Karsch, F.J., Olster, D.H., Ryan, K.D. and Yellon, S.M. (1986) Determinants of puberty in a seasonal breeder. *Recent Progress in Hormone Research* 42, 331-378.

Fowden, A.L. (1989) The role of insulin in prenatal growth. *Journal of Developmental Physiology* 12, 173-182.

Fowden, A.L. and Hay, W.W. (1988) The effects of pancreatectomy on the rates of glucose utilization, oxidation and production in the sheep fetus. *Quarterly Journal of Experimental Physiology* 73, 973-984.

Fowden, A.L., Hughes, P. and Comline, R.S. (1989) The effects of insulin on the growth rate of the sheep fetus during late gestation. *Quarterly Journal of Experimental Physiology* 74, 703-714.

Fowden, A.L., Mao, X.Z. and Comline, R.S. (1986) Effects of pancreatectomy on the growth and metabolite concentrations of the sheep fetus. *Journal of Endocrinology* 110, 225-231.

Gerich, J.J. (1988) Glucose counterregulation and its impact on diabetes mellitus. *Diabetes* 37, 1608-1617.

Gillmer, M.D.G., and Persson, B. (1979) Metabolism during normal and diabetic pregnancy and its effect on neonatal outcome. *Ciba Foundation Symposium* 63, 93-126.

Gluckman, P.D. (1986) The role of pituitary hormones, growth factors and insulin in the regulation of fetal growth. *Oxford Reviews of Reproductive Biology* 8, 1-60.

Gluckman, P.D., Butler, J.H., Comline, R.S. and Fowden, A. (1987) The effects of pancreatectomy on the plasma concentrations of insulin-like growth factors 1 and 2 in the sheep fetus. *Journal of Developmental Physiology* 9, 79-88.

Guesnet, M., Massoud, M.J. and Demarne, Y. (1991) Regulation of adipose tissue metabolism during pregnancy and lactation in the ewe: The role of insulin. *Journal of Animal Science* 69, 2057-2065.

Hallgren, R. and Lundquist, G. (1983) Elevated levels of pancreatic polypeptide are related to impaired glucose handling in inflammatory states. *Scandinavian Journal of Gastroenterology* 18, 561-564.

Harding, J.E., Jones, C.T. and Robinson, J.S. (1985) Studies on experimental growth retardation in sheep. The effects of a small placenta in restricting transport to and growth of the fetus. *Journal of Developmental Physiology* 7, 427-442.

Havel, P.J. and Taborsky, G.J. (1989) The contribution of the autonomic nervous system to changes of glucagon and insulin secretion during hypoglycemic stress. *Endocrine Reviews* 10, 332-350.

Hay, W.W. Jr. (1991) Energy and substrate requirements of the placenta and fetus. *Proceedings of the Nutrition Society* 50, 321-336.

Hay, W.W. and Meznarich, H.K. (1988) Use of fetal streptozotocin injection to determine the role of normal levels of fetal insulin in regulating uteroplacental and umbilical glucose exchange. *Pediatric Research* 24, 312-317.

Hay, W.W. and Meznarich, H.K. (1989) Effect of maternal glucose concentration on uteroplacental glucose consumption and transfer in pregnant sheep (42830). *Proceedings of the Society for Experimental Biology and Medicine* 190, 63-69.

Hay, W.W., Meznarich, H.K., DiGiacomo, J.E., Hirst, K. and Zerbe, G.(1988) Effects of insulin and glucose concentrations on glucose utilization in fetal sheep. *Pediatric Research* 23, 381-388.

Hay, W.W., Sparks, J.W., Wilkening, R.B., Battaglia, F.C. and Meschia, G. (1983) Partition of maternal glucose production between conceptus and maternal tissues in sheep. *American Journal of Physiology* E347- E350.

Hayden, T.J., Thomas, C.R., Smith, S.V. and Forsyth, I.A. (1980) Placental lactogen in the goat in relation to stage of gestation, number of fetuses, metabolites, progesterone and time of day. *Journal of Endocrinology* 86, 279-290.

Hodgson, J.C., Mellor, D.J. and Field, A.C. (1991) Kinetics of lactate and glucose metabolism in the pregnant ewe and conceptus. *Experimental Physiology* 76, 389-398.

Hove, K. and Blom, A.K. (1976) Plasma insulin and growth hormone concentrations in pregnant sheep. *Acta Endocrinologica* 82, 544-552.

Jackson, G.L. and Jansen, H.T. (1991) Persistence of a circannual rhythm of plasma prolactin concentration in ewes exposed to a constant equatorial photoperiod. *Biology of Reproduction* 44, 469-475.

Jones, C.T. (1979) Normal fluctuations in the concentration of corticosteroid and adrenocorticotrophin in the plasma of foetal and pregnant sheep. *Hormone and Metabolic Research* 11, 237-241.

Kalkhoff, R.K., Jacobson, M. and Lemper, D. (1970) Progesterone, pregnancy and the augmented plasma insulin response. *Journal of Clinical Endocrinology and Metabolism* 31, 24-28.

Karsch, F.J., Bittman, E.L., Foster, D.L., Goodman, R.L. Legan, S.J. and Robinson, J.E. (1984) Neuroendocrine basis of seasonal reproduction. *Recent Progress in Hormone Research* 40, 185-233.

Karsch, F.J., Robinson, J.E., Woodfill, C.J.I. and Brown, M.B. (1989) Circannual cycles of luteinizing hormone and prolactin secretion in ewes during prolonged exposure to a fixed photoperiod: Evidence for an endogenous reproductive rhythm. *Biology of Reproduction* 41, 1034-1046.

Leenanuruksa, D. and McDowell, G.H. (1985) Effects of prolonged intravenous infusions of adrenaline on glucose utilization, plasma metabolites, hormones and milk production in lactating sheep. *Australian Journal of Biological Sciences* 38, 197-208.

Leisti, S., Miller, W.L. and Johnson, L.K. (1982) Synthesis of growth hormone, prolactin and proopiomelanocortin by ovine fetal anterior and neurointermediate lobes. *Endocrinology* 111, 1368-1375.

Leury, B.J., Bird, A.R., Chandler, K.D. and Bell, A. W. (1990) Glucose partitioning in the pregnant ewe: effects of undernutrition and exercise. *British Journal of Nutrition* 64, 449-462.

Levidiotis, M.L., Wintour, E.M., McKinley, M.J. and Oldfield, B.J. (1989) Hypothalamic-hypophyseal vascular connections in the fetal sheep. *Neuroendocrinology* 49, 47-50.

Lincoln, G.A. (1990) Correlation with changes in horns and pelage, but not reproduction, of seasonal cycles in the secretion of prolactin in rams of wild, feral and domesticated breeds of sheep. *Journal of Reproduction and Fertility* 90, 285-296.

Lincoln, G.A. (1991) Photoperiod, pineal and seasonality in large mammals. *Advances in Pineal Research* 5, 211-218.

Lincoln, G.A., Klandorf, H. and Anderson, N. (1980) Photoperiodic control of thyroid function and wool and horn growth in rams and the effects of cranial sympathectomy. *Endocrinology* 107, 1543-1548.

Lynch, P. and Russel A.J.F. (1990) The hormonal manipulation of cashmere growth and shedding. *Animal Production* 50, 561 (Abst.).

McMillen, I.C. and Nowak, R. (1989) Maternal pinealectomy abolishes the diurnal rhythm in plasma melatonin concentrations in the fetal sheep and pregnant ewe during late gestation. *Journal of Endocrinology* 120, 459-464.

McMillen, I.C., Thorburn, G.D. and Walker, D.W. (1987) Diurnal variations in plasma concentrations of cortisol, prolactin, growth hormone and glucose in the fetal sheep and pregnant ewe during late gestation. *Journal of Endocrinology* 114, 65-72.

McMillen, I.C. and Walker, D.W. (1991) Effects of different lighting regimes on daily hormonal and behavioural rhythms in the pregnant ewe and sheep fetus. *Journal of Physiology* 442, 465-476.

Mellor, D.J. (1983) Nutritional and placental determinants of foetal growth rate in sheep and consequences for the newborn lamb. *British Veterinary Journal* 139, 307-324.

Mellor, D.J., Flint, D.J., Vernon, R.G. and Forsyth, I.A. (1987) Relationships between plasma hormone concentrations, udder development and the production of early mammary secretions in twin-bearing ewes on different planes of nutrition. *Quarterly Journal of Experimental Physiology* 72, 345-356.

Mellor, D.J. and Matheson, I.C. (1979) Daily changes in the curved crown-rump length of individual sheep fetuses during the last 60 days of pregnancy and effects of different levels of maternal nutrition. *Quarterly Journal of Experimental Physiology* 64, 119-131.

Mellor, D.J. and Murray, L. (1981) Effects of placental weight and maternal nutrition on the growth rates of individual fetuses in single and twin bearing ewes during late pregnancy. *Research in Veterinary Science* 30, 198-204.

Mellor, D.J. and Murray, L. (1982a) Effects of long term undernutrition of the ewe on the growth rates of individual fetuses during late pregnancy. *Research in Veterinary Science* 32, 177-180.

Mellor, D.J. and Murray, L. (1982b) Effects on the rate of increase in fetal girth of refeeding ewes after short periods of severe undernutrition during late pregnancy. *Research in Veterinary Science* 32, 377-382.

Milley, J.R. (1987) Effect of insulin on the distribution of cardiac output in the fetal lamb. *Pediatric Research* 22, 168-172.

Milley, J.R. and Papacostas, J.S. (1989) Effect of insulin on metabolism of fetal sheep hindquarters. *Diabetes* 38, 597-603.

Mueller, P.L., Gluckman, P.D., Kaplan, S.L., Rudolph, A.M. and Grumbach, M.M. (1979) Hormone ontogeny in the ovine fetus. V. Circulating prolactin in mid- and late gestation and in the newborn. *Endocrinology* 105, 129-134.

Newman, C.B., Wardlaw, S.L., Stark, R.I., Daniel, S.S. and Frantz, A.G. (1987) Dopaminergic regulation of a-melanocyte-stimulating hormone and n-acetyl-B-endorphin secretion in the fetal lamb. *Endocrinology* 120, 962-966.

Nowak, R., Young, I.R. and McMillen, I.C. (1990) Emergence of the diurnal rhythm in plasma melatonin concentrations in newborn lambs delivered to intact or pinealectomized ewes. *Journal of Endocrinology* 125, 97-102.

Oddy, V.H., Gooden, J.M., Hough, G.M., Teleni, E. and Annison, E.F. (1985) Partitioning of nutrients in merino ewes. II Glucose utilization by skeletal muscle, the pregnant uterus and the lactating mammary gland in relation to whole body glucose utilization. *Australian Journal of Biological Sciences* 38, 95-108.

Ortavant, R., Bocquier, F., Pelletier, J., Ravault, J.P., Thimonier, J. and Volland-Nail, P. (1988) Seasonality of reproduction in sheep and its control by photoperiod. *Australian Journal of Biological Sciences* 41, 69- 85.

Owens, J.A. (1991) Endocrine and substrate control of fetal growth: Placental and maternal influences and insulin-like growth factors. *Reproduction, Fertility and Development* 3, 501-517.

Owens, J.A., Falconer, J. and Robinson, J.S. (1987a) Effect of restriction of placental growth on oxygen delivery to and consumption by the pregnant uterus and fetus. *Journal of Developmental Physiology* 9, 137- 150.

Owens, J.A., Falconer, J. and Robinson, J.S. (1987b) Effect of restriction of placental growth on fetal and utero-placental metabolism. *Journal of Developmental Physiology* 9, 225-238.

Parkes, M.J. and Bassett, J.M. (1985) Antagonism by growth hormone of insulin action in fetal sheep. *Journal of Endocrinology* 105, 379-382.

Parry, D.M., McMillen, I.C, Robinson, J.S. and Thorburn, G.D. (1979) Immunocytochemical localisation of prolactin and growth hormone in the perinatal sheep pituitary. *Cell and Tissue Research* 197, 501-514.

Pevet, P. (1991) Importance of sex steroids and neuropeptides in the pineal control of seasonal rhythms. *Advances in Pineal Research* 5, 219-224

Phillips, A.F., Carson, B.S., Meschia, G. and Battaglia, F.C. (1978) Insulin secretion in fetal and newborn sheep. *American Journal of Physiology* 235, E467-E474.

Ravault, J.P. and Ortavant, R. (1977) Light control of prolactin secretion in sheep. Evidence for a photoinducible phase during a diurnal rhythm. *Annales de Biologie Animale, Biochimie et Biophysique* 17, 459-473.

Reppert, S.M., Duncan, M.J. and Goldman, B.D. (1985) Photic influences on the developing mammal. *Ciba Foundation Symposium* 117, 116-128.

Robinson, J.J., McDonald, I., McHattie, I. and Pennie, K. (1978) Studies on reproduction in prolific ewes 4. Sequential changes in the maternal body during pregnancy. *Journal of agricultural Science, Cambridge* 91, 291-304.

Robinson, J.S., Hart, I.C., Kingston, E.J., Jones, C.T. and Thorburn, G.D. (1980) Studies on the growth of the fetal sheep. The effects of reduction of placental size on hormone concentration in fetal plasma. *Journal of Developmental Physiology* 2, 239-248.

Rosenfeld, C.R. (1984) Consideration of the uteroplacental circulation in intrauterine growth. *Seminars in Perinatology* 8, 42-51.

Schreiner, R.L., Nolen, P.A., Bonderman, P.W., Moorehead, H.C., Gresham, E.L., Lemons, J.A. and Escobedo, M.B. (1980) Fetal and maternal hormonal response to starvation in the ewe. *Pediatric Research* 14, 103-108.

Schwartz, W.J. and Reppert, S.M. (1985) Neural regulation of the circadian vasopressin rhythm in cerebrospinal fluid: a pre-eminent role for the suprachiasmatic nuclei. *Journal of Neuroscience* 5, 2771-2775.

Seckl, J.R. and Lightman, S.L. (1987) Diurnal rhythm of vasopressin but not of oxytocin in the cerebrospinal fluid of the goat: lack of association with plasma cortisol rhythm. *Journal of Endocrinology* 114, 477-483.

Seron-Ferre, M., Vergara, M., Parraguez, V.H., Riquelme, R. and Llanos, A.J. (1989) Fetal prolactin levels respond to a maternal melatonin implant. *Endocrinology* 125, 400-403.

Shulkes, A. and Hardy, K.J. (1982) Ontogeny of circulating gastrin and pancreatic polypeptide in the foetal sheep. *Acta Endocrinologica* 100, 565-572.

Ssewannyana, E., Lincoln, G.A., Linton, E.A. and Lowry, P.J. (1990) Regulation of the seasonal cycle of B-endorphin and ACTH secretion into the peripheral blood of rams. *Journal of Endocrinology* 124, 443-454.

Stark, R.I. and Daniel, S.S. (1989) Circadian rhythm of vasopressin levels in cerebrospinal fluid of the fetus: Effect of continuous light. *Endocrinology* 124, 3095-3101.

Stetson, M.H., Elliott, J.A. and Goldman, B.D. (1986) Maternal transfer of photoperiodic information influences the photoperiodic response of prepubertal djungarian hamsters. *Biology of Reproduction* 34, 664-669.

Stetson, M.H., Horton, T.H. and Ray, S.L. (1991) Participation of the maternal pineal in communication of photoperiodic information to the fetus: site of action of melatonin. *Advances in Pineal Research* 5, 229-231.

Stevens, D. and Alexander, G. (1986) Lipid deposition after hypophysectomy and growth hormone treatment in the sheep fetus. *Journal of Developmental Physiology* 8, 139-145.

Symonds, M.E., Bryant, M.J. and Lomax, M.A. (1988a) Metabolic adaptation during pregnancy in winter-shorn sheep. *Journal of agricultural Science (Cambridge)* 111, 137-145.

Symonds, M.E., Bryant, M.J., Shepherd, D.A.L. and Lomax, M.A. (1988b) Glucose metabolism in shorn and unshorn pregnant sheep. *British Journal of Nutrition* 60, 249-263.

Symonds, M.E. and Lomax, M.A. (1992) Maternal and environmental influences on thermoregulation in the neonate. *Proceedings of the Nutrition Society* (In press)

Taylor, I.L. (1989) Pancreatic polypeptide family: pancreatic polypeptide, neuropeptide Y, and peptide YY. In: Makhlouf, G.M.(ed.) *Handbook of Physiology. The gastrointestinal system. Neural and endocrine biology.* The American Physiological Society, Washington DC: Section 6, vol II, chapter 21, pp. 474-543.

Thompson, G.E., Bassett, J.M., Samson, D.E. and Slee, J. (1982) The effects of cold exposure of pregnant sheep on foetal plasma nutrients, hormones and birth weight. *British Journal of Nutrition* 48, 59-63.

Thorburn, G.D. and Schneider, W. (1972) The progesterone concentration in the plasma of the goat during the oestrus cycle and pregnancy. *Journal of Endocrinology* 52, 23-30.

Thorburn, G.D. and Hopkins, P.S. (1973) Thyroid function in the foetal lamb. In: Comline, K.S., Cross, K.W., Dawes, G.S. and Nathanielsz, P.W. (eds), *Foetal and Neonatal Physiology, Proceedings of the Sir Joseph Barcroft Centenary Symposium,* Cambridge University Press, Cambridge, pp. 488-507.

Thordarson, G., McDowell, G.H., Smith, S.V., Iley, S. and Forsyth, I.A. (1987) Effects of continuous intravenous infusion of an ovine placental extract enriched in placental lactogen on plasma hormones, metabolites and metabolite biokinetics in non-pregnant sheep. *Journal of Endocrinology* 113, 277-283.

Titlbach, M., Falt, K. and Falkmer, S. (1985) Postnatal maturation of the islets of langerhans in sheep. Light microscopic, immunohistochemical, morpho-metric, and ultrastructural investigations with particular reference to the transient appearance of argyrophil insulin immunoreactive cells. *Diabetes*

Research 2, 5-15.

Vergara, M., Parraquez, V.H., Riquelme, R., Figueroa, J.P., Llanos, A.J. and Seron-Ferre, M. (1989) Ontogeny of the circadian variation of plasma prolactin in sheep. *Journal of Developmental Physiology* 11, 89-95.

Vernon, R.G., Clegg, R.A. and Flint, D.J. (1981) Metabolism of sheep adipose tissue during pregnancy and lactation. *Biochemical Journal* 200, 307-314.

Wallace A.L.C. (1979) Variations in plasma thyroxine concentrations throughout one year in penned sheep on a uniform feed intake. *Australian Journal of Biological Sciences* 32, 371-374.

Wastney, M.E., Arcus, A.C., Bickerstaffe, R. and Wolff, J.E. (1982) Glucose tolerance in ewes and susceptibility to pregnancy toxaemia. *Australian Journal of Biological Sciences* 35, 381-392.

Waters, M.J., Oddy, V.H., McCloghry, C.E., Gluckman, P.D., Duplock, R., Owens, P.C. and Brinsmead, M.W. (1985) An examination of the proposed roles of placental lactogen in the ewe by means of antibody neutralization. *Journal of Endocrinology* 106, 377-386.

Weaver, D.R. and Reppert, S.M. (1986) Maternal melatonin communicates daylength to the fetus in Djungarian Hamsters. *Endocrinology* 119, 2861-2863.

Webster, J.R., Moenter, S.M., Woodfill, C.J.I. and Karsch, F.J. (1991) Role of the thyroid gland in seasonal reproduction. II. Thyroxine allows a season-specific suppression of gonadotropin secretion in sheep. *Endocrinology* 129, 176-183.

Weston, R.H. (1988) Factors limiting the intake of feed by sheep. XIII voluntary roughage consumption in late pregnancy and early lactation in relation to protein nutrition. *Australian Journal of Agricultural Research* 39, 679-689.

Wood, R.I., Ebling, F.J., I'Anson, H., Bucholtz, D.C., Yellon, S.M. and Foster, D.L. (1991a) Prenatal androgens time neuroendocrine sexual maturation. *Endocrinology* 128, 2457-2468.

Wood, R.I., Ebling, F.J., I'Anson, H. and Foster, D.L. (1991b) The timing of neuroendocrine sexual maturity in the lamb by photoperiod. *Biology of Reproduction* 45, 82-88.

Yellon, S.M. and Foster, D.L. (1985) Alternate photoperiods time puberty in the female lamb. *Endocrinology* 116, 2090-2097.

Yellon, S.M. and Longo, L.D. (1987) Melatonin rhythms in fetal and maternal circulation during pregnancy in sheep. *American Journal of Physiology* 252, E799-E802.

Yellon, S.M. and Longo, L.D. (1988) Effect of maternal pinealectomy and reverse photoperiod on the circadian melatonin rhythm in the sheep and fetus during the last trimester of pregnancy. *Biology of Reproduction* 39, 1093-1099.

Zemdegs, I.Z., McMillen, I.C., Walker, D.W., Thorburn, G.D. and Nowak, R. (1988) Diurnal rhythms in plasma melatonin concentrations in the fetal sheep and pregnant ewe during late gestation. *Endocrinology* 123, 284-289.

Chapter 4

Nutrition of the Ewe during Gestation and Lactation

P.V. Rattray

Ruakura Agricultural Centre, Hamilton, New Zealand

Introduction

The potential lamb meat output from a sheep production system based on breeding ewes can be influenced by nutrition during various critical periods of the reproductive cycle or physiological state of the ewe (Rattray, 1987). In a system where the ewe is mated annually the following phases exist:

dry (non-pregnant, non-lactating)	10-16 weeks;
breeding season (pre-mating)	3-6 weeks;
gestation	21 weeks; and
lactation	12-16 weeks.

The potential number of offspring can be influenced during the dry and pre-mating periods when ewe fertility and fecundity can be markedly manipulated by long or short term nutritional effects. Nutrition during pregnancy can also influence the number of offspring carried through to term and the subsequent survival and growth of the neonate. It is during lactation that nutrition of the ewe has the greatest effect on the growth rate and hence final meat production of its offspring. This chapter is devoted to these latter two periods and will highlight recent findings or advances in our knowledge. It is not a purpose of the chapter to tabulate detailed nutrient requirements as these are well known and documented elsewhere.

Early Pregnancy

The timing of the major events in the developing embryo are as follows (Fennessy and Owens, 1985):

P.V. Rattray

Day 0 Ovulation (12 h post-oestrus)
1 Fertilization (up to 24 h post-ovulation)
3 8 cell stage (reaches uterus)
4 32 cell stage

Day 10-14 Migration of embryos
12-14 Growth of chorionic sac
15 Implantation
16 Cotyledon development begins
19-20 Fusion of allantoic and chorionic membranes
28-30 Tissues of cotyledons and caruncles interlock.

Pre-implantation phase

The main consideration during this phase is in reducing the chances of or incidence of losses of fertilized ova. This period is from fertilization until day 15. During this period the embryo migrates to the appropriate uterine horn and implantation is initiated between the cells of the trophoblast and those of the uterine epithelium.

Implantation

In comparison with increased embryonic loss in the pre-implantation phase, when the main symptom is a higher incidence of ewes returning to the ram after the normal interval, losses at this stage result in greater returns to the ram after a longer interval (more than 19 days) or reduced birth weight of survivors relative to litter size (Robinson, 1983; Dingwall *et al.*, 1987; Fennessy and Owens, 1985). This is due to embryonic deaths in the third or fourth week of pregnancy and occurs mainly in ewes carrying multiple embryos and results in reduced placental size of surviving embryos.

Maintenance feeding of the ewe is recommended right through the pre-implantation and implantation periods (Robinson, 1983, 1990a).

Embryonic loss

The feed requirements of the embryo at this early stage are very small and the emphasis is on avoiding extremes in nutrition and ewe body condition as nutritional effects during this time have been associated mainly with embryonic loss.

Some developmental abnormalities have been noted in the absence of embryonic loss, but the carry-over effects on birth weight or the neonate have not been confirmed (Robinson, 1983). Some losses of imperfect embryos during these periods are normal, especially at high levels of multiple ovulation (Cummins *et al.*, 1984), but losses have also been attributed to the effects of very severe under- or over-nutrition (level of feeding and proportion of concentrate in the diet) and to ewes that are either in exceptionally poor body or high condition (Rattray, 1977; Robinson, 1983). Detrimental effects of high feeding levels have also been shown (Gunn *et al.*, 1986) in ewes in good body condition fed high levels in mid-pregnancy (50 days post mating).

Role of progesterone

The key to this early embryo survival appears to be the circulating levels of maternal progesterone. In a number of studies, it has been shown that peripheral progesterone concentration declines with increased level of feeding (Parr *et al.*, 1987; Robinson, 1983, 1990a). Exogenous progesterone administration has improved the embryo survival in animals at risk (McMillan, 1987; Parr *et al.*, 1987; Robinson, 1990a). The role of progesterone appears to involve specific trophoblastic proteins (interferons) and a number of endometrial proteins (Robinson, 1990a and b).

The reduction in progesterone levels under high nutrition is probably due to higher clearance rates of the hormone (Robinson, 1990a). Increased hepatic blood flow and liver size may be implicated, but the possibility of complex interactions and feedback from other hormones or growth factors cannot be discounted.

Specific nutrients, feeds and feed factors

Supplements of trace elements (Cu, Mn, Zn, Fe, Co and Se) have improved lambing rates in deficiency situations (Rattray, 1977; Robinson, 1983), but only for Se is there strong evidence that embryo survival during implantation is affected (Robinson, 1990a). Excess phosphorus can have a similar deleterious effect on embryo mortality. High embryonic losses have been reported when newly mated ewes have been grazing particular legumes such as some varieties of red clover or subterranean clover which have relatively high levels of the isoflavone, formononetin, which is converted by rumen micro-organisms into a phyto-oestrogen. Some crucifer forage crops, such as kale or rape have had similar effects due to the presence of goitrogens, anaemia or copper deficiency (Robinson, 1983). Poisonous plants (e.g. *Veratrum californicum*) can also lead to embryonic death in early pregnancy (Keeler, 1990).

Development of the conceptus

The relative proportions of the conceptus change as gestation progresses (Rattray, 1977). Growth of the fetus(es) is exponential with approximately 85% of the ultimate birth weight accumulating between day 90 of gestation and parturition (Robinson, 1983). On the other hand the placental development is virtually completed by day 100, but amasses only less than 10% of the weight of the fetus(es).

Feed requirements

In quantitative terms the amount of energy and protein stored in the conceptus is minute in the first month, relatively low in the second and third months, and by comparison is massive in the last 2 months. Feed requirements during pregnancy reflect this trend.

Mid Pregnancy

Over this period level of nutrition appears to be relatively unimportant and the grazing ewe has an amazing ability to buffer even moderately severe undernutrition during this period, especially at grazing. Large differences (10-12 kg) have been generated in 50 kg grazing ewes up to day 100 of gestation without any significant carry over effects on lamb birth weight, ewe or lamb survival, ewe milk production and growth rate of the neonatal lamb, provided level of feeding is adequate in late pregnancy (Parr *et al.*, 1986; Rattray *et al.*, 1987). Even substantial weight loss (6 kg or 10%) of grazing ewes during this part of gestation has only affected wool production. However pregnancy *per se* does not lead to obligatory loss of wool production (Williams and Butt, 1989). There have been similar findings with pen-fed sheep (Brink, 1990).

The key to compensation is better feeding in late pregnancy because low feeding levels or maintenance right to term has resulted in reduced birth weights, litter size, survival of fetuses and ewes and reduced ewe fleece weights (Kelly and Ralph, 1988; Smeaton *et al.*, 1985). This latter study showed that very low levels of feeding in mid-pregnancy lowered the proportion of ewes lambing (78 *v.* 85%).

During the middle stages of gestation when the placenta is developing, undernutrition can lead to reduced cotyledon weights and fetal weights by day 90 to 100 of pregnancy, with the placenta being the more sensitive of the two (McCrabb *et al.*, 1986; Robinson, 1983). It seems that in very extreme cases this may not be compensated for by plentiful feeding later and reduced birth weights and lamb viability may result, especially in young ewes, poor conditioned ewes, and ewes carrying multiples (Geenty and Rattray, 1987; Owens, 1985; Robinson, 1983; Vincent *et al.*, 1985).

The placenta is the key organ controlling supply of nutrients to the growing fetus and the weight of placenta and placentome number per fetus has sometimes been significantly related to fetal weight (Fennessy and Owens, 1985; Owens *et al.*, 1986; Owens and Hinch, 1984).

Extremely low levels of feeding from day 30 to 90 have led to fetal losses in twin bearing but not single bearing ewes (Kelly *et al.*, 1989). In small highly fecund ewes live weight loss of 7 kg during days 30 to 100 of gestation has resulted in an increase in the incidence of everted or degenerating cotyledons and cotyledon weight (Owens and Hinch, 1984). While lamb weight was not affected this may have been associated with the higher incidence of embryonic loss over this period compared with low fecundity ewes (18% *v.* 0%). Concurrent reductions in ewe tissue reserves and mammary development may impair the ewe's subsequent rearing ability. Other studies confirm that birth weights may compensate but subsequent ewe milk production and lamb growth may still suffer (Davis *et al.*, 1980). Placental size may also influence subsequent milk production and rearing ability via placental lactogens and is probably responsible for the difference in mammary gland sizes such as occurs between single and twin bearing ewes (Davis *et al.*, 1980; Rattray and Trigg, 1979). Placental mass can decline in late pregnancy and this decline may be greater in ewes that were previously better fed in early-mid pregnancy (McCrabb *et al.*, 1986).

Generally maintenance feeding of ewes is recommended during early-mid pregnancy (Rattray *et al.*, 1987). However some account needs to be taken of ewe condition, age and fecundity and the general feed supply.

Late Pregnancy

This is by far the most important period of pregnancy with fetal mass increasing in a curvilinear manner such that in the last 8, 4 and 2 weeks of gestation the fetus is gaining equivalent to 85, 50 and 25% of its final birth weight (Robinson, 1983). Tables of accretion rates of various nutrients are accessible in the literature or various feeding tables (Robinson, 1983, 1990a), and they invariably follow the pattern of fetal growth.

This is the period when pregnancy nutrition can have the greatest effect on total potential lamb production. There have been a host of different findings in experimentation during this stage of pregnancy. Under controlled feeding levels during late pregnancy birth weight of the lamb can be readily affected especially in ewes that are in poor condition or carrying multiple fetuses (Fennessy and Owens, 1985; Rattray and Trigg, 1979; Rattray *et al.*, 1980; Robinson, 1983; Smeaton *et al.*, 1985). For ewes in good condition and at grazing it is not so clear cut (Fennessy and Owens, 1985; Rattray *et al.*, 1987). In the latter review of pasture allowance studies where grazing 52-55 kg ewes ranged in true live weight change from -8 kg to +8 kg over the last 6 weeks of pregnancy, lamb birth weight was occasionally but not consistently affected by 0.4-0.5 kg. The grazing ewe seems much more resilient and resistant to such effects than her pen-fed counterpart, probably through opportunities for compensatory intake responses. Quite large maternal weight losses can be sustained at this time if nutrition is "steady state". Sudden deprivation however must be avoided.

Shearing of housed ewes in mid to late pregnancy has resulted in increases in birth weight and has been accompanied by variable responses in ewe intake and live weight change (Glanville and Phillips, 1986; Black and Chestnutt, 1990). Multiple and earlier shearings have had greatest effects, and responses of over 1 kg in birth weight have been obtained with twin lambs. Concomitant with these birth weight responses have been changes in glucose metabolism, repartitioning of nutrients, fat mobilization and fatty acid oxidation, respiration rate, ewe rectal temperatures and gestation length (Black and Chestnutt, 1990; Fennessy and Owens, 1985; Symonds *et al.*, 1986, 1988).

In controlled feeding conditions, long-term moderate under feeding may result in a gradual reduction of prenatal growth as gestation approaches, while a severe or sudden restriction can result in a sharp drop or cessation of fetal growth rate (Robinson, 1983). This effect can be temporary, or permanent depending on the period of restriction. Ewes in good body condition appear able to compensate for lowered nutrition to some degree by mobilizing body reserves (Rattray and Trigg, 1979; Rattray *et al.*, 1980).

Neonatal survival

Late pregnancy feeding can influence both perinatal ewe and lamb survival. There are innumerable reports in the literature on the influence of birth weight on lamb survival (Fennessy and Owens, 1985). Very light lambs particularly multiples have a high mortality rate, while at the other end of the scale with very large single lambs there is increased birth trauma and CNS damage, dystocia, prolapsed vagina and increased mortality of both ewes and lambs (Alexandra, 1986).

Maternal nutrition as well as influencing fetal growth also affects fetal energy reserves and colostrum production and composition, thus playing a key role in the survival of the neonate. This topic is well reviewed by Robinson (1990a) and Alexandra (1986), and along with studies by Moore *et al.* (1986) give a good insight into the importance of good maternal feeding in late pregnancy. Lambs with higher birth weights are more resistant to hypothermia and heat stress because they have a lower surface area/volume ratio, higher summit metabolism per unit surface area, higher energy reserves of body lipid and glycogen and also more of the critically important thermogenic brown adipose tissue, than smaller lambs from undernourished ewes. Moore *et al*. (1986) found that lambs from better fed grazing ewes had higher vigour and rectal temperatures than lambs born from lesser fed ewes. This was generally, but not always in the case of rectal temperature, associated with higher birth weights. Small weak lambs have less suckling drive than larger ones. In addition maternal behaviour (desertions) or exhaustion can lead to lamb starvation and increased mortality especially in ewes that lamb multiples and are poorly fed (Alexandra, 1986; Putu *et al*., 1988). Detrimental influences of maternal malnutrition may also affect the birth coat of the lamb and hence its insulation. Cellular hyperplasia is still active in most tissues of the lamb after birth so good post-natal feeding can overcome most prenatal disadvantages (Robinson, 1990a). An exception can be the initiation of secondary wool follicles (Hutchinson and Mellor, 1983), where a reduction in numbers of secondary follicles between day 115 and 135 of gestation can be permanent with repercussions in terms of future wool production.

Undernutrition, maternal metabolism, and pregnancy toxaemia

Because of the high demands of the utero-placenta for glucose (Hough *et al*., 1985; Robinson, 1983), pregnant ewes, especially those carrying multiple fetuses, are metabolically quite fragile (Baird *et al*., 1983). Pregnancy toxaemia ("sleepy sickness" or "twin lamb disease") can be a consequence of energy deficiency in very late pregnancy. The pregnant uterus appears to preferentially take up glucose when compared to some other tissues (Oddy *et al*., 1985). Reduced uterine blood flow and uteroplacental glucose uptake occurs after undernutrition and this is accompanied by fetal hypoglycaemia and hypoxaemia (Chandler *et al*., 1985; Leury *et al*., 1990a; Robinson, 1990a). Fetal hormone responses to partly modulate maternal undernutrition include decreased insulin secretion and increased secretion of GH, ACTH and corticosteroids. Fetal placental gluconeogenesis may compensate in such circumstances (Hay *et al*., 1984; Leury *et al*., 1990b). In experimentally induced pregnancy toxaemia ewes with dead fetuses had higher glucose turnover and hepatic production than in ewes with live fetuses, suggesting the fetus may impair hepatic gluconeogenesis (Wastney *et al*., 1983).

Maternal adipose reserves are mobilized as an energy source in late pregnancy, resulting in increased circulating free fatty acids, ß-hydroxybuturate and ketones (Guada *et al*., 1982). Bickhardt *et al*. (1989) concluded that "the control of glucostasis fails frequently during late pregnancy in ewes and that ketosis of sheep is related more closely to excessive lipid mobilization than to disturbance of glucostasis". Lindsay and Oddy's (1985) review on the origins of the ketonaemia and hypoglycaemia that occur tends to concur with such a conclusion.

Increased litter size has been associated with acetonaemia in prolific ewes - 0, 9, 43 and 73% with twins, triplets, quads and quins respectively (Everts and Kuiper, 1983). Fetal plasma contained significantly smaller amounts and different types of lipoproteins than maternal plasma (Noble and Shand, 1983).

There are a number of blood parameters whose levels may indicate degree of undernutrition, number of fetuses carried or even stage of gestation with varying degrees of success: plasma glucose, plasma free fatty acids, and plasma ß-hydroxybutyrate. Urine ketone body level is also used as an indicator of level of nutrition.

Specific Nutrients

Protein

Detrimental effects of low nutrition on birth weight can be accentuated on diets that are also low in protein (Robinson, 1983, 1990a). Maternal amino-nitrogen levels have been related to crown-rump length of fetuses. Highly prolific ewes are unable to get sufficient amino acids from high quality pasture and such ewes have responded to protein supplements in late pregnancy. Sulphur amino acids may be limiting (Williams *et al.*, 1988). Lamb birth weights have also responded to proteins that do not degrade easily in the rumen (Everts, 1985). Protein insufficiency results in reduced nucleic acid and synthetic activity in various tissues depending on their stage of development (Schaefer *et al.*, 1984). Gluconeogenic amino acids also have a role in modulating glucose metabolism within the utero placenta and fetal urea synthesis in late pregnancy indicates amino acids account for 30-60% of fetal oxygen consumption (Faichney and White, 1987; Lemons and Schreiner, 1984).

Pregnancy itself appears to influence voluntary intake, with a reduction in that of some roughages. The reduction appears to be related to the number of fetuses carried. There appears to be some form of humoral control of intake involving shorter rumination time, lowered rumen volume and faster rate of passage, and even changed digestibility coefficients and hence improved availability of amino acid nitrogen or protein being available for intestinal digestion and absorption, averting possible protein deficiency on diets that appear marginal (Dittrich *et al.*, 1989; Faichney *et al.*, 1988; Robinson, 1990a; Stafford and Leek, 1990). In conjunction, enhanced recycling of urea during late pregnancy may play a role in ewes on low protein diets (Benlamlih and Oukssou, 1986).

Minerals and vitamins

Fetal skeletal growth results in a high demand for major minerals, notably Ca, P and Mg. Mild deficiencies in these elements do not affect fetal growth, as mobilization of maternal skeletal tissues appears to be physiologically normal and caters for such demands (Braithwaite, 1983a, 1983b; Robinson, 1983, 1990a). Generally general undernutrition or protein insufficiency has a greater effect on fetal size than Ca deficiency. A reasonably recent account of the accretion rates of minerals is given by Grace *et al.* (1986). The roles of hormones and vitamin D in Ca metabolism are well documented. It is not necessary either to expand on the essential nature, role and deficiency symptoms of trace elements for normal development and survival of the fetus and neonate (e.g. I, Se, Cu, Co, Zn). Supplementation is efficacious and in some cases trace elements or vitamins are involved in interactions (e.g. Co and Mo; Se and Vitamin E). Vitamin A is readily transported across the placenta to the fetus, and is partly regulated by maternal supply. Supplements to pregnant ewes have led to increases in birth weight, subsequent milk yield, lamb growth and lamb survival (Pdoshibyakin *et al.*, 1988). Vitamin E supplementation to grain diets in late pregnancy has averted muscle myopathy in neonatal lambs (Watson *et al.*, 1988).

Mammogenesis

The mammary secretary tissue of the ewe develops during gestation with virtually all of the lobule-alveolar epithelial cell system being formed prior to parturition (Robinson, 1990a). DNA studies indicate all the development occurs in the last trimester (Treacher, 1983). Milk production is highly correlated with weight of the secretory tissue (Oddy *et al.*, 1984). Undernutrition in late pregnancy retards this development (Mellor and Murray, 1985a; Mellor *et al.*, 1987) and may be caused by nutritionally-induced changes in hormone ratios (GH, insulin). Davis *et al.*, (1980) found that udder volume increased exponentially from day 110 of gestation until lambing, and peak milk production was significantly related to udder volume post-milking. These authors also found that although udder growth was 25% greater in twin-bearing ewes than single-bearing ewes at peak lactation, productivity per unit of udder tissue was similar. The role of placental lactogens as mentioned earlier plays a key role (Treacher, 1983).

Dietary lipid may promote mammogenesis in sheep as mammary parenchymal development has been stimulated by polyunsaturated fatty acids (McFadden *et al.*, 1988).

Colostrum production

Colostrum is a concentrated source of certain nutrients for new born lambs: vitamins A and E, essential fatty acids, proteins and certain trace (Co, Mn, Fe, Cu, Zn, Se) and major minerals (Ca, P, Na). Via immunoglobulins it also confers immunity against infections such as enteropathogenic *E. coli*. The concentrations decline differentially in the first few hours of lactation. Lambs require 180-210 ml colostrum per kg bodyweight in the first 18 hours as an energy source and for such immunity (Altmann and Mukkur, 1983; Mellor and Murray, 1986; Murzaeva, 1983; Thomas *et al.*, 1987).

Normally colostrum is secreted and accumulates in the mammary gland during the last few days of gestation, and is immediately available to the new born after parturition. Undernutrition in late pregnancy reduces colostrum yield and subsequent secretion rates; and the lactose, lipid and protein available in colostrum (Mellor and Murray, 1985a, 1985b; Mellor *et al.*, 1987; Robinson, 1990a). Refeeding previously underfed ewes 5 days prepartum may remedy such yield reductions. Supplements of amino acid have also been beneficial, while protected dietary lipid supplements have also increased the energy content of colostrum (Robinson, 1990a). Levels of Vitamin A and some trace elements especially Cu have also been enhanced by supplementing the ewe in late pregnancy (Fantova, 1989; Gherdan *et al.*, 1984).

Colostrum is an important source of the various immunoglobulins, being 15 to 160 times greater than in milk (Klobasa *et al.*, 1988). Vitamin and mineral supplements have increased immunoglobulins in colostrum and lamb growth (Sapunov *et al.*, 1986). The immunoglobulins pass intact through the lamb's intestine wall, so it is a bit enigmatic that colostrum stimulates intestinal protein synthesis and development more than does milk (Mirand *et al.*, 1990).

Lactogenesis

Severe undernutrition in late pregnancy can delay the onset of lactation (Heap *et al.*, 1986; Mellor *et al.*, 1987), and this effect may be mediated partly through a delay in the prepartum fall of progesterone as well as an insufficient supply of nutrients. Earlier in the chapter the debilitating effects of poor mid-late pregnancy nutrition on maternal reserves and the detrimental effects on subsequent milk production and growth of the offspring for up to 12 weeks was documented. Lamb birth weight and ewe milk yield can be significantly and positively related to ewe weight (Jordan and Mayer, 1989).

Efficiency of Utilization of ME by the Conceptus (k_c) and Energy Requirements for Pregnancy

Use of ME for pregnancy is a relatively inefficient process, due to the high oxygen consumption and heat production of the fetus(es) and placenta, and k_c is generally in the 0.12 to 0.145 range (Geenty and Rattray, 1987; Robinson, 1983). The latter author postulates that k_c is positively related to energy content of the diet.

Energy accretion rates are exponential in late pregnancy, reaching quite substantial levels as parturition approaches, especially in ewes carrying multiples. In such circumstances utilization of maternal energy reserves appears obligatory to sustain fetal growth due to either voluntary intake or feed supply limitations. In fact it is probably physiologically normal for such ewes to be in negative energy balance, as it also is in early lactation (Rattray, 1986). Recommended average feeding levels for ewes during the late pregnancy period (6-8 weeks) are around 1.5 times and 2 times the ewe's maintenance requirement for single- and twin-bearing ewes, respectively, (Geenty and Rattray, 1987). However such recommendations are only guidelines and do not attempt to meet the exact nutrient requirement for fetal growth at each stage of pregnancy (Robinson, 1990a).

Lactation

Level of feeding during lactation has a much greater effect on total lamb production than feeding during pregnancy, by having the largest immediate nutritional effects on actual lamb growth rate.

It may also influence potential future lamb production through long-term effects on the ewe's fertility or fecundity. In most sheep systems, if feed or pasture is in short supply or very expensive, strict rationing throughout pregnancy may be desirable so that ample feeding is possible during lactation (Rattray *et al.*, 1987).

Earlier parts of this chapter dealt in some detail with the numerous ways in which some aspect of pregnancy nutrition can influence potential or actual milk production and these will not be reiterated.

Generous feeding of ewes, especially in early lactation, is recommended. It can often overcome some of the detrimental carry-over effects from poor feeding during pregnancy. Grazing ewes again seem very resilient in this respect. Rattray *et al.* (1982) showed a very wide range of grazing levels during late pregnancy did not consistently influence milk production or lamb growth to weaning even though ewe body weight differences still existed and wool growth was affected. The lowest levels of feeding involved live weight losses in 55 kg ewes of the order of 15 to 20%.

Factors Influencing Milk Production and Composition

Stage of lactation and number of lambs suckled

Milk production of the lactating ewe typically peaks at 2 to 3 kg/day in the second and third weeks of lactation and thereafter declines steadily to approach 1 kg/ewe/day or less by about week 12 of lactation (Rattray *et al.*, 1982; Rattray, 1986; Treacher, 1983). The decline with advancing lactation is associated with post-prandial changes (Bass, 1989) in hormone ratios and concentration (insulin, somatrophin, cortisol, thyroxine and prolactin). Milk production from ewes suckling multiples is greater than that of ewes suckling singles - by 20 to 50% for twins and a further 15 to 20% for triplets - mainly in the early to mid stages of lactation (Hough *et al.*, 1986; Rattray, 1986; Treacher, 1983). Some of this is undoubtedly due to difference in mammary gland size at parturition. Davis *et al.* (1980) found no difference in secretory capacity of mammary tissue between ewes suckling singles or twins (1.9 ml/ml tissue/day) and peak production reflected differences in udder size. The number suckling and hence milk demand, level udder evacuation and the suckling stimulus effects on prolactin and oxytocin levels are generally considered to be more important than the number of lambs born (Loerch *et al.*, 1985; Rattray, 1986; Treacher, 1983). Level of nutrition or intake of the ewes was not strictly controlled in most studies.

Feeding level

There are a host of experiments that demonstrate the importance of nutrition during lactation (Treacher, 1983; Alvarez *et al.*, 1984a,b). With grazing ewes on a number of different pastures Rattray *et al.* (1982) showed that daily milk production and lamb weaning weights increased in a curvilinear manner tending to level off at a pasture allowance of 6 to 8 kg DM/ewe/day. The range in allowance studied over the two year trial was 2 to 10 kg DM/ewe/day. Ewe live weights and wool production were also significantly affected. At the lower levels of feeding ewes lost considerable live weight. Milk fat content may be higher at low feed allowances and positively related to fat mobilization (Geenty and Sykes, 1986).

Diet composition

Responses to energy and protein have been very variable because of the importance of protein/energy ratio in the diet and this increases with increasing milk yield (Treacher, 1983). Milk yield increased in a curvilinear manner from 2.4 to 3.1 kg/day as CP:ME ratio increased from

10.5 to 16.6 g/MJ ME. Responses to dietary protein are very rapid (Treacher, 1983) and mammary amino acid uptake is related to stage of lactation (Fleet and Mepham, 1985).

High protein diets or protein supplements have both increased milk yield, milk protein content, and lamb growth (Bass, 1989; Penning *et al.*, 1988; Pond and Wallace, 1988; Sheehan and Hanrahan, 1989), while feeding of protected methionine to lactating ewes has resulted in wool (length, thickness, and torsion resistance) and milk yield responses, but no change in milk composition (Floris *et al.*, 1988). The efficiency with which truly digested non-ammonia nitrogen is used for milk protein synthesis varies with protein source (Ngongoni *et al.*, 1989) - soyabean meal 0.61, fish meal 0.54 and blood meal 0.29. The difference may reflect methionine content. In cases of protein deficiency during lactation ewes will use body protein reserves, from muscle degradation, and in some circumstance can use urea as a source of N for milk protein synthesis (Farid *et al.*, 1984; Lynch *et al.*, 1988; Vincent and Lindsay, 1985).

Concentrates

Concentrate feeding (grains) has resulted in increased milkfat content and yield and the form the grain is fed (mash, pelleted or whole) may also affect fat levels with higher fat contents from whole grains (Economides *et al.*, 1989).

Pasture

With grazing sheep, milk production increases with the amount of green herbage offered, decreases with increasing pasture maturity and is greater under continuous grazing than under rotational grazing (Cavallero *et al.*, 1988; Foot *et al.*, 1987; Rattray *et al.*, 1987). The latter effect may be due to differences in both proportion of green and/or maturity of herbage on offer.

Minerals and vitamins

Requirements for all minerals increase during lactation, especially those of Ca and P, and these increases reflect milk composition, milk yield and lactation length (Economides, 1986a). Maternal skeletal mineral stores are mobilized in early lactation and replaced in mid-late lactation (Braithwaite, 1983a, 1983b; Chrisp *et al.*, 1989). Some workers attribute this osteoporosis to protein deficiency (Sykes and Geenty, 1986). Regardless of Ca supply in early lactation ewes are unable to absorb enough to meet the lactational demands and the degree of negative Ca and P balance at this time is directly related to milk yield (Rajaratne *et al.*, 1990). This skeletal mobilization may be driven mainly by the need to supply Ca not P. As lactation progresses rate of Ca and P absorption increase, while Ca secretion in milk and bone resorption decreases (Chrisp *et al.*, 1989). Diet can influence whether the Ca balance changes to positive in late lactation or remains negative because availability of Ca from some forages can be quite low. *Medicago sativa* grown in some areas may be low or marginal in Na and Se and lactating ewes have reduced intake and performance when grazing such forages (Held, 1990).

In ewes which have lambs with congenital goitre oral doses of iodine or intramuscular injections of iodized oil have resulted in rapid increases in milk iodine levels (Azuolas and Caple, 1984). The latter method of supplementation leads to long term (16 months) correction of the deficiency. Cobalt supplementation improves the vitamin B_{12} status of lactating ewes, and milk protein content (Peters and Elliott, 1983).

Injecting Vitamin D intravenously into lactating ewes has led to immediate increases in milk levels and hence supplementation of suckling lambs (Hidiroglou and Knipfel, 1984).

Energy requirements of lactation and efficiency of utilization of energy for lactation

Voluntary intake of the ewe is related to ewe live weight at lambing, ewe live-weight change, milk production and number of lambs suckled. Intake of the ewe rises during lactation to peak around week 5 to 6 (Bocquier *et al.*, 1987). This rise in intake lags the pattern of milk production and feed requirements, which tend to peak in weeks 2 or 3 of lactation, hence it is physiologically normal for ewes to mobilize lipid during early lactation (Smith and Walsh, 1984; Vernon and Finley, 1985). This lipolysis is stimulated by noradrenaline levels and the rate depends on adipocyte cell volume. Lipid deposition occurs in late lactation (Smith and Walsh, 1984) and at this time adenosine appears to have an antilipolytic effect on adipose tissue (Vernon and Finley, 1984).

Energy requirements for grazing lactating ewes are given by Geenty and Rattray (1987) and Rattray (1986) and are approximately 2.5 to 3.5 times maintenance requirement, depending on the number of lambs suckled.

ME from most diets is converted to milk energy (k_l) with an efficiency in the 50 to 70% range (Alvarez *et al.*, 1984a, 1984b; Alvarez and Guada, 1987; Oddy *et al.*, 1984; Rattray 1986; Yang *et al.*, 1988) and does not appear to differ with milk yield.

Milk synthesis is an efficient process when compared with other productive efficiencies and this may be attributed to some of the dietary metabolites as well as those available from maternal tissue mobilization that are used by the mammary gland (Rattray, 1986). Geenty and Sykes (1986) obtained estimates of efficiency of utilization of energy above maintenance for milk synthesis (k_l), which included mobilized tissue energy, that decreased from 84% to 51% with increasing tissue mobilization which is rather surprising considering the metabolites available (Rattray, 1986). They did however find a positive relationship between k_l and the proportion of energy derived from body protein. Undernourished ewes appeared to have 10-20% greater energy requirements for lactation. Oddy *et al.* (1984) found that the mammary gland itself used energy to produce milk with an efficiency (90%) close to the theoretical estimate.

The Suckling Lamb

Tactile senses are involved in the newborn lamb's search for the teat and olfactory stimulation from the dam's inguinal wax, wool or milk seem to play a key role (Vince, 1983). *In utero* swallowing of amniotic fluid in late gestation seems to prepare the fetal lamb's intestine for its future milk diet (Trahair *et al.*, 1986).

Milk is essential in the first 3-4 weeks of the lamb's life (Doney *et al.*, 1984; Treacher, 1983) and early live weight gain is highly correlated to lamb milk intake (Geenty and Dyson, 1986). Milk is a source of highly digestible energy and protein (Mayes and Colgrove, 1983) and the ME of a milk diet or milk plus pasture diet is used with a high efficiency (60-70%) compared to the lower efficiencies of completely ruminant diets (Rattray and Jagusch, 1977; Yang *et al.*, 1988). Level of milk intake strongly influences pasture or solid food intake (Doney *et al.*, 1984; Economides, 1986b; Geenty and Dyson, 1986; Joyce and Rattray, 1970; Treacher, 1983) and the development of a fully independent young ruminant.

Acknowledgements

Barry S.F. Pearson for his dedication in ovine pursuits.

References

Alexandra, G. (1986) Physiological and behavioural factors affecting lamb survival under pastoral conditions. In: Alexander, G., Baker, I.D. and Slee, J. (eds.) *Factors Affecting the Survival of Newborn Lambs*. Commission of European Communities, Luxembourg, pp. 99-114.

Altmann, K. and Mukkur, T.K.S. (1983) Passive immunisation of neonatal lambs against infection with enteropathogenic *Escherichia coli* via colostrum of ewes immunised with crude and purified K99 pili. *Research in Veterinary Science* 35, 234-9.

Alvarez, P.J. and Guada, J.A. (1987) Efficiency of metabolizable energy utilization for milk production in Churro ewes. *Investigacion Agraria,*

Produccion y Sanidad Animales 2, 25-35.

Alvarez, P.J., Guada, J.A., Ovejero, F.J. and Zorita, E. (1984a) Effects of roughage: concentrate ratio of the diet and plane of feeding on yield and composition of ewe's milk. *Anales del Instituto Nactional de Investigaciones Agrarias, Ganadera* 21, 69-90.

Alvarez, P.J., Guada, J.A. and Zorita, E. (1984b) Effect of plane of feeding during lactation on milk yield of dairy Churra ewes. *Anales del Instituto Nacional de Investigaciones Agrarias, Ganadera* 21, 47-68.

Azuolas, J.K. and Caple, I.W. (1984) The iodine status of grazing sheep as monitored by concentrations of iodine of milk. *Australian Veterinary Journal* 61, 223-227.

Baird, G.D., Walt, J.G. Van Der and Bergman, E.N. (1983) Whole-body metabolism of glucose and lactate in productive sheep and cows. *British Journal of Nutrition* 50, 249-265.

Bass, J. (1989) Effects of litter size, dietary protein content, ewe genotype and season on milk production and associated endocrine and blood metabolite status of ewes. *Dissertation Abstracts International*, B, Sciences and Engineering 49, 2940-2941.

Benlamlih, S. and Oukssou, M. (1986) Fluid balance and urea recycling during pregnancy and lactation in small ruminants. *Nuclear and related techniques in animal production and health*. Proceedings of a symposium, Vienna, 17-21 March 1986 81-98.

Bickhardt, K., Grocholl, G. and Konig, G. (1989) Studies on glucose metabolism in sheep during different stages of reproduction and in ketotic sheep using the intravenous glucose tolerance test (IVGTT). *Journal of Veterinary Medicine*, Series A 36, 514-529.

Black, H.J. and Chestnutt, D.M.B. (1990) Influence of shearing regime and grass silage quality on the performance of pregnant ewes. *Animal Production* 51, 573-582.

Bocquier, F., Theriez, M. and Brelurut, A. (1987) The voluntary hay intake by ewes during the first weeks of lactation. *Animal Production* 44, 387-394.

Braithwaite, G.D. (1983a) Calcium and phosphorus requirements of the ewe during pregnancy and lactation. 1. Calcium. *British Journal of Nutrition* 50, 711-722.

Braithwaite, G.D. (1983b) Calcium and phosphorus requirements of the ewe during pregnancy and lactation. 2. Phosphorus. *British Journal of Nutrition* 50, 723-736.

Brink, D.R. (1990) Effects of body weight gain in early pregnancy on feed intake, gain, body condition in late pregnancy and lamb weights. *Small Ruminant Research* 3, 421-424.

Cavallero, A., Grignani, C. and Reyneri, A. (1988) Rotational grazing and continuous stocking on the utilization of the indigenous sward with dairy sheep. *Proceedings of the 12th General Meeting of the European Grassland Federation*, Dublin, Ireland, July 4-7, 1988, pp. 168-172.

Chandler, K.D., Leury, B.J., Bird, A.R. and Bell, A.W. (1985) Effects of undernutrition and exercise during late pregnancy on uterine, fetal and uteroplacental metabolism in the ewe. *British Journal of Nutrition* 53, 625-635.

Chrisp, J.S., Sykes, A.R. and Grace, N.D. (1989) Kinetic aspects of calcium betalism in lactating sheep offered herbages with different Ca concentrations and the effect of protein supplementation. *British Journal of Nutrition* 61, 45-58.

Cummins, L.J., Spiker, S.A., Cook, C. and Cox, R.I. (1984) The effects of steroid immunization of ewes and their nutrition on the ovulation rate and associated reproductive wastage. In: Lindsay, D.R. and Pearce, D.T. (eds.) *Reproduction in Sheep*. Cambridge University Press, Cambridge pp. 326-328.

Davis, S.R., Hughson, G.A., Farquhar, P.A. and Rattray, P.V. (1980) The relationship between the degree of udder development and milk production from Coopworth ewes. *Proceedings of the New Zealand Society of Animal Production* 40, 163-165.

Dingwall, W.S., Robinson, J.J., Aitken, R.P. and Fraser, C. (1987) Studies on reproduction in prolific ewes. 9. Embryo survival, early fetal growth and within-litter variation in fetal size. *Journal of agricultural Science,* Cambridge 108, 311-319.

Dittrich, A., Geissler, C. and Hoffman, M. (1989) Influence of pregnancy and lactation on digestion processes in sheep. *Archives of Animal Nutrition* 39, 563-573.

Doney, J.M., Smith, A.D.M., Sim, D.A. and Zygoyannis, D. (1984) Milk and herbage intake of suckled and artificially reared lambs at pasture as influenced by lactation pattern. *Animal Production* 38, 191-199.

Dove, H., Freer, M. and Donnelly, J.R. (1988) Effects of nutrition in early pregnancy on ewe and lamb performance in the subsequent lactation. *Proceedings of the Nutrition Society of Australia* 13, 111.

Economides, S. (1986a) Mineral requirements of dairy sheep. *Nuclear and related techniques in animal production and health.* Proceedings of a symposium, Vienna, 17-21 March 1986, jointly organised by IAEA and FAO pp. 547-557.

Economides, S. (1986b) Comparative studies of sheep and goats: milk yield and composition and growth rate of lambs and kids. *Journal of agricultural Science,* Cambridge 106, 477-484.

Economides, S., Georghiades, E., Koumas, A. and Hadjipanayiotou, M. (1989) The effect of cereal processing on the lactation performance of Chios sheep and Damascus goats and the pre-weaning growth of their offspring. *Animal Feed Science and Technology* 26, 90-104.

Everts, H. (1985) Relationships between the nutrition of the ewe, lamb birth weight and survival in prolific crossbreds. In: Alexander, G., Barker, J.D., and Slee, J. *Factors Affecting the Survival of New born Lambs.* Commission of the European Communities, Luxembourg, pp. 165-176.

Everts, H. and Kuiper, H. (1983) Energy intake and pregnancy toxaemia in prolific ewes. *Proceedings of the Fifth International Conference on Production Disease in Farm Animals,* Uppsala, Sweden, August 10 to 12 1983, pp. 133-136.

Faichney, G.J. and White, G.A. (1987) Effects of maternal nutritional status on fetal and placental growth and on fetal urea synthesis in sheep. *Australian Journal of Biological Sciences* 40, 365-377.

Faichney, G.J., White, G.A. and Donnelly, J.B. (1988) Effect of conceptus growth in the contents of the maternal gastro-intestinal tract in ewes fed at a constant rate throughout gestation. *Journal of agricultural Science,* Cambridge 110, 435-443.

Fantova, M. (1989) Trace elements in sheep colostrum and milk. Sbornik Vysoke Skoly Zemedelske v Praze, Fakulta Agronomicka. Rada B, *Zivocisna Vyroba* 51, 81-86.

Farid, M.F.A., Khamis, H.S., Hassan, N.I., Askar, A. and El-Hofi, A.A. (1984) Effect of feeding urea to lactating ewes on the yield and composition of milk proteins. *World Review of Animal Production* 20, 67-72.

Fennessy, P.F. and Owens, J.L. (1985) Winter nutrition of ewes. *Proceedings of the 15th Seminar of the Sheep and Beef Cattle Society of the New Zealand Veterinary Association* 15, 105-114.

Fleet, I.R. and Mepham, T.B. (1985) Mammary uptake of amino acids and glucose throughout lactation in Friesland sheep. *Journal of Dairy Research* 52, 2.

Floris, B., Bomboi, G. and Sau, F. (1988) Protected methionine in Sarda sheep:

effect on lactation and wool growth. *Bollettino Societa Italiana Biologia Sperimentale* 64, 1143-1149.

Foot, J.Z., Maxwell, T.J. and Heazlewood, P.G. (1987) Effects of pasture availability on herbage intake by autumn-lambing ewes. In: Wheeler, J.L., Pearson, C.J., Robards, G.E. (eds.) *Temperate pastures: their production, use and management*, Commonwealth Scientific and Industrial Organization, East Melbourne, Victoria.

Geenty, K.G. and Dyson, C.B. (1986) The effects of various factors on the relationship between lamb growth rate and ewe milk production. *Proceedings of the New Zealand Society of Animal Production* 46, 265-269.

Geenty, K.G. and Rattray, P.V. (1987) The energy requirements of grazing sheep and cattle. In: Nicol, A.M. (ed.) *Livestock feeding on pasture*. New Zealand Society of Animal Production Occasional Publication 10, pp. 39-53.

Geenty, K.G. and Sykes, A.R. (1986) Effect of herbage allowance during pregnancy and lactation on feed intake, milk production, body composition and energy utilization of ewes at pasture. *Journal of agricultural Science*, Cambridge 106, 351-367.

Gherdan, A., Trif, A., Malaesteanu, S. and Kalciov, P. (1984) Carotene and vitamin A in pregnant ewes. *Lucari Stiintifice, Institul Agronomic Timisoara, Medicina Veterinara* 19, 53-64.

Glanville, J.R.D. and Phillips, C.J.C. (1986) The effect of winter shearing Welsh Mountain ewes in the hill environment. *British Society of Animal Production*. Paper No. 77, 2 pp.

Grace, N.D., Watkinson, J.H. and Martinson, P.L. (1986) Accumulation of minerals by the fetus(es) and conceptus of single- and twin-bearing ewes. *New Zealand Journal of Agricultural Research* 29, 207-222.

Guada, J.A., Gonzalez, J.S. and Carriedo, J.A. (1982) Effect of pregnancy and plane of feeding on the level of non-esterified fatty acids in plasma of sheep. *Anales de la Facultad de Veterinaria de Leon* 28, 117-129.

Gunn, R.G., Russel, A.J.F. and Barthram, E. (1986) A note on the effect of nutrition during mid pregnancy on lamb production of primiparous ewes in high body condition at mating. *Animal Production* 43, 175-177.

Hay, W.W. Jr, Sparks, J.W., Wilkens, R.B., Battaglia, F.C. and Meschia, G. (1984) Fetal glucose uptake and utilization as functions of maternal glucose concentration. *American Journal of Physiology* 246, E237-E242.

Heap, R.B., Fleet, I.R., Hamon, M., Booth, J.M. and Chaplin, V.M. (1986) Hormone changes in milk at the onset of lactogenesis and parturition in Friesland sheep. *Journal of agricultural Science*, Cambridge 106, 265-269.

Held, J.E. (1990) The selenium content of alfalfa grown in fourteen Wisconsin counties. II. Sodium requirement of lactating ewes fed Wisconsin feeds. Dissertation Abstracts International. *Sciences and Engineering* 50, 4291B.

Hidiroglou, M. and Knipfel, J.E. (1984) Plasma and milk concentrations of vitamin D3 and 25-hydroxy vitamin D3 following intravenous injection of vitamin D3 or 25-hydroxy vitamin D3. *Canadian Journal of Comparative Medicine* 48, 78-80.

Hough, G.M., McDowell, G.H., Annison, E.F. and Williams, A.J. (1985) Glucose metabolism in pregnant and lactating ewes. *Proceedings of the Nutrition Society of Australia* 10, 153.

Hough, G.M., Williams, A.J., Annison, E.F. and Murison, R.D., McDowell, G. H. (1986) Influence of genotype, number of lambs suckled and nutrition on milk production and composition in Merino ewes. *Proceedings of the Australian Society of Animal Production* 16, 416.

Hutchinson, G. and Mellor, D.J. (1983) Effects of maternal nutrition on the initiation of secondary wool follicles in fetal sheep. *Journal of Comparative Pathology* 93, 577-583.

Jordan, D.J. and Mayer, D.G. (1989) Effects of udder damage and nutritional plane on milk yield, lamb survival and lamb growth of Merinos. *Australian Journal of Experimental Agriculture* 29, 315-320.

Joyce, J.P. and Rattray, P.V. (1970) The intake and utilization of milk and grass by lambs. *Proceedings of the New Zealand Society of Animal Production* 30, 94-105.

Keeler, R.F. (1990) Early embryonic death in lambs induced by *Veratrum californicum*. *Cornell Veterinarian* 80, 2, 203-207.

Kelly, R.W. and Ralph, I.G. (1988) Lamb and wool production from ewes fed differentially during pregnancy. *Proceedings of the Australian Society of Animal Production* 17, 218-21.

Kelly, R.W., Wilkins, J.F. and Newnham, J.P. (1989) Fetal mortality from day 30 of pregnancy in Merino ewes offered different levels of nutrition. *Australian Journal of Experimental Agriculture* 29, 339-342.

Klobasa, F., Herbort, B. and Kallweit, E. (1988) Immunoglobulin supply of newborn lambs through colostrum. Immunglobulinversorgung neugeborener Schaflammer durch die Kolostralmilch, German Federal Republic; *Deutsche Veterinarmedizinische Gessellschaft*, Giessen 55-60.

Lemons, J.A. and Schreiner, R.L. (1984) Metabolic balance of the ovine fetus during the fed and fasted states. *Annals of Nutrition and Metabolism* 28, 268-280.

Leury, B.J., Bird, A.R. and Chandler, K.D., Bell, A.W. (1990a) Glucose partitioning in the pregnant ewe: effects of undernutrition and exercise. *British Journal of Nutrition* 64, 449-462.

Leury, B.J., Chandler, K.D., Bird, A.R. and Bell, A.W. (1990b) Effects of maternal undernutrition and exercise on glucose kinetics in fetal sheep. *British Journal of Nutrition* 64, 463-472.

Lindsay, D.B. and Oddy, V.H. (1985) Pregnancy toxaemia in sheep - a review. In: Cumming, D.B. (ed.) *Recent advances in animal nutrition in Australia 1985*. University of New England, Armidale. Paper No.30.

Loerch, S.C., McClure, K.E. and Parker, C.F. (1985) Effects of number of lambs suckled and supplemental protein source on lactating ewe performance. *Journal of Animal Science* 60, 6-13.

Lynch, G.P., Elsasser, T.H., Rumsey, T.S., Jackson, C. Jr and Douglass, L.W. (1988) Nitrogen metabolism by lactating ewes and their lambs. *Journal of Animal Science* 66, 12, 3285-3294.

Mayes, R.W. and Colgrove, P.M. (1983) The digestibilities of milk fat and milk protein in lambs receiving fresh herbage. *Proceedings of the Nutrition Society* 42, 126A.

McCrabb, G.J., Hosking, B.J. and Egan, A.R. (1986) Placental size and fetal growth in relation to maternal undernutrition during mid-pregnancy. *Proceedings of the Nutrition Society of Australia* 11, 147.

McFadden, T.B., Daniel, T.E. and Akers, R.M. (1988) Effects of feeding protected unsaturated fat, growth hormone treatment and feeding level on circulating hormones and receptor concentrations in prepubertal ewe lambs. Journal of Dairy Science 71, Suppl. 1, 230 (Abstr.).

McMillan, W.H. (1987) Post mating progesterone implants in ewes and hoggets. *Proceedings of the New Zealand Society of Animal Production* 47, 151-153.

Mellor, D.J., Flint, D.J., Vernon, R.G. and Forsyth, I.A. (1987) Relationships between plasma hormone concentrations, udder development and the production of early mammary secretions in twin-bearing ewes on different planes of nutrition. *Quarterly Journal of Experimental Physiology and Cognate Medical Sciences* 72, 345-56.

Mellor, D.J. and Murray, L. (1985a) Effects of maternal nutrition on udder development during late pregnancy and on colostrum production in Scottish Blackface ewes with twin lambs. *Research in Veterinary Science* 39, 230-234.

Mellor, D.J. and Murray, L. (1985b) Effects of maternal nutrition on the availability of energy in the body reserves of fetuses at term and in colostrum from Scottish Black-face ewes with twin lambs. *Research in Veterinary Science* 39, 235-240.

Mellor, D.J. and Murray, L. (1986) Making the most of colostrum at lambing. *Veterinary Record* 118, 351-353.

Mirand, P.P., Mosoni, L., Levieux, D., Attaix, D. and Bayle, G., Bonnet, Y. (1990) Effect of colostrum feeding on protein metabolism in the small intestine of newborn lambs. *Biology of the Neonate* 57, 30-36.

Moore, R.W., Millar, C.M. and Lynch, P.R. (1986) The effect of pre-natal nutrition and type of birth and rearing of lambs on vigour, temperature and weight at birth, and weight and survival at weaning. *Proceedings of the New Zealand Society of Animal Production* 46, 259-262.

Murzaeva, A.N. (1983) Minerals in milk of ewes. *Ovtsevodstvo.* 10, 34.

Ngongoni, N.T., Robinson, J.J., Aitken, R.P. and Fraser, C. (1989) Efficiency of utilization during pregnancy and lactation in the ewe of the protein reaching the abomasum and truly digested in the small intestine. *Animal Production* 49, 249-265.

Noble, R.C. and Shand, J.H. (1983) A comparative study of the distribution and fatty acid composition of the lipoproteins in the fetal and maternal plasma of sheep. *Biology of the Neonate* 44, 10-20.

Oddy, V.H., Gooden, J.M. and Annison, E.F. (1984) Partitioning of nutrients in Merino ewes. I. Contribution of skeletal muscle, the pregnant uterus and the lactating mammary gland to total energy expenditure. *Australian Journal of Biological Sciences* 37, 375-388.

Oddy, V.H., Gooden, J.M., Hough, G.M., Teleni, E. and Annison, E.F. (1985) Partitioning of nutrients in Merino ewes. II. Glucose utilization by skeletal muscle, the pregnant uterus and the lactating mammary gland in relation to whole body glucose utilization. *Australian Journal of Biological Sciences* 38, 95-108.

Owens, J.L. (1985) Mid-pregnancy feeding of triplet-bearing ewes. New Zealand, Ministry of Agriculture and Fisheries, Agricultural Research Division, *Annual Report 1983-84*, 261.

Owens, J.L. and Hinch, G.N. (1984) Factors influencing placental and fetal development and lamb birth weights. *New Zealand Ministry of Agriculture and Fisheries, Agricultural Research Division, Annual Report 1982/83*, p.259.

Owens, J.L., Kyle, B. and Fennessy, P.F. (1986) Observations on the effect of litter size, pregnancy nutrition and fat genotype on ewe and fetal parameters. *Proceedings of the New Zealand Society of Animal Production* 46, 41-44.

Parr, R.A., Davis, I.F., Fairclough, R.J. and Miles, M.A. (1987) Overfeeding during early pregnancy reduces peripheral progesterone concentration and pregnancy rate in sheep. *Journal of Reproduction and Fertility* 80, 317-320.

Parr, R.A., Williams, A.H., Campbell, I.P., Witcombe, G.F. and Roberts, A.M. (1986) Low nutrition of ewes in early pregnancy and the residual effect on the offspring. *Journal of agricultural Science*, Cambridge 106, 81-87.

Pdoshibyakin, A.E., Sapunov, A.G., Golovskoi, I.P. and Chebotarev, I.I. (1988) Use of a complex mixture of vitamin A and trace elements as a preventive against vitamin A deficiency in sheep. *Doklady Vsesoyuznoi Akademii Sel'skokhozyaistvennykh Nauk* 2, 32-34.

Penning, P.D., Orr, R.J. and Treacher, T.T. (1988) Responses of lactating ewes, offered fresh herbage indoors and when grazing, to supplements containing differing protein concentrations. *Animal Production* 46, 403-415.

Peters, J.P. and Elliott, J.M. (1983) Effect of vitamin B12 status on performance of the lactating ewe and gluconeogenesis from propionate. *Journal of Dairy Science* 66, 1917-1925.

Pond, W.G. and Wallace, M.H. (1988) Inability of Finnish Landrace crossbred ewes fed corn silage-based diets during lactation to respond to increments of supplemental protein. *Nutrition Reports International* 37, 149-156.

Putu, I.G., Poindron, P. and Lindsay, D.R. (1988) A high level of nutrition during late pregnancy improves subsequent maternal behaviour of Merino ewes. *Proceedings of the Australian Society of Animal Production* 17, 294-297.

Rajaratne, A.A.J., Scott, D., Buchan, W. and Duncan, A. (1990) Effect of variation in dietary protein or mineral supply on calcium and phosphorus metabolism in lactating ewes. *British Journal of Nutrition* 64, 147-160.

Rattray, P.V. (1977) Nutrition and reproductive efficiency. In: Cole, H.H. and Cupps, P.T. (eds.) *Reproduction in Domestic Animals*. Academic Press, Inc., New York, San Francisco, London. pp. 553-575.

Rattray, P.V. (1986) Feed requirements for maintenance, gain and production. In: Wickham, G.A. and MacDonald, S.M. (eds.) *Sheep Production Vol.2: Feeding Growth and Health*. R.A. Richards in association with New Zealand Institute of Agricultural Science, pp. 75-109.

Rattray, P.V. (1987) Sheep production from managed grasslands. In: Snaydon, R.W. (ed.) *Managed Grasslands, B. Analytical Studies*. Elsevier Science Publishers B.V. Amsterdam, pp. 113-122.

Rattray, P.V. and Jagusch, K.T. (1977) Energy cost of protein deposition in the pre-ruminant and young ruminant lamb. *Proceedings of the New Zealand Society of Animal Production* 37, 167-172.

Rattray, P.V., Jagusch, K.T., Duganzich, D.M., Maclean, K.S. and Lynch, R.J. (1982) Influence of feeding post-lambing on ewe and lamb performance at grazing. Proceedings of the New Zealand Society of Animal Production 42, 179-182.

Rattray, P.V., Thompson, K.F., Hawker, H. and Sumner, R.M.W. (1987) Pastures for sheep production. In: Nicol, A.M. (ed.) *Livestock feeding on pasture*. New Zealand Society of Animal Production, Occasional Publication No. 10, pp. 89-103.

Rattray, P.V. and Trigg, T.E. (1979) Minimal feeding on pregnant ewes. *Proceedings of the New Zealand Society of Animal Production* 39, 242-250.

Rattray, P.V., Trigg, T.E. and Urlich, C.F. (1980) Energy exchanges in pregnant ewes. In: Mount, L.E. (ed.) *Energy Metabolism*. Butterworths, London, pp. 325-328.

Robinson, J.J. (1983) Nutrition of the pregnant ewe. In: Haresign, W. (ed.) *Sheep Production*. Butterworths, London, pp. 111-131.

Robinson, J.J. (1990a) Nutrition in the reproduction of farm animals. *Nutrition Research Reviews* 3, 253-276.

Robinson, J.J. (1990b) The pastoral animal industries in the 21st century. *Proceedings of the New Zealand Society of Animal Production* 50, 345-359.

Sapunov, A.G., Podshibyakin, A.E., Chebotarev, I.I. and Golovskoi, I.P. (1986) Biological value of ewes' colostrum and milk, and survival of newborn lambs. *Veterinariya*, Moscow 9, 65-66.

Schaefer, A.L., Krishnamurti, C.R., Heindze, A.M. and Gopinath, R. (1984) Effect of maternal starvation on fetal tissue nucleic acid, plasma amino acid and growth hormone concentration in sheep. *Growth* 48, 404-414.

Sheehan, W. and Hanrahan, J.P. (1989) A comparison of soyabean meal and fish meal as protein supplements for the lactating ewe. *Irish Journal of Agricultural Research* 28, 133-140.

Smeaton, D.C., Wadams, T.K. and Hockey, H.U.P. (1985) Effects of very low nutrition during pregnancy on live weight and survival of ewes and lambs. *Proceedings of the New Zealand Society of Animal Production* 45, 151-154.

Smith, R.W. and Walsh, A. (1984) Effect of lactation on the metabolism of sheep adipose tissue. *Research in Veterinary Science* 37, 320-323.

Stafford, K.J. and Leek, B.F. (1990) Rumination in pregnant ewes. *Journal of*

Veterinary Medicine Series A 37, 154-160.

Sykes, A.R. and Geenty, K.G. (1986) Calcium and phosphorus balances of lactating ewes at pasture. *Journal of agricultural Science*, Cambridge 106, 369-375.

Symonds, M.E., Bryant, M.J. and Lomax, M.A. (1986) The effect of shearing on the energy metabolism of the pregnant ewe. *British Journal of Nutrition* 56, 635-43.

Symonds, M.E., Bryant, M.J., Shepherd, D.A.L. and Lomax, M.A. (1988) Glucose metabolism in shorn and unshorn pregnant sheep. *British Journal of Nutrition* 60, 2.

Thomas, V.M., Hanford, K.J. and Kott, R.W. (1987) Effects of lasalocid in late gestation and other factors on blood metabolite profiles of Finn-Targhee ewes and their lambs and colostrum composition. *Nutrition Reports International* 36, 1257-1265.

Trahair, J.F., Harding, R., Bocking, A.D., Silver, M. and Robinson, P.M. (1986) The role of ingestion in the development of the small intestine in fetal sheep. *Quarterly Journal of Experimental Physiology* 71, 99-104.

Treacher, T.T. (1983) Nutrient requirements for lactation in the ewe. In: Haresign, W. *(ed.) Sheep Production*. Butterworths, London pp. 133-153.

Vernon, R.G., Clegg, R.A. and Flint, D.J. (1985) Adaptations of adipose tissue metabolism and number of insulin receptors in pregnant sheep. *Comparative Biochemistry and Physiology*, B. 81, 909-913.

Vernon, R.G. and Finley, E. (1984) Modulation of noradrenaline-stimulated lipolysis by an adenosine analog in adipose tissue from control and lactating sheep. *Proceedings of the Nutrition Society* 43, 37A.

Vernon, R.G. and Finley, E. (1985) Regulation of lipolysis during pregnancy and lactation in sheep. Response to noradrenaline and adenosine. *Biochemical Journal* 230, 651-656.

Vince, M.A. (1983) Sensory factors involved in the newly born lamb's initial search for the teat. *Journal of Physiology* 343, 2.

Vincent, I.C., Williams, H.L. and Hill, R. (1985) The influence of a low-nutrient intake after mating on gestation and perinatal survival of lambs. *British Veterinary Journal* 141, 611-617.

Vincent, R. and Lindsay, D.B. (1985) Effect of pregnancy and lactation on muscle protein metabolism in sheep. *Proceedings of the Nutrition Society* 44, 77A.

Wastney, M.E., Wolff, J.E. and Bickerstaffe, R. (1983) Glucose turnover and hepatocyte glucose production of starved and toxaemia pregnant sheep. *Australian Journal of Biological Sciences* 36, 271-284.

Watson, M.J., Judson, G.J., Harrigan, K.E. and Caple, I.W. (1988) Vitamin E deficiency and myopathy in neonatal lambs of ewes fed wheat-based diets for two months. *Proceedings of the Nutrition Society of Australia* 13, 93.

Williams, A.J. and Butt, J. (1989) Wool growth of pregnant Merino ewes fed to maintain maternal liveweight. *Australian Journal of Experimental Agriculture* 29, 503-507.

Williams, A.J., Murison, R. and Padgett, J. (1988) Metabolism of sulfur-containing amino acids by pregnant Merino ewes. *Australian Journal of Biological Sciences* 41, 247-259.

Yang, S., Peng, D., Zhang, W., Zhang, Z., Gao, T., Shi, B., Liu, J., Liu, S., Kou, Y., Zhang, L. and Pan, L. (1988) Energy and protein requirements of Hu-sheep. *Scientia Agricultura Sinica* 21, 73-80.

Chapter 5

Lentivirus Diseases of Sheep and Goats: Maedi-Visna and Caprine Arthritis-Encephalitis

G. Pétursson, V. Andrésdóttir, Ó.S. Andrésson,
G. Georgsson, P.A. Pálsson, B. Rafnar and
S. Torsteinsdóttir

Institute for Experimental Pathology, University of Iceland, Keldur, P.O. Box 8540, IS-128 Reykjavík, Iceland

Introduction

The virus family *Retroviridae* consists of three subfamilies: *Oncovirinae*, *Lentivirinae* and *Spumavirinae*. The lentiviruses are non-oncogenic, lytic in cell cultures and cause cell fusion resulting in multinucleated giant cells or syncytia. They are exogenic and cause life-long systemic infections, usually with slow development of disease, characterized by chronic inflammation and leading in a progressive manner to severe disability and death in most cases. Although the host responds with development of serological and cell-mediated immune responses to the virus antigens, it is unable to clear itself of the infection. The name lentivirus is derived from Sigurdsson's (1954) ideas of a special group of infectious diseases which he called slow infections. The basic genomic structure of lentiviruses is similar to that of oncoviruses but more complex with a number of regulatory genes, which influence the replication and the expression of the virus genome which is also influenced by the differentiation and the physiological state of the host cells.

Lentiviruses have been found in sheep, goats, horses, cattle, cats, monkeys and man (Pétursson *et al.*, 1989). The realization that the causative agents of AIDS (HIV-1 and HIV-2) belong to the lentivirus group (Gonda *et al.*, 1986) has greatly intensified the research effort in this field with a resulting unprecedented plethora of publications. Although the research effort is by far most concentrated on the human lentiviruses, interest in animal lentivirus research is growing because of the potential comparative value of animal models for AIDS.

The first lentivirus was isolated in Iceland, in 1957 (Sigurdsson *et al.*, 1960) from the brain of sheep with a demyelinating leucoencephalomyelitis called visna (wasting in Icelandic). Subsequently a closely related or almost identical virus was obtained from the lungs of sheep with a chronic interstitial pneumonia called maedi, an Icelandic word for dyspnoea (Sigurdardóttir and Thormar, 1964). Since then ovine lentiviruses have been isolated from sheep flocks in many countries (Houwers, 1990). Although there are considerable genetic differences between the various strains of ovine lentiviruses they will be referred to collectively as maedi-visna viruses (MVV) in this chapter. In the USA, the sheep lentivirus isolates are referred to as ovine progressive pneumonia virus (OPPV).

Ovine lentiviruses cause systemic infections. Pathological changes are mainly found in the lungs, the central nervous system, the mammary glands and sometimes the joints. The first description of lentivirus-induced histopathology is probably that of Mitchell (1915) who described interstitial pneumonia in South African sheep. Another disease, jaagsiekte or sheep pulmonary adenomatosis, with an entirely different histopathology and now known to be caused by an oncovirus (Sharp and Angus, 1990) was prevalent in South Africa at that time. For decades these two different pulmonary diseases of sheep were confused, since they often occurred in the same animal.

The pulmonary form of MVV-infection has been known under different names in various countries: Graff-Reinet disease in South Africa, progressive pneumonia in Montana, USA, where sheepmen called affected animals "lungers", "heavers" or "blowers", *zwoegerziekte* in Holland and *la bouhite* in France.

The isolation and characterization of a lentivirus from goats is a relatively recent development (Crawford *et al.*, 1980a). Previously maedi-like lesions of goats were described in India (Rajya and Singh, 1964). Later German authors described non-suppurative encephalomyelitis and chronic polyarthritis in goats with virological and serological studies indicating a relationship to visna of sheep (Stavrou *et al.*, 1969; Dahme *et al.*, 1973; Weinhold *et al.*, 1973). This condition was referred to as goat visna by European authors (Weinhold and Triemer, 1978; Sundquist *et al.*, 1981; Fankhauser and Theus, 1983).

Cork *et al.* (1974a) described an infectious leucoencephalomyelitis of goats and following the isolation of a lentivirus from goats in the USA (Crawford *et al.*, 1980a) the use of the term caprine arthritis-encephalitis virus (CAEV) has become established in the literature. Lentivirus-induced arthritis in goats is referred to as "big knee" or "dicke knie" by goat farmers and a chronic indurative mastitis due to lentivirus infection as "hard bag".

Lentivirus infections of sheep and goats appear to be widespread in many countries (Pálsson, 1976; Houwers, 1990; Adams *et al.*, 1984) although the proportion of infected animals varies widely among flocks.

Australia, New Zealand and Finland are reported to be free of MVV virus infection in sheep and ovine lentiviral infection was eradicated from Iceland by a radical stamping out over a 20 year period (Pálsson, 1976). Today Iceland remains the only country in the world from which a lentivirus infection has been completely eradicated. Caprine arthritis-encephalitis apparently has a world-wide distribution especially where dairy goats of European origin are present (Adams *et al.*, 1984). It has also been found in countries which are considered free of sheep lentiviruses (Houwers, 1990) such as Australia and New Zealand.

Aetiology

Virus ultrastructure

The ultrastructure of lentiviruses is similar to that of C-type retroviruses and the virions are assembled at cellular membranes in a similar way. They can be distinguished, however, from C-type virions by the presence of a cone-shaped core in the fully formed virion and the core shell is formed closer to the cell membrane during the budding process and appears thicker at this stage than seen in C-type viruses. The appearance of budding is very similar to that seen in HTLV-I, HTLV-II and bovine leukaemia viruses but they have a differently shaped core in the mature virion (Gelderblom *et al.*, 1990). The diameter of the virion is about 110-130 nm and the surface is studded with projections or knobs about 9-10 nm in length, composed of the envelope glycoprotein. These knobs are easily lost from the virions especially during purification procedures and even from free virus particles after several days in cell cultures (Gelderblom *et al.*, 1990).

Molecular structure of virions

Like other retroviruses the lentiviruses are composed of approximately 60% protein, 35% lipid, 3% carbohydrate and 1% RNA (Gelderblom *et al.*, 1990). The core contains two copies of the RNA genome associated with a few molecules of the enzyme reverse transcriptase (RT), a small nucleocapsid protein (NC) and is surrounded by a shell composed of the capsid protein (CA). Together with the so-called lateral bodies, the core is in turn surrounded by a layer of matrix protein (MA). Finally the virion has an outer envelope composed of a lipid bilayer containing the transmembrane protein (TM) to which are attached the surface knobs already mentioned containing the outer envelope glycoprotein (SU). The composition of the lateral bodies is unknown but they may contain "non-structural" regulatory proteins. Since the outer envelope of the virus is derived from the cell membrane it is thought to contain host derived constituents, not only lipids but also MHC class I and class II determinants (Gelderblom *et al.*, 1987).

Virus genome

The lentiviruses have a genome organization basically similar to other retroviruses (Sonigo *et al.*, 1985; Narayan and Clements, 1989). The basic order of virus structural genes is 5' - *gag* - *pol* - *env* - 3'. The *gag* open frame codes for three structural proteins of the virus core MA, CA and NC; the *pol* gene codes for viral protease (PR), reverse transcriptase (RT) (which also has ribonuclease H activity) and the viral integrase (IN), and the *env* gene contains the genetic information for the surface envelope glycoprotein (SU) and the transmembrane glycoprotein (TM).

In contrast to most oncoviruses, the lentiviruses produce several regulatory proteins which control the rate of viral genome expression and virus replication. However, similar complex regulation of lentivirus genome expression is also seen in the oncovirus subgroup consisting of the human HTLV-I, HTLV-II and the bovine BLV (Cullen, 1991).

These additional regulating genes have been studied most extensively in HIV which produces six regulatory proteins (*vif, vpr, vpu, tat, rev,* and *nef*) and at least three such genes have been identified in MVV and CAEV. The open reading frames for two of them are located between the *pol* and *env* gene, the so-called *Q* gene whose function is still unknown and the *tat* which transactivates viral gene transcription (Davis and Clements, 1988). The *rev* gene is bipartite, coded for by a small region at the 5' end of the *env*-gene and a longer region at the 3' end of the *env*-gene. The *rev* gene of MVV has been reported to have a transactivating effect on the transcription of MVV genes (Gourdou *et al.*, 1989; Sargan and Bennet, 1989). It has genetic homologies with the *rev* genes of primate lentiviruses, whose function is to activate the cytoplasmic expression of large viral mRNAs coding for the viral structural proteins.

As in other retroviruses the DNA genomes of lentiviruses are flanked by regulatory sequences, the so-called long terminal repeats (LTRs). The 5' end of the viral RNA starts with a short region named R followed by a unique sequence called U5 whereas at the 3' end there is a unique sequence (U3) followed by the R region. The R region contains the signal and the site for the addition of a poly-A tail, characteristic of messenger RNA in eucaryotic cells. Thus the general organization of the viral RNA molecule is R-U5-*gag-pol-env*-U3-RAAAA......

When the viral RNA enters a permissive cell it is transcribed into double stranded DNA in which the viral genes are flanked by identical terminal repeats as follows:

U3-R-U5-*gag-pol-env*-U3-R-U5. The U3 region contains promoter sequences and viral specific enhancer sequences that regulate the initiation and the rate of RNA synthesis in the infected cells, including the TATA box recognized by the cellular RNA polymerase II which catalyses the synthesis of viral RNA from the viral DNA.

The U3 region of sheep and goat lentiviruses also contains sequences known to respond to cellular factors controlling gene expression. These sequences include several AP-1 sites and one AP-4 site, named after HeLa cell activator proteins (Hess *et al.*, 1989).

Virus replication

Sheep and goat lentiviruses generally replicate to high titres in permissive cell cultures of the fibroblastic type, such as choroid plexus and synovial cell cultures. Ovine and caprine cells of the monocyte/macrophage lineage also support productive infection in culture. Some ovine field strains replicate poorly in fibroblastic cell cultures (Quérat *et al.*, 1984) and this seems to be a more general finding with CAEV (Narayan *et al.*, 1988). Based on those characteristics isolates of the small ruminant lentiviruses have been divided into two types: 1) lytic isolates such as the Icelandic MVV and 2) non-lytic, persistent isolates typical for CAEV and some sheep isolates of the American ovine progressive pneumonia (Quérat *et al.*, 1984). A general feature of the cytopathic effect of these viruses is the production of multi-nucleated syncytia or giant cells by cell fusion and formation of star-shaped or spider-like cells with a rounded refractile cell body and many long filiform cytoplasmic extensions.

The initial step in virus replication is the attachment of virus to the cell membrane. The principal receptor molecule for HIV is known to be the CD4 surface molecule of the so-called helper-inducer lymphocyte (Dalgleish *et al.*, 1984) but also present on macrophages. Much less is known about the virus receptor or receptors for the small ruminant lentiviruses, although there are findings implicating the class II histocompatibility antigens on cells of the immune system (R.G. Dalziel, personal communication). Other cell surface molecules may also serve as virus receptors for MVV and CAEV.

Once attached to the cell the virion envelope probably fuses with the cell membrane. It is then decoated and the RNA genome is released and transcribed into double-stranded DNA with the help of the reverse transcriptase (RT) contained in the virion core. The RT is an RNA-dependent DNA polymerase which needs a short double-stranded stretch of RNA to start. Immediately downstream of the U5 region is a short primer binding site (PBS) with a base sequence complementary to the 3' sequence of $tRNA_{1,2\,lys}$ and this marks the start of the synthesis of the minus-strand viral DNA (Sonigo *et al.*, 1985; Saltarelli *et al.*, 1990). When this strand is completed, the plus-strand of the viral DNA is synthesized. There are two purine-rich sequences or polypurine tracts (PPT) thought to be initiation sites for the synthesis of the plus-strand viral DNA. One is located towards the 3' end of the genome immediately upstream of the U3 region and the other is located close to the middle of the genome at the end of the *pol* gene. This second PPT is located in the single-strand region of the unintegrated linear viral DNA described by Harris *et al.* (1981) and it has been shown in the case of MVV that plus-strand viral DNA synthesis is initiated at both these PPTs (Blum *et al.*, 1985). Thus it appears that the synthesis of the plus-strand is discontinuous.

In contrast to the oncogenic retroviruses, where viral DNA synthesis takes place in the cytoplasm, lentiviral DNA is replicated in the nucleus of the cell (Haase *et al.*, 1982). It has been proposed that this explains why lentiviruses can replicate in non-dividing cells (Thormar, 1963). In lentivirus-infected cells most of the viral DNA is in the linear form and apparently very few viral DNA molecules integrate into the host cell DNA, possibly in the order of 1-2 molecules per cell on the average (Ó.S. Andrésson, unpublished results).

On the other hand 100-200 copies of free linear-double stranded DNA molecules are found in lentivirus-infected cells (Clements *et al.*, 1979). It has been suggested that this high number of free viral DNA molecules in infected cells may be associated with the cytolytic effects of lentiviruses usually not seen in oncovirus-infected cells.

Transcription of viral RNA and gene expression *in vitro*

Like other lentiviruses and in contrast to most oncoviruses MVV and CAEV exhibit a complex transcriptional pattern (Vigne *et al.*, 1987). Northern blot analysis of MVV-infected cells shows at least six species of mRNA. The largest 9.4 kilobase RNA represents the genomic RNA and seems to be the template for translation of the *gag* and *pol* genes. There are in addition three singlyspliced mRNAs varying in size from 3.7 to 5 kilobases. The smallest of these corresponds to the *env* gene whereas the other two include one or both of the small open reading frames between the *pol* and the *env* gene. In addition two doubly-spliced small mRNAs are found, 1.7 kilobases and 1.4 kilobases in size. They contain sequences from the region between the *pol* and *env* genes and from the 3' terminal part of the *env* gene. It should be added that all these subgenomic mRNAs contain a short 5' leader of ca. 300 nucleotides and they all contain sequences derived from the 3' end of the *env* gene and the U3 and R regions (Davis *et al.*, 1987; Mazarin *et al.*, 1988).

It has been shown that there is a temporal control of transcription of the MVV genes during the lytic infection in permissive cell cultures (Vigne *et al.*, 1987). Thus 24 hours after infection there is a predominant expression of the small mRNAs, 1.7 and 1.4 kilobases, whereas later (72 hours post infection) the four larger mRNAs are also transcribed. The early small mRNAs are derived from genes controlling the rate of transcription from the viral genome, such as the *tat* and the *rev* genes. There is evidence that the 1.7 kilobase mRNA is derived from the *tat* gene and the 1.4 kilobase mRNA from the bipartite *rev* gene (Mazarin *et al.*, 1990).

In vivo restriction of virus replication and gene expression

In contrast to the highly productive viral infection in permissive cells in culture, the replication of lentiviruses in the tissues of the animal is highly focal and unproductive (Haase, 1989). Very few cells are found to contain the viral DNA and many of those produce little or no viral RNA and even fewer detectable viral antigens. This restriction seems to depend on the physiological and developmental state of the cells. Thus when cells from tissues of lentivirus-infected sheep or goats are grown in tissue culture, this restriction is relieved with resulting productive infection (Pétursson *et al.*, 1976). It has also been shown that when infected monocytes with low production of virus differentiate into macrophages the host cell restriction is lifted leading to increased viral production (Gendelman *et al.*, 1986). Thus it appears that the expression of viral genes and the replication of these lentiviruses is strictly controlled by a complex interaction of viral and host cell genes. The restriction of viral replication in the host is considered a major factor in the "slowness" of lentiviral diseases.

Clinical Disease

The most common clinical manifestation of lentivirus infections of sheep is a chronic pneumonia with signs of progressive respiratory failure (Pálsson, 1976). When the sheep are driven, the affected animals lag behind, they exhibit laboured breathing and, after exertion, the respiration becomes rapid and shallow. As the disease progresses the respiration at rest becomes gradually more difficult with dilated nostrils and flank breathing aided by accessory respiratory muscles accompanied by characteristic rhythmic jerks of the head. In contrast to jaagsiekte, coughing is not a prominent sign and not productive when it occurs. Because of the long silent preclinical phase of the infection, clinical signs are rarely seen in animals under 3-4 years of age. As the disease progresses the sheep loose condition and become emaciated, affected ewes often give birth to small and weak lambs and milk production is apparently decreased. In advanced cases the sheep are anaemic and sometimes a mild lymphocytosis is observed in the blood.

Another manifestation of lentivirus infection of sheep is visna, where clinical signs from the central nervous system predominate. One of the earliest signs is that affected sheep lag behind the flock when driven uphill or over uneven ground and stumble for no apparent reason. There is slight aberration of gait notably affecting the hind quarters. There is a gradually increasing weakness of the hindlegs progressing to paraplegia or almost total paralysis, so the animal is unable to rise. A fine tremor or twitching of lips and facial muscles is sometimes seen. A general loss of condition is often seen and spinal taps reveal an increased number of mononuclear cells in the cerebrospinal fluid. Clinical visna is apparently rarely seen in ovine lentivirus infections outside of Iceland, where it was particularly common in certain flocks in the south-west area of the country (Pálsson, 1976).

The third type of clinical disease in sheep induced by lentivirus infection is chronic, indurative mastitis with diffuse or nodular hardening of the udder and reduction in milk production (Van der Molen *et al.*, 1985; Cutlip *et al.*, 1985a).

Arthritis due to sheep lentiviral infection seems to be found mainly in certain breeds infected with the American progressive pneumonia virus (Cutlip *et al.*, 1985b). It primarily affects the carpal joints with swelling of the joints and the sheep become lame and emaciated.

In contrast to sheep, arthritis is the predominant disease manifestation of lentivirus infection of goats (Crawford *et al.*, 1980b). The disease develops slowly and usually appears in adult animals 2-9 years of age. In some animals it progresses rapidly but in others the course is chronic with episodes of more acute manifestations followed by remissions. The joints most affected are the carpal ones but the tarsal, stifle, fetlock and even the atlanto-occipital ones are affected to a lesser degree. Animals with advanced arthritis may show deformities of the joints and flexion contractures with loss of function of the limbs.

Neurological disease in goats is in contrast to visna of sheep mainly seen in young animals 2-6 months old and follows a rapid course with similar symptoms of paresis and paralysis as seen in sheep (Cork, 1990). Goats that are less severely affected may survive and they usually develop arthritis later on. Sometimes encephalitis occurs in adult animals.

Although mild, chronic interstitial pneumonia may be found associated with lentiviral infections in goats (Robinson and Ellis, 1984) and even results in dyspnoea, it seems to be a much less prominent feature of lentiviral infection of goats than usually observed in sheep. Chronic indurative mastitis is seen in goats infected with CAEV and may affect the milk yield (Zwahlen *et al.*, 1983).

Pathology

The pathological changes in lentiviral infections of sheep and goats are characterized by inflammation of the chronic type, i.e. with infiltrates of mononuclear cells, macrophages, lymphocytes and plasma cells.

The target organs mainly affected include the lungs, mediastinal lymph nodes, the central nervous system, the mammary glands and the joints. Even if clinical signs primarily indicate one organ, histopathological examination often shows that other organs are also affected. Since detailed accounts of pathological changes have been published (Georgsson *et al.*, 1976; Georgsson, 1990; Cork *et al.*, 1974b; Cork, 1990), only the principal features of the lesions will be described here. In advanced cases of maedi, the lungs are much heavier than normal and do not collapse upon removal. Microscopically there is thickening of the interalveolar septa due to infiltration of mononuclear cells, hyperplasia of smooth muscle cells and some fibrosis. The alveoli are found to be almost obliterated in severe cases. Peribronchial and perivascular lymphoid follicles with active germinal centres are a prominent feature of lentiviral pneumonia. Almost diagnostic is the typical enlargement of the mediastinal lymph node with hyperplastic changes.

In the central nervous system the lesions consist of multifocal mononuclear inflammatory changes most pronounced in subependymal and perivascular locations. Glial nodules are also seen and in the most severe cases necrotic changes. The inflammation affects the meninges to a variable degree. The grey matter is relatively spared and destruction of neurons is not usual. The choroid plexus is often heavily affected with formation of lymphoid follicles with germinal centres.

Demyelination is commonly seen, both secondary or concomitant with axon destruction, but also in severe or advanced cases primary demyelination with relative sparing of axons is observed (Georgsson *et al.*, 1982). In the spinal cord the lesions are primarily around the central canal but sometimes demyelinated areas are seen in the tracts which closely resemble plaques in multiple sclerosis of humans.

The arthritic changes in goats and sheep are usually associated with synovial lined cavities, joints, tendon sheaths and bursae (Cheevers and McGuire, 1988).

The affected joints are enlarged due to thickening of the subcutis and distension of joint capsules with increased synovial fluid. There is hyperplasia of the synovia with villous protrusions and fibrin tags. This is accompanied by erosion of articular cartilage and subchondral bone destruction. Microscopically there is mononuclear cell infiltration of subsynovial structures with synovial cell proliferation. There is perivascular infiltration of macrophages, lymphocytes and plasma cells and a tendency to lymphoid follicle formation in active cases. In advanced cases the lesions are more degenerative than inflammatory.

Lentiviral mastitis is characterized by non-suppurative inflammation with diffuse infiltrates of mononuclear cells of the parenchyma (Narayan and Cork, 1985). Lymphoid follicles are also seen in the mammary glands adjacent to ducts and within interlobular septa.

Immune Response

In experimental and natural infections with MVV of sheep and CAEV of goats, antibodies specific for the major viral antigens develop. They can be detected by various methods: complement-fixation (Gudnadóttir and Kristinsdóttir, 1967), gel immunodiffusion (Cutlip *et al.*, 1977; Robinson, 1981), immunofluorescence (De Boer, 1970), ELISA tests (Houwers *et al.*, 1982; Houwers and Schaake, 1987; Archambault *et al.*, 1988) and Western blotting (Houwers and Nauta, 1989; Zanoni *et al.*, 1989). Analysis by the last method demonstrates antibody response to the viral structural proteins: the CA (p25), the MA (p16) and the NC (p14) antigens coded for by the *gag* gene. There is a major envelope glycoprotein line on Western blots with an apparent molecular weight variously reported (gp110-135). The calculated molecular weight of the TM and the SU visna virus envelope proteins before glycosylation is 38 and 64 kilodaltons respectively (Sonigo *et al.*, 1985). Additional bands may appear on the Western blot but have not been definitely identified. The product of the rev regulatory gene has also been shown to react with immune sera from MVV-infected sheep (Mazarin *et al.*, 1990). Virus-neutralizing antibodies are also found in lentiviral infection of small ruminants but the titre seems to depend on the virus strain and neutralizing antibodies have been reported to be absent or of low titre in many cases of CAEV infection in goats (McGuire *et al.*, 1990).

In experimental infection with MVV of sheep neutralizing antibodies develop later than other types of antibodies (Pétursson *et al.*, 1976). They are very strain-specific, especially in early stages of infection, but become more broadly reactive later in the course of the disease. Neutralizing antibodies to MVV have been reported to be of low affinity so that the virus binds more readily to cells than to the antibodies (Kennedy-Stoskopf and Narayan, 1986). It has also been shown that the envelope epitopes for cell fusion and virus neutralization are separate so that antifusion antibodies are distinct from virus-neutralizing antibodies (Crane *et al.*, 1988).

Neither neutralizing nor other antibodies have shown to be of protective value for the host. It has even been suggested that antibodies may enhance infection, especially of cells carrying Fc-receptors (Jolly *et al.*, 1989). There is little if any documented evidence that the immune response can ever clear the host of infection with lentiviruses although a modulating effect on the course of the disease progression cannot be ruled out.

Cell-mediated immune response is thought to be more likely to play an important role in host defence against lentivirus than antibodies. Such immunity to MVV has been demonstrated by lymphocyte blast transformation but the response is irregular, varies from sheep to sheep and with time in individual animals (Sihvonen, 1981). It is also possible that such a response may be directly detrimental and contribute to immunopathological production of lesions as discussed below under pathogenesis.

Pathogenesis

Target cells

Cells of the monocyte/macrophage lineage have been shown to be infected in all lentivirus infections of animals and man (Narayan and Clements, 1989). In addition other cell types have been found to express virus antigens by immunohistochemical methods. They include lymphocytes, plasma cells, endothelial cells, fibroblasts, pericytes, choroidal epithelial cells and meningeal cells in MVV infection of sheep (Georgsson *et al.*, 1989). In CAEV infection of goats, most cells with viral transcripts were found in inflamed areas of the central nervous system, lungs, joints and mammary glands and most were identified as macrophages. Viral RNA was also found in endothelial cells, synovial cells and in epithelial cells of intestinal crypts, renal tubules and thyroid follicles (Zink *et al.*, 1990). On the whole only a few scattered positive cells are seen in the tissues of infected animals and they are mainly present in areas of inflammation. Although CD4 positive T cells do not seem to be an important target for goat and sheep lentiviruses, there is evidence that at least some lymphocytes can carry and express the viral genome (S. Torsteinsdóttir, unpublished results).

Immunopathogenesis

There is evidence that both excessive and deficient immune responses play a role in the pathogenesis of lentiviral diseases of man and animals. The pronounced immune deficiency in the final stage of HIV infection (AIDS), at least in part due to the disappearance of CD4 positive T cells, is the best example of a deficient response.

On the other hand it has been shown that early lesions in experimental MVV infection of sheep were almost abolished by immunosuppressive treatment with antithymocyte serum and cyclophosphamide (Nathanson *et al.*, 1976). Hyper-immunization with ovine and caprine lentiviruses has also been reported to increase the severity of the disease (Nathanson *et al.*, 1981; McGuire *et al.*, 1986). It is likely that the immune attack is directed against virus induced antigens rather than host antigens (Panitch *et al.*, 1976) although autoimmune reactions have not been completely excluded. Since surprisingly few cells in the lesions express viral antigens as revealed by immunostaining (Georgsson *et al.*, 1989), it appears that tissue damage may be in large part due to amplification of the immune response to viral antigens with a great influx of macrophages and lymphocytes and secretion of cytokines and other soluble factors resulting in non-specific (innocent bystander) tissue damage.

The severe immunodeficiency with associated opportunistic infections and tumours characteristic of human AIDS is not seen in ovine and caprine

lentivirus infections. There are, however, reports indicating mild disturbances of certain immune functions associated with lentiviral infection of sheep
(Ellis and DeMartini, 1985; Myer *et al.*, 1988; Bird *et al.*, 1990). These
disturbances of immune functions have been attributed to viral infection of
macrophages rather than to a depletion of CD4 positive lymphocytes since
studies of CD4/CD8 ratios in peripheral blood have failed to reveal any
abnormalities similar to those seen in AIDS (Bird *et al.*, 1990; S.
Torsteinsdóttir, unpublished results).

Role of viral genetic (antigenic) variation

The RT enzyme which catalyses the transcription of lentiviral RNA into
DNA, is known to be rather inaccurate, resulting in a high mutation
frequency of these viruses. Some regions of the genome are better
conserved than others, presumably because genetic variation at these sites
is less compatible with survival of the virus. Thus the *pol* gene and to a
certain extent the *gag* gene are better conserved than parts of the *env* gene
(Pyper *et al.*, 1984) which have been reported to be hypervariable (Braun
et al., 1987).

Since neutralizing antibodies seem to be directed primarily against
epitopes on the SU envelope glycoprotein, neutralization tests are well
suited to pick up virus variants.

In fact it has been shown that during the course of infection in individual
animals antigenic variants can be isolated that are less readily neutralized
by early serum samples than the infecting virus (Gudnadóttir, 1974). It has
been proposed that this antigenic variation plays a role in the persistence of
the virus and in the progression of lesions (Narayan *et al.*, 1978). Other
reports, however, indicate that the original infecting virus type persists in
the animal coexisting with the new variants and that progression of
experimental disease could not be linked with the appearance of new
variants (Thormar *et al.*, 1963). The high mutation rate must, however, be
of importance for the evolution and the epidemiology of lentiviruses and
may pose problems in attempts to develop vaccines and drugs to control
these diseases.

When strains of ovine lentivirus from different parts of the world are
compared, considerable biological differences are revealed (McGuire *et al.*,
1990). Some strains are more lytic in tissue culture and grow to higher
titres than others and differ in their ability to induce neutralizing antibodies.
Arthritis in sheep has only been reported in the field associated with
American OPPV infection (Cutlip *et al.*, 1985b). MVV strains from
Iceland, Scotland and South Africa show about 20% difference in genome
base sequences and the American OPPV is even more divergent in genetic
make-up. Differences between the Icelandic MVV strain 1514 and the South
African SA-OMVV strain virus proteins vary from 8.5 to 35% mismatched

amino acids (Quérat *et al.*, 1990). The goat virus CAEV also shows considerable strain variation (McGuire *et al.*, 1990). The structural genes of sheep and goat viruses show a considerable homology which explains the serological cross reactivity between these viruses. However, the *env* genes of these viruses show only 60% amino acid identity (Saltarelli *et al.*, 1990).

Although genetic differences of the various sheep and goat breeds seem to play a role in susceptibility to lentiviral infection and influence the disease manifestation (Cheevers *et al.*, 1988; Ruff and Lazary, 1988; Dawson, 1988), the genetic make-up of virus strains is also of importance. When Icelandic lentivirus strains isolated from the pulmonary form (maedi) are compared with strains isolated from the CNS of sheep with neurological disease (visna) by restriction enzyme analysis, they are found to differ by about 7% in base sequence (V. Andrésdóttir, unpublished results).

We have now obtained infectious pathogenic molecular DNA clones both from visna and maedi which may serve to elucidate which viral genes may influence organ tropism or affect other biological characteristics of different viral strains (Ó.S. Andrésson *et al.*, unpublished results).

Diagnosis

Although clinical symptoms may be highly suggestive and histopathological lesions practically diagnostic in typical or advanced cases of lentivirus disease in sheep and goats, serological and virological methods are necessary to confirm the diagnosis, especially in early phases of the infection and in animals which are infected but show little or no pathological changes.

Of the serological tests the agar gel immunodiffusion test and the ELISA are most suitable for practical application (Dawson *et al.*, 1982) but the Western blot is helpful in confirming doubtful results in the other tests (Zanoni *et al.*, 1989). Virus isolation from autopsy material (spleen, lymph nodes, CNS, joints, lungs) can be used to confirm the diagnosis. In the living animal virus may be isolated from buffy coat cells by cocultivation with permissive cells such as choroid plexus cells. Attempts to isolate virus from the buffy coat are not always positive in infected animals so negative results are not reliable (Pétursson *et al.*, 1990).

Amplification of the viral genetic material by means of the polymerase chain reaction (PCR) has been used to demonstrate viral DNA of MVV and CAEV in tissue culture at early stages of infection (Zanoni *et al.*, 1990). This method is expected to become a useful diagnostic method when it has been perfected and suitable primers have been found. Such primers may be chosen from highly conserved regions of the genome for general use or for initial grouping whereas primers from more variable parts of the genome can be used for more exact typing of virus strains.

Economic Importance

The economic consequences of lentiviral infection for sheep and goat breeding vary depending on several factors, the virus strain, the animal breed, husbandry and management (Houwers, 1990).

The most dramatic example of direct losses due to lentiviral infection comes from Iceland where the importation of silent carrier sheep in 1933 resulted in heavy losses of up to 30% per flock per year and sheep breeding became virtually uneconomic (Pálsson, 1976).

On the other hand it has been reported (Snowder *et al.*, 1990) that subclinical ovine progressive pneumonia virus infection did not influence ewe wool and lamb production in several sheep breeds in an experimental station in Idaho. The incidence of clinical disease will depend on how widespread the infection is in a flock and it will take years to build up following the introduction of the infection. Gradually losses will increase due to culling of sick animals either with overt clinical signs of pneumonia, arthritis, mastitis or animals in poor body condition and ewes which produce poorly growing lambs. Indirect effects include lower market value of breeding animals from infected flocks with increasing recognition by prospective buyers of the potential dangers of these infections (Houwers, 1990).

Peterhans *et al.* (1988) report 10% reduction in milk yield in infected goats in Switzerland and report adverse effects on exportation of goats due to a high prevalence (60% of CAEV-infected goats in that country). It was estimated that lentivirus infection of Dutch sheep caused 10-20% animal losses (De Boer *et al.*, 1979) and Dohoo *et al.* (1987) estimate 33% reduction in conception rate and 3-6% reduction in birth weight of lambs due to MVV infection of sheep. The economic consequences of lentiviral infections of sheep and goats are difficult to evaluate in exact financial terms. They may be predicted to become of increasing concern with the apparent tendency towards intensification of the sheep industry and further information from well controlled studies of the effects of lentivirus infection on sheep and goat productivity.

Epidemiology

There is ample evidence of horizontal transmission of MVV infection in sheep (Sigurdsson *et al.*, 1953; Pálsson, 1976; Houwers, 1990). Thus when virus-carrying sheep are introduced into a virgin population the infection will spread gradually between adult animals. The most convincing evidence of this mode of transmission was obtained in Iceland following the importation of Karakul sheep in 1933. Because of the rather special husbandry conditions in Iceland, the long winter housing period and the autumn roundups where thousands of sheep from different flocks were gathered in pens to be sorted out, the infection spread quickly within flocks and gradually between flocks. Close contact favours the spread of infection which is assumed to be mainly by the respiratory route (Pálsson, 1976).

Where jaagsiekte or pulmonary adenomatosis is also endemic, the copious production of bronchial secretion with coughing seems to favour the spreading of MVV (Dawson *et al.*, 1985).

Another important mode of transmission is from ewe to lamb, apparently mainly by colostrum and milk. When lambs are separated from their mothers immediately after birth and reared in isolation they remain with few exceptions free of the infection (Houwers *et al.*, 1983). This implies that infection of the fetus *in utero* must be very rare although some workers report evidence of transplacental infection (Cutlip *et al.*, 1981; Hoff-Jørgensen, 1977). Although the evidence strongly points to the lactogenic mode of spread from ewe to lamb, other means of transmission such as by the respiratory route may also occur. In infected flocks the progeny of seropositive ewes are much more likely to show evidence of infection than lambs of seronegative mothers (Houwers, 1990).

For the CAEV the importance of lactogenic transmission has been demonstrated by experimental transmission of the infection with colostrum and milk from infected goats to kids (Adams *et al.*, 1983) and even to lambs (Oliver *et al.*, 1985). There is also evidence that horizontal spread in goat herds is favoured by close contact (Adams *et al.*, 1983; East *et al.*, 1987). There is no evidence that male goats or sheep play a particular role in spreading the infection to ewes by the semen, although the presence of virus or virus-infected cells in semen cannot be excluded.

It has been shown experimentally that goats can be infected by the MVV sheep virus and vice versa sheep by experimental injection of CAEV (Banks *et al.*, 1983). Under field conditions, however, this does not appear to take place. Thus in Australia where CAEV infection of goats is present, the sheep population is free of lentivirus (Smith *et al.*, 1985).

Control Measures

The most radical way of eliminating lentivirus infection of sheep and goats is to slaughter all animals in a flock or within a district, replacing them with animals from an unaffected population. In this way, MVV was eliminated from Iceland in about 20 years (1944-1965). This drastic and expensive programme was justified because of the exceptionally heavy losses due to the disease in that country and because adequate serological diagnostic methods were not available at that time (Pálsson, 1976). Today Iceland remains the only country where a lentiviral infection has been totally eradicated. Such methods are impractical in most other countries and less drastic methods are now available. They all require extensive, repeated serological testing and restrictions of movement of animals between flocks

(Houwers, 1990). One way of eliminating the infection is culling of seropositive animals with repeated testing over several years. This is a slow process and mainly suitable where the proportion of infected animals is low. By also culling the progeny of seropositive ewes the process can be speeded up to some extent (Houwers et al., 1984). In a heavily infected flock, the main route of infection appears to be from mother ewe to lamb after birth. Attempts to clean up infected herds by removal of lambs at birth and raising them artificially in isolation from the flock have been successful in the Netherlands and in the USA (Houwers et al., 1987; Cutlip et al., 1988). Similar methods of control are suitable for elimination of CAEV from goat herds (Robinson and Ellis, 1986; McGuire et al., 1990).

Methods of elimination of infection from infected flocks are difficult, expensive and time-consuming. It is therefore of great importance to prevent introduction of infection into lentivirus-free populations. There are numerous examples of introduction of lentiviral diseases into countries previously free from these diseases. This is invariably associated with importation of infected animals. In order to prevent this, import controls are necessary. Animals should only be bought from flocks free of infection. Serological testing of individual animals is not enough, the whole flock of origin must be tested repeatedly, since animals in early stages of infection may be serologically negative.

Examples of introductions of MVV into countries that were apparently free of the infection are numerous. They include Iceland, Denmark, Norway and Great Britain (Houwers, 1990).

The outlook for prevention by vaccination is not very promising. Published and unpublished attempts to vaccinate against MVV and CAEV have failed to obtain any protection (Cutlip et al., 1987) and have even lead to more severe disease in vaccinated animals (Nathanson et al., 1981; McGuire et al., 1986). Although some drugs may inhibit lentiviral replication in cultured cells (Frank et al., 1987), there are no indications that drug treatment will have any practical value in the near future.

References

Adams, D.S., Klevjer-Anderson, P., Carlson, J.L. and McGuire, T.C. (1983) Transmission and control of caprine arthritis-encephalitis virus. *American Journal of Veterinary Research* 44, 1670-1675.

Adams, D.S., Oliver, R.E., Ameghino, E., DeMartini, J.C., Verwoerd, W., Houwers, D.J., Waghela, S., Gorham, J.R., Hyllseth, B., Dawson, M. and Trigo, F.J. (1984) Global survey of serological evidence of caprine arthritis-encephalitis virus infection. *Veterinary Record* 115, 493-495.

Archambault, D., East, N., Perk, K. and Dahlberg, J.E. (1988) Development of an enzyme-linked immunosorbent assay for caprine arthritis-encephalitis virus. *Journal of Clinical Microbiology* 26, 971-975.

Banks, K.L., Adams, D.S., McGuire, T.C. and Carlson, J. (1983) Experimental infection of sheep by caprine arthritis-encephalitis virus and goats by progressive pneumonia virus. *American Journal of Veterinary Research* 44, 2307-2311.

Bird, P., Allen, D., Reyburn, H., Watt, N. and McConnell, I. (1990) Immunological studies in sheep naturally infected with maedi-visna virus. *European Federation of Immunological Societies. 10th meeting*. Edinburgh. Abstract.

Blum, H.E., Harris, J.D., Ventura, P., Walker, D., Staskus, K., Retzel, E. and Haase, A.T. (1985) Synthesis in cell culture of the gapped linear duplex DNA of the slow virus visna. *Virology* 142, 270-277.

Braun, M.J., Clements, J.E. and Gonda, M.A. (1987) The visna virus genome: evidence for a hypervariable site in the *env* gene and sequence homology among lentivirus envelope proteins. *Journal of Virology* 61, 4046-4054.

Cheevers, W.P. and McGuire, T.C. (1988) The lentiviruses: maedi/visna, caprine arthritis-encephalitis, and equine infectious anemia. *Advances in Virus Research* 34, 189-215.

Cheevers, W.P., Knowles, D.P., McGuire, T.C., Cunningham, D.R., Adams, D.S. and Gorham, J.R. (1988) Chronic disease in goats orally infected with two isolates of the caprine arthritis-encephalitis lentivirus. *Laboratory Investigation* 58, 510-517.

Clements, J.E., Narayan, O., Griffin, D.E. and Johnson, R.T. (1979) The synthesis and structure of visna virus DNA. *Virology* 93, 377-387.

Cork, L.C. (1990) Pathology and epidemiology of lentiviral infection of goats. In: Pétursson, G. and Hoff-Jørgensen (eds.) *Maedi-Visna and Related Diseases*. Developments in Veterinary Virology (series eds. Becker, Y. and Hadar, J.). Kluwer Academic Publishers, Boston/Dordrecht/London, pp. 119-127.

Cork, L.C., Hadlow, W.J., Crawford, T.B., Gorham, J.R. and Piper, R.C. (1974a) Infectious leukoencephalomyelitis of young goats. *Journal of Infectious Diseases* 129, 134-141.

Cork, L.C., Hadlow, W.J., Gorham, J.R., Piper, R.C. and Crawford, T.B. (1974b) Pathology of viral leukoencephalomyelitis of goats. *Acta Neuropathologica* (Berlin) 29, 281-291.

Crane, S.E., Clements, J.E. and Narayan, O. (1988) Separate epitopes in the envelope of visna virus are responsible for fusion and neutralization: Biological implications for anti-fusion antibodies in limiting virus replication. *Journal of Virology* 62, 2680-2685.

Crawford, T.B., Adams, D.S., Cheevers, W.P. and Cork, L.C. (1980a) Chronic arthritis in goats caused by a retrovirus. *Science* 207, 997-999.

Crawford, T.B., Adams, D.S., Sande, R.D., Gorham, J.R. and Henson, J.B. (1980b) The connective tissue component of the caprine arthritis-encephalitis syndrome. *American Journal of Pathology* 100, 443-454.

Cullen, B.R. (1991) Human immunodeficiency virus as a prototype complex retrovirus. *Journal of Virology* 65, 1053-1056.

Cutlip, R.C., Jackson, T.A. and Laird, G.A. (1977) Immunodiffusion test for ovine progressive pneumonia. *American Journal of Veterinary Research* 38, 1081-1084.

Cutlip, R.C., Lehmkuhl, H.D. and Jackson, T.A. (1981) Intrauterine transmission of ovine progressive pneumonia virus. *American Journal of Veterinary Research* 42, 1795-1797.

Cutlip, R.C., Lehmkuhl, H.D., Brogden, K.A. and Bolin, S.R. (1985a) Mastitis associated with ovine progressive pneumonia. *American Journal of Veterinary Research* 46, 326-328.

Cutlip, R.C., Lehmkuhl, H.D., Brogden, K.A. and Schmerr, M.J.F. (1987) Failure of experimental vaccines to protect against infection with ovine progressive pneumonia (maedi-visna) virus. *Veterinary Microbiology* 13, 201-204.

Cutlip, R.C., Lehmkuhl, H.D., Schmerr, M.J.F. and Brogden, K.A. (1988) Ovine progressive pneumonia (maedi-visna) in sheep. *Veterinary Microbiology* 17, 237-250.

Cutlip, R.C., Lehmkuhl, H.D., Wood, R.L. and Brog den, K.A. (1985b) Arthritis associated with ovine progressive pneumonia. *American Journal of Veterinary Research* 46, 65-68.

Dahme, E., Stavrou, D., Deutschländer, N., Arnold, W. and Kaiser, E. (1973) Klinik und Pathologie einer übertragbaren granulomatösen Meningo-encephalo-myelitis (gMEM) bei der Hausziege. *Acta Neuropathologica (Berlin)* 23, 59-76.

Dalgleish, A.G., Beverly, P.C.L., Clapham, P.R., Crawford, D.H., Greaves, M.F. and Weiss, R.A. (1984) The CD4 (T4) antigen is an essential component of the receptor for the AIDS retrovirus. *Nature* 312, 763-766.

Davis, J.L. and Clements, J.E. (1988) Characterization of a cDNA clone encoding the visna virus trans-activating protein. *Proceedings of the National Academy of Sciences*, USA 86, 414-418.

Davis, J.L., Molineaux, S. and Clements, J.E. (1987) Visna virus exhibits a complex transcriptional pattern: One aspect of gene expression shared with the acquired immunodeficiency syndrome retrovirus. *Journal of Virology* 61, 1325-1331.

Dawson, M. (1988) Lentivirus diseases of domesticated an imals. *Journal of Comparative Pathology* 99, 401-414.

Dawson, M., Biront, P. and Houwers, D.J. (1982) Comparison of serological tests used in three state veterinary laboratories to identify maedi-visna infection. *Veterinary Record* 111, 432-434.

Dawson, M., Venables, C. and Jenkins, C.E. (1985) Experimental infection of a natural case of sheep pulmonary adenomatosis with maedi-visna virus. *Veterinary Record* 116, 588-589.

De Boer, G.F. (1970) Antibody formation in zwoegerziekte, a slow infection in sheep. *Journal of Immunology* 104, 414-422.

De Boer, G.F., Terpstra, C. and Houwers, D.J. (1979) Studies in epidemiology of maedi/visna in sheep. *Research in Veterinary Science* 26, 202-208.

Dohoo, I.R., Heaney, D.P., Stevenson, R.G., Samagh, B.S. and Rhodes, C.S. (1987) The effect of maedi-visna virus infection on productivity in ewes. *Preventive Veterinary Medicine* 4, 471-484.

East, N.E., Rowe, J.D., Madewell, G.R. and Floyd, K. (1987) Serologic prevalence of caprine arthritis-encephalitis virus in California goat dairies. *Journal of the American Veterinary Medical Association* 190, 182-186.

Ellis, J.E. and DeMartini, J.C. (1985) Ovine interleukin-2: partial purification and assay in normal sheep with ovine progressive pneumonia. *Veterinary Immunology and Immunopathology* 8, 15-25.

Fankhauser R. and Theus, T. (1983) Visna bei der Ziege. *Schweizer Archiv für Tierheilkunde* 125, 387-390.

Frank, K.B., McKernan, P.A., Smith, R.A. and Smee, D.F. (1987) Visna virus as an in vitro model for human immunodeficiency virus and inhibition by ribavirin, phosphonoformate, and 2', 3',-dideoxynucleosides. *Antimicrobial Agents and Chemotherapy* 31, 1369-1374.

Gelderblom, H.R., Özel, M., Gheysen, D., Reupke, H., Winkel, T., Herz, U., Grund, C. and Pauli, G. (1990) Morphogenesis and fine structure of lenti-viruses. In: Schellekens, H. and Horzinek, M.C. (eds.). *Animal Models in AIDS*. Elsevier, Amsterdam/New York/Oxford, pp. 1-26.

Gelderblom, H.R., Reupke, H., Winkel, T., Kunze, R. and Pauli, G. (1987) MHC-antigens: Constituents of the envelopes of human and simian immuno-deficiency viruses. *Zeitschrift für Naturforschung* 42c, 1328-1334.

Gendelman, H.E., Narayan, O., Kennedy-Stoskopf, S., Kennedy, P.G.E., Ghotbi, Z., Clements, J.E., Stanley, J. and Pezeshkpour, G. (1986) Tropism of sheep lentiviruses for monocytes: Susceptibility to infection and virus gene expression increase during maturation of monocytes to macrophages. *Journal of Virology* 58, 67-74.

Georgsson, G. (1990) Maedi-Visna. Pathology and Pathogenesis. In: Pétursson, G. and Hoff-Jørgensen, R.H. (eds.). *Maedi-Visna and Related Diseases.* Developments in Veterinary Virology (series eds. Becker, Y. and Hadar, J.). Kluwer Academic Publishers, Boston/Dordrecht/London, pp. 19-54.

Georgsson, G., Houwers, D.J., Pálsson, P.A. and Pétursson, G. (1989) Expression of viral antigens in the central nervous system of visna-infected sheep: An immunohistochemical study in experimental visna induced by virus strains of increased neurovirulence. *Acta Neuropathologica* (Berlin) 77, 299-306.

Georgsson, G., Martin, J.R., Klein, J., Pálsson, P.A., Nathanson, N. and Pétursson, G. (1982). Primary demyelination in visna. *Acta Neuropathologica* (Berlin) 57, 171-178.

Georgsson, G., Nathanson, N., Pálsson, P.A. and Pétursson, G. (1976) The pathology of visna and maedi in sheep. In: Kimberlin, R.H. (ed.) *Slow Virus Diseases of Animals and Man.* North-Holland Publishing Company, Amsterdam/Oxford/New York, pp. 61-96.

Gonda, M.A., Braun, M.J., Clements, J.E., Pyper, J.M., Casey, J.W., Wong-Staal, F., Gallo, R.C. and Gilden, R.V. (1986) HTLV-III shares sequence homology with a family of pathogenic lentiviruses. *Proceedings of the National Academy of Sciences*, USA 83, 4007-4011.

Gourdou, I., Mazarin, V., Quérat, G., Sauze, N. and Vigne, R. (1989) The open reading frame S of visna virus genome is a trans-activating gene. *Virology* 171, 170-178.

Gudnadóttir, M. (1974) Visna-maedi in sheep. *Progress in Medical Virology* 18, 336-349.

Gudnadóttir, M. and Kristinsdóttir, K. (1967) Complement-fixing antibodies in sera of sheep affected with visna and maedi. *Journal of Immunology* 98, 663-667.

Haase, A.T. (1989) Pathogenesis of lentivirus infection. *Nature* 322, 130-136.

Haase, A.T., Stowring, L., Harris, J.D., Traynor, B., Ventura, P., Peluso, R. and Brahic, M. (1982) Visna DNA synthesis and the tempo of infection in vitro. *Virology* 119, 399-410.

Harris, J.D., Scott, J.V., Traynor, B., Brahic, M., Stowring, L., Ventura, P., Haase, A.T. and Peluso, R. (1981) Visna virus DNA: discovery of a novel gapped structure. *Virology* 113, 573-583.

Hess, J.L, Small, J.A. and Clements, J.E. (1989) Sequences in the visna virus long terminal repeat that control transcriptional activity and respond to viral trans-activation: Involvement of AP-1 sites in basal activity and trans-activation. *Journal of Virology* 63, 3001-3015.

Hoff-Jørgensen, R. (1977) Slow virus infections with particular reference to maedi-visna and enzootic bovine leukemia. *Veterinary Science Communications* 1, 251-263.

Houwers, D.J. (1990) Economic importance, epidemiology and control. In: Pétursson, G. and Hoff-Jørgensen, R. (eds.), *Maedi-Visna and Related Diseases.* Developments in Veterinary Virology (Series eds. Becker, Y. and Hader, J.). Kluwer Academic Publishers, Boston/Dordrecht/London pp. 83-117.

Houwers, D.J. and Nauta, I.M. (1989) Immunoblot analysis of the antibody response to ovine lentivirus infections. *Veterinary Microbiology* 19, 127-139.

Houwers, D.J. and Schaake, J. (1987) An improved ELISA for the detection of antibodies to ovine and caprine lentiviruses, employing monoclonal antibodies in a one-step assay. *Journal of Immunological Methods* 98, 151-154.

Houwers, D.J., Gielkens, A.L.J. and Schaake, J. (1982) An indirect enzyme-linked immunosorbent assay (ELISA) for the detection of antibodies to maedi-visna virus. *Veterinary Microbiology* 7, 209-219.

Houwers, D.J., König, C.D.W., Bakker, J., de Boer, M.J., Pekelder, J.J., Sol, J., Vellema, P. and de Vries, G. (1987) Maedi-visna control in sheep III. Results of a voluntary control program in the Netherlands over a period of four years. *The Veterinary Quarterly* 9, Suppl. 1, 29-36.

Houwers, D.J., König, C.D.W., de Boer, G.F. and Schaake, J. (1983) Maedi-visna control in sheep I. Artificial rearing of colostrum deprived lambs. *Veterinary Microbiology* 8, 179-185.

Houwers, D.J., Schaake, J. and de Boer, G.F. (1984) Maedi-visna control in sheep II. Half-yearly serological testing with culling of positive ewes and progeny. *Veterinary Microbiology* 9, 445-451.

Jolly, P.E., Huso, D., Hart, G. and Narayan, O. (1989) Modulation of lentivirus replication by antibodies. Non-neutralizing antibodies to caprine arthritis-encephalitis virus enhance early stages of infection in macrophages, but do not cause increased production of virions. *Journal of General Virology* 70, 2221-2226.

Kennedy-Stoskopf, S. and Narayan, O. (1986) Neutralizing antibodies to visna lentivirus: mechanism of action and possible role in virus persistence. *Journal of Virology* 59, 37-44.

Mazarin, V., Gourdou, I., Quérat, G., Sauze, N. and Vigne, R. (1988) Genetic structure and function of an early transcript of visna virus. *Journal of Virology* 62, 4813-4818.

Mazarin, V., Gourdou, I., Quérat, G., Sauze, N., Audoly, G., Vitu, C., Russo, P., Rousselat, C., Filippi, P. and Vigne, R. (1990) Subcellular localization of rev-gene product in visna virus-infected cells. *Virology* 178, 305-310.

McGuire, T.C., Adams, D.S., Johnson, G.C., Klevjer-Anderson, P., Barbee, D.D. and Gorham, J.R. (1986) Acute arthritis in caprine arthritis-encephalitis virus challenge exposure of vaccinated or persistently infected goats. *American Journal of Veterinary Research* 47, 537-540.

McGuire, T.C., O'Rourke, K.I., Knowles, D.P. and Cheevers, W.P. (1990) Caprine arthritis encephalitis lentivirus. Transmission and disease. *Current Topics in Microbiology and Immunology* 160, 61-75.

Mitchell, D.T. (1915) Investigations into Jagziekte or Chronic Catarrhal-Pneumonia of Sheep. *Third and fourth reports of the Director of Veterinary Research*, Union of South Africa pp. 585-614.

Myer, M.S., Huchzermeyer, H.F.A.K., York, D.F., Hunter, P., Verwoerd, D.W. and Garnett, H.M. (1988) The possible involvement of immuno-suppression caused by a lentivirus in the aetiology of jaagsiekte and pasteurellosis in sheep. *Onderstepoort Journal of Veterinary Research* 55, 127-133.

Narayan, O. and Clements, J.E. (1989) Biology and pathogenesis of lentiviruses. *Journal of General Virology* 70, 1617-1639.

Narayan, O. and Cork, L.C. (1985) Lentiviral diseases of sheep and goats: chronic pneumonia, leukencephalomyelitis and arthritis. *Review of Infectious Diseases* 7, 89-98.

Narayan, O., Griffin, D.E. and Clements, J.E. (1978) Visna mutation during 'slow infection': temporal development and characterization of mutants of visna virus recovered from sheep. *Journal of General Virology* 41, 343-352.

Naryan, O., Zink, M.C., Huso, D., Sheffer, D., Crane, S., Kennedy-Stoskopf, S., Jolly, P.E. and Clements, J.E. (1988) Lentiviruses of animals are biological models of the human immunodeficiency viruses. *Microbial*

Pathogenesis 5, 149-157.

Nathanson, N., Martin, J.R., Georgsson, G., Pálsson, P.A., Lutley, R.E. and Pétursson, G. (1981) The effect of post-infection immunization on the severity of experimental visna. *Journal of Comparative Pathology* 91, 185-191.

Nathanson, N., Panitch, H., Pálsson, P.A., Pétursson, G. and Georgsson, G. (1976) Pathogenesis of visna II. Effect of immunosuppression upon early central nervous system lesions. *Laboratory Investigation* 35, 444-451.

Oliver, R., Cathcart, A., McNiven, R., Poole, W. and Robati, G. (1985) Infection of lambs with caprine arthritis-encephalitis virus by feeding milk from infected goats. *Veterinary Record* 116, 83.

Pálsson, P.A. (1976) Maedi and visna in sheep. In: Kimberlin (ed.), *Slow Virus Diseases of Animals and Man*. North-Holland Publishing Company, Amsterdam, Oxford, pp. 17-43.

Panitch, H., Pétursson, G., Georgsson, G., Pálsson, P.A. and Nathanson, N. (1976) Pathogenesis of visna III. Immune responses to central nervous system antigens in experimental allergic encephalomyelitis and visna. *Laboratory Investigation* 35, 452-460.

Peterhans, E., Zanoni, R., Krieg, T. and Balcer, T. (1988) Lentiviren bei Schaf und Ziege: Eine Literaturübersicht. *Schweizer Archiv für Tierheilkunde* 130, 681-700.

Pétursson, G., Georgsson, G. and Pálsson, P.A. (1990) Maedi-visna virus. In: Dinter, Z. and Morein, B. (eds.). *Virus Infections of Vertebrates. Vol. 3. Virus Infections of Ruminants*. Elsevier Science Publishers B.V., Amsterdam, pp. 431-440.

Pétursson, G., Nathansson, N., Georgsson, G., Panitch, H. and Pálsson, P.A. (1976) Pathogenesis of visna. I. Sequential virologic, serologic and pathologic studies. *Laboratory Investigation* 35, 402-412.

Pétursson, G., Pálsson, P.A. and Georgsson, G. (1989) Maedi-visna in sheep: Host-virus interactions and utilization as a model. *Intervirology* 30, 36-44

Pyper, J.M., Clements, J.E., Molineaux, S.M. and Narayan, O. (1984) Genetic variation among lentiviruses: homology between visna virus and caprine arthritis-encephalitis virus is confined to the 5' *gag-pol* region and a small portion of the *env* gene. *Journal of Virology* 51, 713-721.

Quérat, G., Andoly, G., Sonigo, P. and Vigne, R. (1990) Nucleotide sequence analysis of SA-OMVV, a visna-related ovine lentivirus: phylogenetic history of lentiviruses. *Virology* 175, 434-447.

Quérat, G., Barban, V., Sauze, N., Filippi, R., Vigne, R., Ruse, R. and Vitu, C. (1984) Highly lytic and persistent lentiviruses naturally present in sheep with progressive pneumonia are genetically distinct. *Journal of Virology* 52, 672-679.

Rajya, B.S. and Singh, C.M. (1964) The pathology of pneumonia and associated respiratory disease of sheep and goats. I. Occurrence of jaagziekte and maedi in sheep and goats in India. *American Journal of Veterinary Research* 25, 61-67.

Robinson, W.F. (1981) Chronic interstitial pneumonia in association with a granulomatous encephalitis in a goat. *Australian Veterinary Journal* 57, 127-131.

Robinson, W.F. and Ellis, T.M. (1984) The pathological features of an interstitial pneumonia of goats. *Journal of Comparative Pathology* 94, 55-64.

Robinson, W.F. and Ellis, T.M. (1986) Caprine arthritis-encephalitis virus infection: from recognition to eradication. *Australian Veterinary Journal* 63, 237-241.

Ruff, G. and Lazary, S. (1988) Evidence for linkage between the caprine leucocyte antigen (CLA) system and susceptibility to CAE virus-induced arthritis in goats. *Immunogenetics* 28, 303-309.

Saltarelli, M., Quérat, G., Konings, D.A.M., Vigne, R. and Clements, J.E. (1990) Nucleotide sequence and transcriptional analysis of molecular clones of CAEV which generate infectious virus. Virology 179, 347-364.

Sargan, D.R. and Bennet, I.D. (1989) A transcriptional map of visna virus: Definition of the second intron structure suggests a rev-like product. Journal of General Virology 70, 1995-2006.

Sharp, J.M. and Angus, K.W. (1990) Sheep pulmonary adenomatosis: Studies on its aetiology. In: Pétursson, G. and Hoff-Jórgensen, R. (eds.). Maedi-Visna and Related Diseases. Developments in Veterinary Virology (Series eds. Becker, Y. and Hadar, J.). Kluwer Academic Publishers, Boston/Dordrecht/London, pp. 177-185.

Sigurdardóttir, B. and Thormar, H. (1964) Isolation of a viral agent from the lungs of sheep affected with maedi. Journal of Infectious Diseases 114, 55-60.

Sigurdsson, B. (1954) Observations on three slow infections of sheep. British Veterinary Journal 110, 255-270.

Sigurdsson, B., Pálsson, P.A. and Tryggvadóttir, A. (1953) Transmission experiments with maedi. Journal of Infectious Diseases 93, 166-175.

Sigurdsson, B., Thormar, H. and Pálsson, P.A. (1960) Cultivation of visna virus in tissue culture. Archiv für die Gesamte Virusforschung 10, 368-381.

Sihvonen, L. (1981) Experimental maedi virus infection in sheep. In vivo and in vitro studies. Thesis, College of Veterinary Medicine, Helsinki, Finland. 33 pp.

Smith, V.W., Dickson, J., Coakley, W. and Carman, H. (1985) Response of Merino sheep to inoculation with a caprine retrovirus. Veterinary Record 117, 61-63.

Snowder, G.D., Gates, N.L., Glimp, H.A. and Gorham, J.R. (1990) Prevalence and effect of subclinical ovine progressive pneumonia virus infection on ewe wool and lamb production. Journal of the American Veterinary Medical Association 197, 475-479.

Sonigo, P., Alizon, M., Staskus, K., Klatzmann, D., Cole, S., Danos, O., Retzel, E., Tiollais, P., Haase, A. and Wain-Hobson, S. (1985) Nucleotide sequence of the visna lentivirus: Relationship to the AIDS virus. Cell 42, 369-382.

Stavrou, D., Deutschländer, N. and Dahme, E. (1969) Granulomatous encephalitis in goats. Journal of Comparative Pathology 79, 393-396.

Sundquist, B., Joensson, L., Jacobsson, S.-O. and Hammarberg, K.-E. (1981) Visna virus meningo-encephalomyelitis in goats. Acta Veterinaria Scandinavica 22, 315-330.

Thormar, H. (1963) The growth cycle of visna virus in monolayer culture of sheep cells. Virology 19, 273-278.

Van der Molen, E.J., Vecht, U. and Houwers, D.J. (1985) A chronic indurative mastitis in sheep associated with maedi/visna infection. Veterinary Quarterly 7, 112-119.

Vigne, R., Barban, V., Quérat, G., Mazarin, V., Gourdou, I. and Sauze, N. (1987) Transcription of visna virus during its lytic cycle: Evidence for a sequential early and late gene expression. Virology 161, 218-227.

Weinhold, E., Müller, A. and Leuchte, S. (1973) Visna-Virus-ähnliche Partikel in der Kultur von Plexus chorioideus-Zellen einer Ziege mit Visna-Symptomen. Zentralblatt für Veterinäre Medizin B 21, 32-36.

Weinhold, E. and Triemer, B. (1978) Visna bei der Ziege. Zentralblatt für Veterinäre Medizin B 25, 525-538.

Zanoni, R., Krieg, A. and Peterhans, E. (1989) Detection of antibodies to caprine arthritis-encephalitis virus by protein G enzyme-linked immunosorbent assay and immunoblotting. Journal of Clinical Microbiology 27, 580-582.

Zanoni, R., Pauli, U. and Peterhans, E. (1990) Detection of caprine arthritis-encephalitis and maedi-visna viruses using the polymerase chain reaction. *Experientia* 46, 316-318.

Zink, M.C., Yager, J.A. and Myers, J.D. (1990) Pathogenesis of caprine arthritis-encephalitis virus. Cellular localization of viral transcripts in tissues of infected goats. *American Journal of Pathology* 136, 843-853.

Zwahlen, R., Aeschbacher, M., Balcer, T., Stucki, M., Wyder-Walther, M., Weiss, M. and Steck, F. (1983) Lentivirusinfektionen bei Ziegen mit Carpitis und interstitieller Mastitis. *Schweizer Archiv für Tierheilkunde* 125, 281-299.

Chapter 6

Scrapie in Sheep and Goats

Nora Hunter

Institute for Animal Health, AFRC/MRC Neuropathogenesis Unit, Ogston Building, West Mains Road, Edinburgh EH11 1JZ, UK

Introduction

Scrapie is a transmissible degenerative disease which affects the central nervous system of sheep and goats. It can be transmitted from affected to healthy sheep both vertically and horizontally, by inoculation or by feeding of diseased tissues, however the susceptibility of goats to scrapie is not so well understood. There is experimental evidence suggesting that there are different strains of scrapie and that it behaves biologically as a virus, however the infectious entity has never been identified. There are several views on its aetiology: some citing the unusual physico-chemical resistances of scrapie (e.g. Brown *et al.*, 1990) suggest that it may be made up mostly or completely of protein (the prion hypothesis) (Prusiner, 1982) or largely of protein with a small nucleic acid component (the virino hypothesis) (Dickinson and Outram, 1988) while still others, because sheep scrapie tends to "run in families", suggest that it is simply a genetic disease (Parry, 1984).

There are several similar diseases of mammals including man (Creutzfeldt-Jakob Disease - CJD, Gerstmann-Straussler-Scheinker syndrome - GSS) and domesticated or captive animals (including Bovine Spongiform Encephalopathy of cattle - BSE). These disorders are often grouped together under the general term Transmissible Spongiform Encephalopathies (TSE) and they have in common many features including long, largely asymptomatic, incubation periods which may last for months or years (Dickinson, 1976). In experimental scrapie models, it has proved possible to lengthen incubation period with various treatments (Pocchiari *et al.*, 1989). However, although there are a few anecdotal stories of recovery of individual animals from scrapie-like symptoms (e.g. Pattison and Millson, 1961), it is generally true that once signs of TSE develop, it is not possible to effect a cure.

Scrapie is a difficult disease to work with - there is no preclinical

131

diagnostic test (hence no way of knowing whether an apparently healthy animal may be incubating the disease); there is no immune or inflammatory response in the sick animal and the only way to detect scrapie is by bioassay (transmission to other animals) as the infectious agent has not been characterized. Diagnosis of the disease is based on clinical signs, post-mortem brain pathology and post-mortem biochemistry.

Clinical symptoms vary but typical signs in sheep were described by Dickinson (1976) as starting insiduously with mildly impaired social behaviour followed by locomotor incoordination or ataxia with a fine trembling pruritis is often seen as an apparently intense itching is relieved by rubbing against fences or trees and even biting of the affected area. The raw wounds and loss of fleece which develop can be quite extensive. The duration of clear signs of illness lasts from 2 weeks to 6 months depending on the sheep flock involved - these differences both in symptoms and clinical phase in different flocks may be simply breed dependent or may indicate the existence in the field (as in the laboratory: Bruce *et al.*, 1991) of different scrapie strains. Goat scrapie symptoms are generally the same as in sheep (Hadlow, 1960) but outbreaks vary - three naturally infected Nubian x Toggenburg goats described by Hadlow *et al.* (1980) suffered from severe incoordination but did not show signs of scratching or rubbing. There is also experimental evidence of two different forms of scrapie in goats. These were called "drowsy" and "scratching" because of the differences in major symptoms and were the first indication that there might be strain variation in scrapie (Pattison and Millson, 1961).

Some of the neurological lesions of scrapie seen in post-mortem examination of the central nervous system (CNS) are also found in the other TSEs: common features in the brain include neuronal degeneration and the formation of vacuoles, proliferation of astroglial cells but no demyelination or other overt inflammatory response. One major study of experimental scrapie in sheep and goats (Zlotnik, 1962) described brain areas with "almost complete destruction of the brain substance", certain foci having a "network of fibres and strands with only a few neurones remaining". These features develop in the later stages of the incubation period, and it is very difficult to detect (by histopathology) those animals with scrapie which are not visibly affected by the disease. This has greatly hindered the effectiveness of approaches for the control and management of scrapie.

The only biochemical markers which can be used to indicate the presence of scrapie like agents are scrapie-associated fibrils (SAF) (Merz *et al.*, 1984) which are made up of modified forms of a host protein called PrP. These and other PrP aggregates (amyloid plaques) accumulate in the brain and some peripheral tissues (spleen, lymph nodes) (Farquhar *et al.*, 1990) during disease development but the process involved is not understood. The normal function of PrP, a relatively abundant protein in mammalian brain, is not known.

The advice to farmers facing an outbreak of scrapie in their sheep has remained unchanged for a very long time. From the time of the first recording of scrapie in Britain in the eighteenth century it has been noticed that the disease "runs in families". The most common route of transmission is from ewe to lamb and therefore the recommendation is to cull all animals in an affected line whether they appear sick or healthy. This drastic and often expensive measure will not always succeed in eradicating scrapie - contagious spread undoubtedly exists in sheep (Dickinson *et al.*, 1974) and not every susceptible animal will necessarily succumb in any one year (Gordon, 1966). Contact transmission of scrapie in goats was reported by Chelle in 1942: a case of scrapie occurred in a goat kept in close contact with affected sheep and indeed, it has been thought to be usual for goats to contract scrapie from association with scrapie sheep. There is not enough information available however, to say whether scrapie "runs in families" in goats.

Experimental Scrapie in Sheep

Early experimental studies of transmission of scrapie using sheep often gave confusing and contradictory results. One of those most often quoted is the "twenty four breed" experiment described by W.S. Gordon in 1966. Twenty four different breeds of sheep (approximately 40 animals per breed) were injected subcutaneously (s.c.) with the source of scrapie known as SSBP/1 (Dickinson, 1976) and the animals were observed for two years. Seventy-eight per cent of the Herdwick group succumbed in this time and each breed was ranked in decreasing order of susceptibility ending with Dorset Downs which were all apparently resistant. These breed differences in susceptibility were not always reproducible; in a different group of Herdwicks only 30% were susceptible and, in the original group of "resistant" Dorset Downs, some animals did eventually develop SSBP/1 scrapie at much later dates (Dickinson, 1976). It became clear that there was as much variation in the "take" of scrapie inoculation between different flocks of the same breed as there was between different breeds and so lines of sheep were bred to have more predictable response to induced scrapie. There are three such flocks in Britain - Cheviots and Herdwicks (both at NPU, Edinburgh) and Swaledales (at Redesdale, Northumberland).

NPU Cheviots

The Cheviot flock selection was started in the 1950's and the sheep were split into two lines (positive and negative) on the basis of their response to subcutaneous (s.c.) injection with SSBP/1 (sheep scrapie brain pool 1) (Dickinson, 1976). The survival time of the sheep is mainly determined by a single gene - *Sip*, with two alleles sA and pA; sA is dominant (Dickinson and Outram, 1988). When injected s.c., the positive line (*Sip* sAsA or *Sip* sApA) developed scrapie in an average of 300 days whereas the negative line (*Sip* pApA) survived a natural lifespan. Negative line sheep were not thought to be completely resistant as some would contract scrapie in around 1,000 days following intracerebral (i.c.) injection. Positive line sheep died within 200 days of i.c. injection of SSBP/1 (Dickinson *et al.*, 1968). Isolates (and perhaps natural outbreaks) of scrapie were classified according to their relative effects on sheep of the different *Sip* genotypes. Most isolates (A group) produced the disease in carriers of the sA allele faster than in pApA sheep, but with one isolate (CH1641) (C group) the ranking of *Sip* type (in respect of survival time) is not clear cut and may even be reversed (Foster and Dickinson, 1988a) - see Table 6.1.

Table 6.1 Incubation periods of scrapie in NPU Cheviot sheep.

Scrapie source	Route of injection	Incubation period (days ± sem)	
		Positive line *Sip* sAsA/sApA	Negative line *Sip* pApA
SSBP/1	i.c.[*]	197 ± 7	917 ± 90
	s.c.[*]	313 ± 9	SURVIVE
CH1641	i.c.[*]	595 ± 122	360 ± 15

Data adapted from Dickinson *et al.* (1968) & Foster and Dickinson (1988a).
[*]i.c. = intracerebral, s.c = subcutaneous

NPU Herdwicks

Also in 1961, W.S. Gordon at Compton began a similar selection process using SSBP/1 and Herdwick sheep (Nussbaum *et al.*, 1975). Positive and negative lines were produced in the same way as the Cheviot lines and susceptibility differences were shown again to be under the control of a single gene with two alleles: that conferring high susceptibility to SSBP/1 was dominant. There is no formal proof that this gene is *Sip* but the results of crossing experiments suggest that it is (Foster and Hunter, 1991). There are very few of these selected Herdwicks remaining now - about thirty in number.

Swaledales

During the course of the development of the Herdwick lines, it became apparent that resistance to experimental scrapie could be identified within one generation and this discovery was used to breed a low susceptibility Swaledale flock (Hoare *et al.*, 1977; Davies and Kimberlin, 1985). Sheep were inoculated subcutaneously with natural scrapie isolates and only survivors were used in subsequent breeding. The progeny of these survivors had a greatly reduced susceptibility to experimental challenge with scrapie and were used to form a nucleus flock of scrapie "resistant" (or negative line) Swaledales now at the MAFF Experimental Husbandry Farm, Redesdale, Northumberland. It is likely, although formal proof is lacking, that these sheep are also of the genotype *Sip* pApA in which replication or the deleterious effects of the scrapie pathogen may be sufficiently inhibited to allow survival.

Experimental Scrapie in Goats

The transmission of SSBP/1 scrapie to all of a group of crossbred goats by i.c. injection (Pattison *et al.*, 1959) was followed by its passage from goats to goats again with 100% incidence. Pattison and Millson (1962) reviewed 24 experiments carried out between 1956 and 1962 involving 170 goats injected i.c. with scrapie brain homogenates - all animals succumbed with incubation periods ranging from 7 to 22 months (mean 12 months). This finding of 100% susceptibility with goats of various types has been reported several times but experiments have usually been carried out with goats of uncontrolled genetic background: for example, the crossbreds in Pattison's studies at Comptom had a mixture of Anglo-Nubian, Toggenburg, Saanen and British Alpine ancestry (Dickinson, 1976). Comparison with genetic analysis of mice and sheep would suggest that differences in susceptibility are likely. Until this is demonstrated, however, the existence of a goat homologue of *Sip* cannot be assumed.

Biochemical and Genetic Studies

The Herdwick sheep lines have been studied in an attempt to find genetic or biochemical differences between them which could be linked to the differences in susceptibility or which could be used to differentiate the sick from the healthy animal. No correlation was found with susceptibility and phenotypes of albumin, pre-albumin, esterase, haemoglobin, transferrin, reduced glutathione or a-mannosidase (Collis and Millson, 1975; Collis *et al.*, 1977). Increased serum levels of IgG were found in positive line Herdwick sheep with natural scrapie but this finding was too variable to be useful (Collis *et al.*, 1979) and was not confirmed in Suffolk sheep with natural scrapie (Strain *et al.*, 1984).

A number of lymphocyte antigens were assessed by Cullen *et al.* (1984) for linkage with susceptibility differences in the Herdwicks but without success. This is interesting in the light of the report by Millot *et al.* (1988) of a linkage of OLA (Ovine Lymphocyte Antigen) haplotypes in Ile de France sheep with natural scrapie. Cullen (1989) has disputed this linkage on the grounds of a possible founder effect and the study clearly needs to be extended to unrelated sheep.

A few biochemical studies of goats affected by scrapie have been reported. Slater (1965), because scrapie infectivity co-purifies with membranes, suspected that it might be possible to detect changes in mitochondrial or lysosomal enzymes during scrapie infection. Activities of cytochrome oxidase and succinic dehydrogenase were measured in brain tissue of goats experimentally infected with the "drowsy" form of scrapie, and found to be slightly elevated compared with controls injected with saline. This was not found in mice affected by ME7 scrapie however.

PrP protein and its link with scrapie susceptibility

PrP protein was discovered originally as a component of scrapie-associated fibrils (SAF) which can be extracted from brain tissue of all mammals affected by TSEs (Oesch *et al.*, 1985; Hope *et al.*, 1986). Although it was at first thought to be a component of the infectious agent, it was quickly discovered to be a host encoded protein the product of a single copy gene expressed mostly in the brain (Basler *et al.*, 1986). It is now thought to be present on the outside of neuronal membranes and is also found in other tissues, however its normal function remains unknown although it has some sequence homology with ARIA, a protein which affects chicken acetyl choline receptors (Harris *et al.*, 1989).

The PrP gene in transgenic mice

Linkage of alleles of the PrP gene with the gene controlling susceptibility differences was first demonstrated in the mouse (Carlson *et al.*, 1986) and this has since been extended to sheep and man (Hunter *et al.*, 1989; Foster and Hunter, 1991; Hsiao *et al.*, 1989). Indeed transgenic studies suggest that the PrP gene not only controls host range of TSEs but also, in a mutant form linked to incidence of the GSS in humans, will independently produce spongiform disease pathology (Hsiao *et al.*, 1990).

The hamster PrP gene (HaPrP) inserted into the mouse genome was found to change the response of the mice to a hamster-passaged strain of scrapie (Prusiner *et al.*, 1990). Transgenic mouse lines became susceptible to hamster scrapie with incubation periods of 48-277 days related inversely to the amount of HaPrP protein produced. Non-transgenic (or normal) mice were much more resistant to hamster scrapie with only a few succumbing more than 400 days after inoculation. The HaPrP transgenic mice replicated and reproduced the type of scrapie inoculated into them - injection with hamster scrapie resulting in high titres of hamster scrapie while a mouse-passaged strain produced high titres of mouse scrapie. Pathology typical of hamster scrapie was found in terminal transgenic mouse brain and included large deposits or plaques of PrP protein which reacted with hamster PrP specific monoclonal antibodies. This work suggests that in some way PrP protein controls the host range of scrapie in these animals. HaPrP transgenic animals remain healthy unless inoculated with scrapie. However this is not true with the mice transgenic for the variant of human PrP linked to the incidence of GSS - these animals spontaneously develop scrapie-like symptoms and die with spongiform brain pathology at around 166 days of age (Hsiao *et al.*, 1990). The precise reason for this result is the subject of great debate but at the very least it emphasizes the central importance of PrP in these diseases.

The PrP gene in sheep and goats

Studies of PrP protein in ruminants are at an earlier stage compared with the more easily controlled laboratory mouse work. Sheep and goats with experimental and natural scrapie, however also produce abnormal PrP deposits (see Figure 6.1). Disease specific PrP is differentiated from the normal protein by its relative resistance to proteases. In Figure 6.1, scrapie goat PrP protein is shown. This exhibits the resistance of all the TSE scrapie PrP protein preparations to Proteinase K. As in all other cases tested so far, goat normal PrP is completely degraded by these treatments.

Sheep

In NPU Cheviot sheep, an association between restriction fragment length polymorphisms (RFLPs) of the PrP gene and alleles of *Sip* has been demonstrated (Figure 6.2). An *Eco* RI fragment of 6.8 kb (e1) is associated with *Sip* sA and a fragment of 4.0 kb (e3) with *Sip* pA. Following subcutaneous inoculation with SSBP/1, animals of PrP genotype e1e1 have incubation periods averaging 192 days, those with PrP genotype e1e3, 324 days and those with PrP e3e3 are survivors (Table 6.2; Foster and Hunter, 1991). PrP e1e1 behave as *Sip* sAsA; e1e3 as *Sip* sApA and e3e3 as *Sip* pApA. (The exact incubation periods differ from those shown in Table 6.1 as it details a different more recent experiment.) Goldmann *et al.* (1990) have also demonstrated an amino acid polymorphism of the PrP protein itself in Suffolk sheep and this may be linked to alleles of *Sip*.

Figure 6.1 Infection-specific PrP protein extracted from brain tissue of scrapie affected goats. Lanes 1 and 2 - experimental scrapie; Lanes 3 and 4 - natural scrapie. This is an immunoblot and the protein in Lanes 2 and 4 has been treated with Proteinase K. (Photo courtesy of Dr James Hope, NPU, Edinburgh.)

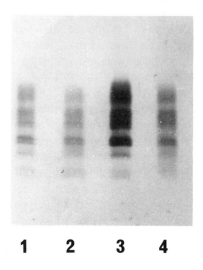

1   2   3   4

Figure 6.2 PrP gene *Eco*RI fragments e1 (6.8 kb) and e3 (4.0 kb) in NPU Cheviot sheep. Lane 1 - negative line (*Sip* pApA); Lanes 2 and 3 - positive line (*Sip* sApA and *Sip* sAsA). This is a Southern blot probed with a sheep PrP gene clone.

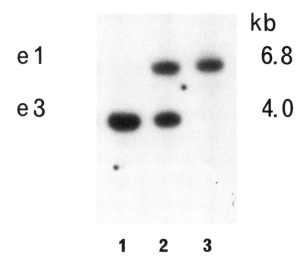

1   2   3

Table 6.2 Incubation period of SSBP/1 injected subcutaneously into NPU Cheviot sheep.

NPU Cheviot line	Positive line		Negative line
PrP gene *Eco*RI RFLP genotype	e1e1	e1e3	e3e3
Incubation period	192 ± 11* n = 11**	323 ± 16* n = 10**	SURVIVE
Putative *Sip* gene	sAsA	sApA	pApA

* days ± SE
** n = number of sheep
e1 - 6.8 kb
e3 - 4.0 kb
Data taken from Foster and Hunter (1991). The incubation period data is different from Table 6.1 - they represent different experiments.

Table 6.3 PrP RFLP genotypes in negative line Swaledale sheep.

*Eco*RI genotype	e1e1	e1e3	e3e3	(n)*
Frequency (%)	3	23	74	(79)

* n = numbers of sheep
e1 - 6.8kb
e3 - 4.0kb

In the scrapie "resistant" Swaledale flock, there is also a high frequency of PrP e3 (see Table 6.3) which in NPU Cheviots is associated with *Sip* pA or low susceptibility. As the Swaledales were selected, not with SSBP/1 but with SW73 and SW75 (natural Swaledale scrapie sources), this result suggests that the linkage of the PrP alleles with susceptibility differences might have more general application.

Mice transgenic for the sheep PrP protein gene are now being produced (Baybutt, personal communication; Westaway, personal communication) and it will be of great interest to see whether such animals become more susceptible to sheep scrapie.

Goats

The goat PrP gene has very high homology with both the sheep and the bovine PrP genes and the mRNA's from the three species are approximately the same size (4.6 kb) around twice the size of the rodent mRNA (Goldmann *et al.*, 1990). Although goat DNA also shows a 6.8 kb *Eco* RI restriction fragment, no polymorphism has been seen with this enzyme in 22 Anglo-Nubians and in 30 British Dairy Goats (Benson and Hunter, unpublished). There *is* a polymorphism with *Hin* dIII, but no linkage with susceptibility so far (Hunter *et al.*, 1989).

The Natural Disease

Natural transmission

The route of transmission of natural scrapie in sheep or goats is not known although extensive studies of the spread of infection within naturally or experimentally infected animals have been made in order to suggest possibilities.

A flock of Suffolk sheep with high incidence of natural scrapie has been used to study the body distribution of scrapie (Hadlow *et al.*, 1982). Time course assays of infectivity (demonstrated by mouse bioassay) in various tissues were carried out on sheep known to be at high risk of developing scrapie. Scrapie was detected at low titre in lymphatic tissues and in intestine in clinically normal 10-14 month lambs: the central nervous system was not found to be involved until about 25 months; nine control animals developed scrapie at approximately 3.5 years of age. Because of the early involvement of the alimentary tract in this form of Suffolk sheep scrapie, an oral route of transmission was postulated. However when milk, colostrum and faeces were assayed (Hourrigan *et al.*, 1979), they were found negative. Pattison *et al.* (1972 and 1974) had already discovered that it was possible to transmit scrapie to Herdwick sheep and to goats by feeding placental material from scrapie affected Swaledale ewes and this, coupled with the familial appearance of the disease, gave rise to the idea of oral transmission of scrapie via infected placentae contaminating pasture. Interestingly, infection-specific PrP protein has been found, by Western blotting, in the placental tissue of a naturally infected scrapie sheep (Ikegami *et al.*, 1991). However this single result was not compared with control placentae. Contamination of pasture as a source of infection has some supporting evidence from Icelandic scrapie (Palsson, 1979) and from the work of Grieg (1940a, 1950) and could be quite long lived because of scrapie's infamous resistance to inactivation.

Skin abrasions are also possible natural entry points for scrapie in sheep or goats and could be the route by which contact transmission occurs.

Scarification, or infection through skin abrasions, has not been exploited experimentally although Dickinson (1976) quoting Wilson's unpublished work from the 1950s implied that it is quite efficient.

Contact transmission

Published experiments on contact transmission whether sheep to sheep, sheep to goat or goat to goat are often contradictory. One of the main problems with such work is the uncertainty of whether the indicator animals (those exposed to scrapie animals) are genuinely scrapie free prior to exposure and would not go on to develop the disease anyway.

In 1968, Brotherston *et al.* described two experiments where sheep and goats were kept in close contact with naturally infected scrapie sheep for a number of years. Out of 17 test goats, 10 scrapie cases occurred after contact of up to 3.5 years with the scrapie sheep. Of the test sheep, 7 animals survived long enough (> 2 years) to develop scrapie and out of these 3 developed the disease. Prolonged contact from the time of birth was thought necessary in order to develop scrapie in this way as those animals born in the test area were judged more likely to show signs of contact transmission. However Harcourt (1974) reported two scrapie cases in goats with very little exposure to any sheep, let alone scrapie sheep.

The experimental disease appears to be more difficult to transmit by contact alone. Pattison *et al.* (1972) reported no such transmissions in 18 years of mixing scrapie affected sheep with other sheep and goats.

Hadlow *et al.* (1974) reported that experimentally infected goats had "limited and inconstant" amounts of scrapie (detected by mouse bioassay) in salivary glands and intestinal tract and attributed to this their difficulty in demonstrating goat to goat transmission. They believed at that time that scrapie was unlikely to be established in goat herds completely isolated from affected sheep but in a following study of naturally infected goats (Hadlow *et al.*, 1980) infectivity was found in the alimentary tracts of three animals, i.e. in tonsil, ileum, proximal and distal colon. Because of this, it was suggested that goat natural scrapie (as opposed to experimental scrapie) was essentially like sheep scrapie and could be maintained by contagion in a herd of goats kept apart from sheep.

Hourrigan *et al.* (1979) carried out a large scale field trial exposing sheep and goats in field conditions to more than 500 sheep taken from flocks known to have a scrapie problem. Lateral (or contact) transmission appeared at a very low rate (about 3%) but the incidence of disease amongst animals born at the trial site was much higher. Early exposure and close contact were again implicated as important for this route of transmission.

Intermediate host?

The mystery of scrapie appearing suddenly in flocks of sheep or herds of goats which have previously been scrapie-free has prompted the idea that there may be a reservoir of infection in the field in some other animal. Fitzsimmonds and Pattison (1968) were unsuccessful, however, in detecting infectivity in three species of nematode taken from scrapie affected and normal Herdwick sheep. Preparations from these parasites were inoculated into mice, rats, sheep and goats with negative results. Hourrigan *et al.* (1979), however, did report a successful tranmission from nematode samples to one cage of mice.

A recent short report from Kolomietz and colleagues (1990) suggests that the wild mouse could harbour infection. In a study of mice taken from farms in Russia with a scrapie problem, they carried out pathological studies and found evidence of scrapie-specific PrP protein deposits in 4% of cases. If this work is confirmed and extended it could provide a neat explanation of how scrapie gets around. Dickinson (1976) noted that scrapie administered orally to the laboratory mouse passes through the animal with little loss of infectivity. Wild mice could therefore contribute to pasture contamination and spread of infected material.

Carrier status

Data on the "carrier status" of sheep and goats is lacking as there is neither a convenient assay for infection nor has there been (until recently) an assay for *Sip* genotype. Such carrier animals, if they exist, could act as sources of infection to other animals.

Very low doses of infection may result in the host becoming a carrier. One of the earliest studies of scrapie in mice (Chandler, 1963) showed that incubation period was dependent on dose, i.e. incubation period lengthened as the infectivity in the inoculum was reduced. At very low doses scrapie replication may be at such a low level that disease does not develop with the animal's normal lifespan (Dickinson *et al.*, 1975). Some Suffolk sheep used in the trials described by Hourrigan *et al.* (1979) were at high risk of scrapie and did not show symptoms for 5 to nearly 11 years. It is possible that they could be shedding infectivity during this time.

In the experimental mouse model 87V scrapie in IM mice, an intraperitoneal dose of 1% brain homogenate results in long term persistence of high scrapie titres in the spleen without replication in the brain or development of scrapie symptoms (Collis and Kimberlin, 1985). This has suggested the spleen and other areas of the lymphoreticular system as tissues for examination for infectivity or abnormal PrP protein deposits in apparently healthy animals.

Infection specific PrP protein has been demonstrated in both brain and spleen of experimentally infected mice (Farquhar *et al.*, 1990). In a mouse/scrapie model which gives an incubation period of 170 days, abnormal proteinase K resistant PrP was detected as early as 32 days (21 day assays were negative). This finding, if extended to lymph nodes or other more amenable tissues for biopsy, could form the basis of a pre-clinical diagnostic test for scrapie.

Skarpheoinsson *et al.* (1990) found SAF by electron microscopy and infection-specific PrP protein by Western blotting in the brains of some apparently healthy sheep. This abnormal form of PrP protein was also detected in brain, spleen and lymph nodes early in the incubation period of experimentally infected sheep *and* in spleen and lymph nodes of a small number of apparently normal control animals (Ikegami *et al.*, 1991).

The issue of carrier status - whether an animal is silently "carrying" the disease or whether simply at a symptomless stage in its scrapie incubation period - has not been resolved. Carrying out a large series of mouse transmission assays on unaffected animals will only be useful if a positive result is produced. Negative transmission (as in Hadlow *et al.*, 1982) could mean that there is no infectivity, that the titre is too low to give a short incubation period in the test mice or that the strain of scrapie is not one which transmits easily to rodents. It will only be possible to be absolutely sure about the presence of the infectious agent when it has been positively identified and a direct biochemical test for scrapie is available.

Genetics of Susceptibility to the Natural Disease

The relevance of the selected sheep lines to the natural disease has been underlined by the fact that outbreaks of natural scrapie have occurred in the NPU Cheviot and Herdwick positive lines but not in the negative lines (Dickinson, 1974) nor in the Swaledale nucleus flock (A.J. Chalmers, personal communication). The *Sip* gene is also thought to control the host response to natural scrapie. Crossbreeding experiments described by Foster and Dickinson (1988b) involved negative line Cheviots (*Sip* pApA) and Suffolks from a flock bred for high incidence of natural scrapie (Dickinson *et al.*, 1974). Progeny were observed for development of scrapie both naturally and following experimental challenge with both SSBP/1 and SUF81 (a homogenate of five natural scrapie brains from the Suffolk flock). The incidence of scrapie was close to the expected ratio for the segregation of a single gene with one dominant allele giving good evidence that the *Sip* gene affects the host response to both experimental and natural scrapie in these two breeds of sheep.

However, molecular genetic results obtained from NPU Cheviot sheep are open to doubts because of the potential problems of inbreeding. Great care has been taken to reduce this to a minimum within this flock but nevertheless it is necessary to test the PrP gene RFLPs for association with incidence of scrapie in the "real world" of the farm with unselected and unrelated sheep.

A study of naturally infected scrapie sheep from all over Britain was undertaken in a twenty-eight month period which ended in November 1990 (Hunter *et al.*, 1991). One hundred and thirteen sheep of 29 different breeds and cross breeds were studied. Approximately 30% came from Scotland, 60% from England and 10% from Wales. The average age of the animals was 3.2 years which agrees well with previous age studies (e.g. Dickinson *et al.*, 1964) and all diagnoses of scrapie were confirmed by histopathology.

The DNA of these sheep was examined for the PrP gene *Eco* RI fragments described earlier: e1 (associated with *Sip* sA in NPU Cheviots) and e3 (associated with *Sip* pA) and for another fragment e2 (5.2 kb) whose association with *Sip* alleles is as yet uncertain. The results are shown in Table 6.4. Approximately 60% of these natural scrapie sheep were PrP e1e1, 25% were PrP e1e3 and 12% were PrP e3e3. The remaining (4) animals were PrP e2 carriers. The PrP RFLP type e1 had a frequency therefore of approximately 73% in these sheep; e3, 24% and e2, 3%.

Table 6.4 PrP gene RFLP genotypes in sheep affected by natural scrapie.

*Eco*RI genotype	e1e1	e1e2	e1e3	e2e2	e2e3	e3e3	Total
n*	68	1	28	2	1	13	113
%	60	1	25	2	1	12	
RFLP type frequency		e1			e2	e3	
%		73			3	24	

n = numbers of sheep of 29 different breeds and crossbreeds
e1 - 6.8 kb
e2 - 5.2 kb
e3 - 4.0 kb

To assess the significance of these results, further studies were carried out to compare directly scrapie cases with age-matched unaffected flock mates. The results from about 250 sheep have indicated that affected flocks may be dividing into two groups. In one of these, scrapie cases all carried e1 with high percentages of e1e1 homozygotes (found in three flocks) and, in the other (one flock), none of the scrapie cases was e1e1 but all carried e3 in approximately equal numbers of homozygotes and heterozygotes (Hunter *et al.*, 1991). Whether or not this can be regarded as indirect evidence for different field strains of scrapie is difficult to say. If only one strain of scrapie affects British sheep then the PrP gene e1 fragment will mark scrapie susceptibility in 86% of cases; if there are several strains, e1 may mark susceptibility to one (more common) strain and e3 to another (less common) strain. This work is at an early stage and there is a need to assess the evidence for different field strains.

Primary transmissions from naturally scrapie affected sheep

Dickinson (1976) described the first attempt to consider the epidemiology of scrapie in terms of the strains of agent isolated in primary transmissions to laboratory mice. Twenty-six source cases from 13 breeds and cross-breeds and from a goat were injected intracerebrally into a panel of mice of differing *Sinc* genotype (the mouse homologue of *Sip*). There was one strain (ME7) which was isolated from over half the sheep cases and in several breeds; there were also up to ten other isolates. Table 6.5 shows incubation time data on the strain ME7 in sheep, goats (Foster and Dickinson, 1988a) and mice (Bruce *et al.*, 1991). Although the incubation periods can be quite long, ME7, isolated in mice, was able to replicate and cause scrapie in the positive and negative line Cheviot sheep and in goats. ME7 may therefore represent a common type of British sheep scrapie.

Table 6.5 Incubation period of ME7 scrapie in sheep, goats and mice.

Test animal	Genotype		Incubation period[*]
NPU Cheviot sheep*	*Sip* sAsA/sApA		1081 ± 361 n=7
	Sip pApA		1179 ± 246 n=3
Goats		-	412 ± 25 n=3
Mice	C57BL	*Sinc* s7s7	168 ± 2 n=19
	VM/DK	*Sinc* p7p7	346 ± 3 n=84

[*] days +/- sem, i.c. route
[a] Data from Foster and Dickinson (1988a)
[b] Data from Bruce *et al.* (1991)
Sinc is the mouse homologue of *Sip*

Icelandic sheep may suffer from a different type of scrapie, or at least one which is not readily transmissible to mice. Fraser (1983) carried out a study in which samples from 5 scrapie sheep from 3 different areas of Iceland were injected into a panel of mice. There were only 49 transmissions out of several hundred recipients with incubation periods ranging from 312 to 918 days.

Carp and Callahan (1991) describe the isolation of 10 passage lines in mice from 5 USA Suffolk sheep suffering from natural scrapie. They found these scrapie isolates showed some differences in characteristics (e.g. incubation times and amount of weight gained by the recipient mice) and attributed this to strain differences in the types of scrapie isolated from the sheep.

There is also experimental evidence for at least two types of scrapie in NPU Cheviot sheep. SSBP/1 and CH1641 have quite different incubation periods in these sheep as shown in Table 6.1.

Is scrapie a genetic disease?

Whether or not a lamb contracts scrapie depends more on the present or future scrapie status of its mother than on anything else. This familial (running in families) nature of scrapie has led some to suggest that scrapie is purely a genetic disease. A primary advocate of this view was Parry (1960) who believed that scrapie was caused by a defective gene. He invoked the term "provirus" (first used by Darlington in the 1940s) to describe this gene and to account for its being both genetic and transmissible. The transmissible agent was suggested to be "formed *de novo* in each affected animal by the metabolic activity" of the responsible gene (Parry, 1979). This work has been elegantly disputed by Dickinson *et al.* (1965), amongst others, but has recently gained credence from the spontaneous scrapie-like disease developed by transgenic mice carrying multicopies of a mutant human PrP gene genetically linked to incidence of GSS (described above) (Hsiao *et al.*, 1990). Brown *et al.* (1991) describe a "genetic and transmissible" hypothesis to apply to the TSEs based on the idea of abnormal infection-specific PrP molecules acting as "seeds" for the conversion of normal PrP protein polypeptides to more of the abnormal form. This "seed" PrP could enter the body by experimental inoculation, by natural routes such as the oral route or via skin abrasions or could be inherited in the form of a mutant gene. It is still difficult to explain the apparent existence of 15-20 different scrapie strains (Dickinson *et al.*, 1984) using such a hypothesis, however.

Therapeutic Intervention

To date, nothing has been found which will prevent the death of an individual person or animal once the symptoms of TSE are evident. There have been, however, a number of studies in which lengthening of incubation period of experimentally induced scrapie in laboratory rodents has been achieved.

The organic polyanion dextran sulphate (DS500) prolongs the incubation period of scrapie in mice. The effective titre was found to be reduced by 90% when DS500 was injected 72 hours before or 7 hours after an intraperitoneal injection of ME7 scrapie (Farquhar and Dickinson, 1986).

Another potential "antiscrapie" drug has been described recently by Pocchiari *et al.* (1989). This is Amphotericin B (AmB), a systemic anti-fungal antibiotic. AmB delays the replication of 263K inoculated into hamsters but although it is thought to damage to cell membranes, it is not known how AmB has this effect on scrapie.

Lengthening of incubation period could appear to the sheep or goat owner as a cure if no scrapie symptoms develop within an animal's lifetime. However such treatment (which is not available outside the laboratory) may only increase the number of "carrier" animals and build up a reserve of infection.

Prospects for the Future

Genetic selection of sheep for low or high susceptibility to scrapie may be greatly facilitated by the availability of a rapid, laboratory test for *Sip* genotype. The PrP gene linkage test has yet to be fully developed but it is potentially useful to pedigree livestock breeders. For example, the selection of animals by commercially based criteria such as fecundity, milk or wool quality or appearance, could inadvertently lead to an increased susceptibility to natural scrapie in a selection line and might not be apparent for several years. A DNA probe test may act as a warning indicator of the direction of the selection process. Selection for scrapie "resistance" by this means or any other might not ensure a scrapie free flock, since our studies have suggested that animals "resistant" to one type of scrapie strain can be susceptible to another (Table 6.1). Selection strategies need to be tested in parallel with more fundamental studies on the mechanism of *Sip* action and on the dynamics of spread of infection in *Sip* pApA, *Sip* sApA and *Sip* sAsA sheep. However maintenance of mixed populations of *Sip* alleles in a selection line may be useful in preventing high incidence outbreaks of scrapie.

Goat PrP genetic studies are at an early stage and are hindered by the lack of genetically homogeneous lines of goats with different and predictable responses to scrapie. However the same could be said of cattle and mutations have also been found in the bovine PrP gene (Goldmann *et al.*, 1991) although linkage with susceptibility to BSE will take some time to assess. Perhaps a study of goat scrapie cases may provide similar polymorphisms which could be used both in management of the disease in goats and to extend the limited and sometimes confusing information on goat scrapie.

References

Basler, K., Oesch, B., Scott, M., Westaway, D., Walchli, M., Groth, D.F., McKinley, M.P., Prusiner, S.B. and Weissmann, C. (1986) Scrapie and cellular PrP isoforms are encoded by the same chromosomal gene. *Cell* 46, 417-428.

Brotherston, J.G., Renwick, C.C., Stamp, J.T. and Zlotnik, I. (1968) Spread of scrapie by contact to goats and sheep. *Journal of Comparative Pathology* 78, 9-17.

Brown, P., Liberski, P.P., Wolff, A. and Gajdusek, D.C. (1990) Resistance of scrapie infectivity to steam autoclaving after formaldehyde fixation and limited survival after ashing at 360°C: practical and theoretical implications. *Journal of Infectious Diseases* 161, 467-472.

Brown, P., Goldfarb, L.G. and Gajdusek, D.C. (1991) The new biology of spongiform encephalopathy: infectious amyloidoses with a genetic twist. *Lancet* 337, 1019-1022.

Bruce, M.E., McConnell, I., Fraser, H. and Dickinson, A.G. (1991) The disease characteristics of different strains of scrapie in *Sinc* congenic mouse lines: implications for the nature of the agent and host control of pathogenesis. *Journal of General Virology* 72, 595-603.

Carlson, G.A., Kingsbury, D.T., Goodman, P.A., Coleman, S., Marshall, S.T., DeArmond, S., Westaway, D. and Prusiner, S.B. (1986) Linkage of prion protein and scrapie incubation time genes. *Cell* 46, 503-511.

Carp, R.I. and Callahan, S.M. (1991) Variation in the characteristics of 10 mouse-passaged scrapie lines derived from five scrapie-positive sheep. *Journal of General Virology* 72, 293-298.

Chandler, R.L. (1963) Experimental scrapie in the mouse. *Research in Veterinary Science* 4, 276-285.

Chelle, P.L. (1942) Un cas de tremblante chez la chèvre. *Bulletin Academie Veterinaire de France* 15, 294-295.

Collis, S.C. and Kimberlin, R.H. (1985) Long term persistence of scrapie infection in mouse spleens in the absence of clinical disease. *FEMS Microbiology Letters* 29, 111-114.

Collis, S.C. and Millson, G.C. (1975) Transferrin polymorphism in Herdwick sheep. *Animal Blood Groups and Biochemical Genetics* 6, 117-120.

Collis, S.C., Millson, G.C. and Kimberlin, R.H. (1977) Genetic markers in Herdwick sheep: no correlation with susceptibility or resistance to experimental scrapie. *Animal Blood Groups and Biochemical Genetics* 8, 79-83.

Collis, S.C., Kimberlin, R.H. and Millson, G.C. (1979) Immunoglobulin G concentrations in the sera of Herdwick sheep with natural scrapie. *Journal of Comparative Pathology* 89, 389-396.

Cullen, P.R. (1989) Scrapie and the sheep MHC: claims of linkage refuted. *Immunogenetics* 29, 414-416.

Cullen, P.R., Brownlie, J. and Kimberlin, R.H. (1984) Sheep lymphocyte antigens and scrapie. *Journal of Comparative Pathology* 94, 405-415.

Davies, D.C. and Kimberlin, R.H. (1985) Selection of Swaledale sheep of reduced susceptibility to experimental scrapie. *Veterinary Record* 116, 211-214.

Dickinson, A.G. (1974) Natural infection, 'spontaneous generation' and scrapie. *Nature* 252, 179-180.

Dickinson, A.G. (1976) Scrapie in sheep and goats. In: Kimberlin, R.H.(ed), *Slow virus diseases of animals and man*. North-Holland Publishing Company, Amsterdam, Oxford. pp. 209-241.

Dickinson, A.G. and Outram, G.W. (1988) Genetic aspects of unconventional virus infections: the basis of the virino hypothesis. In: CIBA symposium no. 135. *Novel infectious agents and the central nervous system*. Wiley, Chichester. pp. 63-83.

Dickinson, A.G., Young, G.B., Stamp, J.T. and Renwick, C.C. (1964) A note on the distribution of scrapie in sheep of different ages. *Animal Production* 6, 375-377.

Dickinson, A.G., Young, G.B., Stamp, J.T. and Renwick, C.C. (1965) An analysis of natural scrapie in Suffolk sheep. *Heredity* 20, 485-503.

Dickinson, A.G., Stamp, J.T., Renwick, C.C. and Rennie, J.C. (1968) Some factors controlling the incidence of scrapie in Cheviot sheep injected with a Cheviot-passaged scrapie agent. *Journal of Comparative Pathology* 78, 313-321.

Dickinson, A.G., Stamp, J.T. and Renwick, C.C. (1974) Maternal and lateral transmission of scrapie in sheep. *Journal of Comparative Pathology* 84, 19-25.

Dickinson, A.G., Fraser, H., Outram, G.W. (1975) Scrapie incubation time can exceed natural lifespan. *Nature* 256,732-733.

Dickinson, A.G., Bruce, M.E., Outram, G.W. and Kimberlin, R.H. (1984) Scrapie strain differences: the implications of stability and mutations. In: Tateishi, J. (ed.) *Proceedings of Workshop on Slow Transmissible Diseases*. Japanese Ministry of Health and Welfare, Tokyo. pp. 105-118.

Farquhar, C.F. and Dickinson, A.G. (1986) Prolongation of scrapie incubation period by an injection of dextran sulphate 500 within the month before or after infection. *Journal of General Virology* 67, 463-473.

Farquhar, C.F., Somerville, R.A., McBride, P., Bruce, M.E. and Hope, J. (1990) PrP detection in peripheral organs of scrapie infected mice. *VIIIth International Congress of Virology* Berlin, 26-31 August 1990. Abstract P28-013, p. 284.

Fitzsimmonds, W.M. and Pattison, I.H. (1968) Unsuccessful attempts to transmit scrapie by nematode parasites. *Research in Veterinary Science* 9, 281-283.

Foster, J.D. and Dickinson, A.G. (1988a) The unusual properties of CH1641, a sheep passaged isolate of scrapie. *Veterinary Record* 123, 5-8.

Foster, J.D. and Dickinson,A.G. (1988b) Genetic control of scrapie in Cheviot and Suffolk sheep. *Veterinary Record* 123, 159.

Foster, J.D. and Hunter, N. (1991) Partial dominance of the sA allele of the *Sip* gene in the control of experimental scrapie in Cheviot sheep. *Veterinary Record* 128, 548-549.

Fraser, H. (1983) A survey of primary transmission of the scrapie agent. In: Court, L.A. and Cathala, F.(eds.) *Virus non conventionnnels et affections du systeme nerveux central*. Masson, Paris. pp. 34-46.

Goldmann, W., Hunter, N., Foster, J.D., Salbaum, J.M., Beyreuther, K. and Hope, J. (1990) Two alleles of a neural protein gene linked to scrapie in sheep. *Proceedings of the National Academy of Sciences USA* 87, 2476-2480.

Goldmann, W., Hunter, N., Martin, T., Dawson, M. (1991) Different forms of the bovine PrP gene have five or six copies of a short, G-C-rich element within the protein-coding exon. *Journal of General Virology* 72, 201-204.

Gordon, W.S. (1966) Variation in susceptibility of sheep to scrapie and genetic implications. In: ARS 91-53 US Department of Agriculture *Report of scrapie seminar 1964* pp. 53-67.

Greig, J.R. (1940a) Scrapie. *Transactions of the Highland Agricultural Society of Scotland* 52, 71-90.

Greig, J.R. (1940b) Observations on the transmission of the disease by mediate contact. *Veterinary Journal* 96, 203-206.

Greig, J.R. (1950) Scrapie in sheep. *Journal of Comparative Pathology* 60,263-266.

Hadlow, W.J. (1960) Pathology of scrapie in goats. In: ARS 91-22 US Department of Agriculture. *Report of scrapie seminar 1959* pp. 23-26.

Hadlow, W.J., Eklund, C.M., Kennedy, R.C. Jackson, T.A., Whitford, H.W. and Boyle, C.C. (1974) Course of experimental infection in the goat. *Journal of Infectious Diseases* 129, 559-567.

Hadlow, W.J., Kennedy, R.C., Race, R.E. and Eklund, C.M. (1980) Virologic and neurohistologic findings in dairy goats affected with natural scrapie. *Veterinary Pathology* 17, 187-199.

Hadlow, W.J., Kennedy, R.C. and Race, R.E. (1982) Natural infection of Suffolk sheep with scrapie virus. *Journal of Infectious Diseases* 146, 657-664.

Harcourt, R.A. (1974) Naturally-occurring scrapie in goats. *Veterinary Record* June 1st, 1974, p. 504.

Harris, D.A., Falls, D.L., Walsh, W. and Fischbach, G.D. (1989) Molecular cloning of an acetylcholine receptor-inducing protein. *Society of Neurosciences* 15, 164 (Abstract).

Hoare, M., Davies, D.C. and Pattison, I.H. (1977) Experimental production of scrapie-resistant Swaledale sheep. *Veterinary Record* 101, 482-484.

Hope, J., Morton, L.J.D., Farquhar, C.F., Multhaup, G., Beyreuther, K. and Kimberlin, R.H. (1986) The major polypeptide of scrapie-associated fibrils (SAF) has the same size, charge distribution and N-terminal sequence as predicted for the normal brain protein (PrP) *EMBO Journal* 5, 2591-2597.

Hourrigan, J., Klingsporm, A., Clark, W.W. and deCamp, M. (1979) Epidemiology of scrapie in the United States. In: Prusiner, S.B. and Hadlow, W.J. (eds) *Slow transmissible diseases of the nervous system.* Vol. 1. Academic Press, New York. pp. 331-356.

Hsiao, K., Baker, H.F., Crow, T.J., Poulter, M., Owen, F., Terwilliger, J.D., Westaway, D., Ott, J. and Prusiner, S.B. (1989) Linkage of a prion protein missense variant to Gerstmann-Straussler syndrome. *Nature* 338, 342-345.

Hsiao, K.K., Scott, M., Foster, D., Groth, D.F., DeArmond, S.J. and Prusiner, S.B. (1990) Spontaneous neurodegeneration in transgenic mice with mutant prion protein. *Science* 250, 1587-1590.

Hunter, N., Foster, J.D., Dickinson, A.G. and Hope, J. (1989) Linkage of the gene for the scrapie-associated fibril protein (PrP) to the *Sip* gene in Cheviot sheep. *Veterinary Record* 124, 364-366.

Hunter, N., Foster, J.D., Benson, G. and Hope, J. (1991) Restiction fragment polymorphisms of the scrapie-associated fibril protein (PrP) gene and their association with susceptibility to natural scrapie in British sheep. *Journal of General Virology* 72, 1287-1292.

Ikegami, Y., Ito, M., Isomura, H., Momotani, E., Sasaki, K., Muramatsu, Y., Ishiguro, N. and Shinagawa, M. (1991) Pre-clinical and clinical diagnosis if scrapie by detection of PrP protein in tissues of sheep. *Veterinary Record* 128, 271-275.

Kolomietz, N.D., Zukovsky, V.G., Roichel, V.M., Kolomietz, A.G. and Grachev. (1990) Scrapie - naturally slow focal infection. *VIIIth International Congress of Virology,* Berlin, August 26-31, 1990. Abstract W28-005, p. 59.

Merz, P.A., Rohwer, R.G., Kascsak, R., Wisniewski,H.M., Somerville,R.A., Gibbs, C.J. and Gajdusek, D.C. (1984) Infection-specific particle from the unconventional slow virus diseases. *Science* 225, 437-440.

Millot, P., Chatelain, J., Dautheville, C., Salmon, D. and Cathala, F. (1988) Sheep major histocompatibility (OLA) complex: linkage between a scrapie susceptibility/resistance locus and the OLA complex in Ile-de-France sheep progenies. *Immunogenetics* 27, 1-11.

Nussbaum, R.E., Henderson, W.M., Pattison,I.H., Elcock, N.V. and Davies, D.C. (1975) The establishment of sheep flocks of predictable susceptibility to experimental scrapie. *Research in Veterinary Science* 18, 49-58.

Oesch, B., Westaway, D., Walchli,M., McKinley, M.P., Kent, S.B.H., Aebersold, R., Barry, R.A., Tempst, P., Teplow, D.B., Hood, L.E., Prusiner, S.B. and Wiessmann, C. (1985) A cellular gene encodes scrapie PrP 27-30 protein. *Cell* 40, 735-746.

Palsson, P.A. (1979) Rida (scrapie) in Iceland and its epidemiology. In: Prusiner, S.B. and Hadlow,W.J. (eds) *Slow transmissible diseases of the nervous system.* Vol. 1. Academic Press, New York. pp. 357-366.

Parry, H.B. (1960) Scrapie: a transmissible hereditary disease of sheep. *Nature* 185, 441-443.

Parry, H.B. (1979) Elimination of natural scrapie in sheep by sire genotype selection. *Nature* 277, 127-129.

Parry, H.B. (1984) *Scrapie.* Academic Press, London.

Pattison, I.H. and Millson, G.C. (1961) Scrapie produced experimentally in goats with special reference to the clinical syndrome. *Journal of Comparative Pathology* 71, 101-108.

Pattison, I.H. and Millson, G.C. (1962) Distribution of the scrapie agent in the tissues of experimentally inoculated goats. *Journal of Comparative Pathology* 72, 233-244.

Pattison, I.H., Gordon, W.S. and Millson, G.C. (1959) Experimental production of scrapie in goats. *Journal of Comparative Pathology* 69, 300-312.

Pattison, I.H., Hoare, M.N., Jebbett, J.N. and Watson, W.A. (1972) Spread of scrapie to sheep and goats by oral dosing with foetal membranes from scrapie affected sheep. *Veterinary Record* 90, 465-468.

Pattison, I.H., Hoare, M.N., Jebbett, J.N. and Watson, W.A. (1974) Further observations on the production of scrapie in sheep by oral dosing with foetal membranes from scrapie-affected sheep. *British Veterinary Journal* 130, 65-67.

Pocchiari, M., Casaccia, P. and Ladogana, A. (1989) Amphotericin B: a novel class of antiscrapie drugs. *Journal of Infectious Diseases* 160, 795-802.

Prusiner, S.B. (1982) Novel proteinaceous infectious particles cause scrapie. *Science* 216, 136-144.

Prusiner, S.B., Scott, M., Foster, D., Pan, K-M., Groth, D., Mirenda, C., Torchia, M., Yang, S-L., Serban, D., Carlson, G.A., Hoppe, P.C., Westaway, D. and DeArmond, S.J. (1990) Transgenetic studies implicate interactions between homologous PrP isoforms in scrapie prion replication. *Cell* 63, 673-686.

Skarpheoinsson, S., Johannsdottir, R., Siguroarson, S. and Georgsson, G. (1990) Detection of scrapie associated fibrils (SAF) and protease resistant proteins (PrP) in preclinical scrapie of sheep. *VIIIth International Congress of Virology* Berlin, 26-31 August, 1990. Abstract P28-007, p. 283.

Slater, J.S. (1965) Succinic dehydrogenase, cytochrome oxidase and acid phosphatase activities in the brains of scrapie-infected goats and mice. *Research in Veterinary Science* 6, 155-161.

Strain, G.M., Barta, O., Olcott, B.M. and Braun, W.F. (1984) Serum and cerebrospinal fluid concentrations of immunoglobulin G in Suffolk sheep with scrapie. *American Journal of Veterinary Research* 45, 1812-1813.

Zlotnik, I. (1962) The pathology of scrapie: a comparative study of lesions in the brain of sheep and goats. *Acta Neuropathologica,* Supplement 1, 61-70.

Chapter 7

Foraging Strategy: From Monoculture to Mosaic

I.J. Gordon[1] and A.W. Illius[2]

[1] *Macaulay Land Use Research Institute, Pentlandfield, Roslin, Midlothian, EH25 9RF, Scotland, UK*
[2] *Institute of Cell, Animal and Population Biology, Division of Biological Sciences, University of Edinburgh, West Mains Road, Edinburgh, EH9 3JG, Scotland, UK*

Introduction

Sheep and goats form the backbone of most of the world's ruminant livestock enterprises (Fitzhugh *et al.*, 1978) and the majority of these enterprises are based upon small farmers who graze their animals on rangelands. Rangelands consist of mosaics of vegetation communities containing a complex of plant species. These communities result from differences in the underlying soil, rainfall, aspect and also through the impact of humans and their livestock. To consume a diet which has an adequate level of nutrients to meet its requirements for maintenance, growth and reproduction the animal is faced with a series of short-term tactical decisions about what diet to select, how long to search between bites and the resulting rate of food intake. In the longer term, strategic decisions concern the length of time to spend feeding and where to feed, given topographic influences on energy expenditure and distance travelled between foraging sites, water and shelter (Figure 7.1). This suite of decision-making processes is defined as the foraging strategy of the animal. The degree of complexity of the decisions will reflect the heterogeneity of the environment in which the animal is foraging. Animals living in simple monocultures, typical of temperate sown swards, have limited heterogeneity to deal with in everyday foraging choices. In contrast because rangelands comprise a range of vegetation communities varying in species diversity and structural heterogeneity the foraging animal has many more foraging choices to make. If we are to the make most efficient use of the plant and animal resources available on rangelands it is essential to improve our understanding of the foraging strategies of the herbivores which utilize these ecosystems.

Figure 7.1 Schematic presentation of interactions between the tactical and strategic components of foraging strategy and external factors.

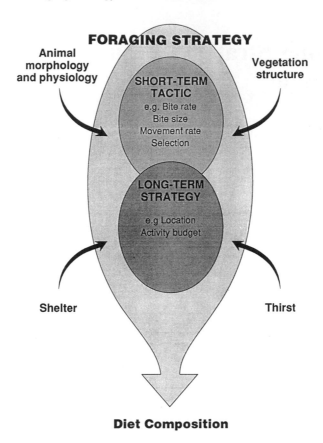

Diet Composition

To date, most studies of the foraging strategy of sheep and goats have been descriptive catalogues of the diets of animals at the taxonomic or plant part level (e.g. Van Dyne *et al.*, 1980). These studies are of limited predictive use outside the circumstances and site where they were conducted and as such do not advance our general understanding of foraging strategy and diet selection. Lately, however, a more predictive and theoretical framework for the study of ungulate foraging strategies has been developed (Belovsky, 1978, 1986; Skogland, 1980, 1984; Hanley, 1982; Owen-Smith and Novellie, 1982; Illius and Gordon, 1987, 1990). This review draws recent research on the foraging behaviour of sheep and goats together within this framework and describes how the results may be used in developing ecosystem models which integrate plant/herbivore interactions. This will allow us to predict both the influence of vegetation structure on the productivity of the animal component of the system and the impact of the herbivore in its environment.

The Comparative Biology of Sheep and Goats

Domestic sheep (*Ovis aries*) and goats (*Capra hircus*) are part of the Order *Artiodactyla* (MacDonald, 1984). Within this order they form part of the Tribe *Caprinae*. Sheep and goats evolved under similar environmental conditions (Ryder, 1984; Mason 1984) and the wild species within the *Caprinae* all have a foraging strategy defined as that of mixed/intermediate feeders (Hofmann, 1989), feeding on a mixture of plant types including grass, forbs and browse.

The wild congeners of sheep tend to consume a greater proportion of grass in their diets than do the wild congeners of goats (Roberts, 1977; Browning and Monson, 1990). Domestic sheep and goats have been domesticated by humans for millennia and still differ in their foraging strategies. On rangelands, there is strong evidence for differences in diet selection in sheep and goats (Clark *et al.*, 1982; Lu, 1988; Bullock, 1985; Malechek and Narjisse, 1990; Norton *et al.*, 1990). Goats are more catholic in their feeding habits and tend to consume grass and forb species when available and more browse than sheep during the dry season or winter when grass and forbs are limited. This difference in foraging strategy of sheep and goats is associated with differences in their physiology and anatomy.

The size and shape of the animal's mouth affects its rate of forage intake and its ability to select discrete food items from a heterogeneous array of plant material (Meyer *et al.*, 1957; Grant *et al.*, 1985, 1987; Illius and Gordon, 1987; Taylor *et al.*, 1987). More selective feeding habits are associated with a narrower and more pointed dental arcade which is better able to select food items from surrounding material of lower quality (Gordon and Illius, 1988). Goats have an incisor arcade structure similar to that of browsers, whereas that of sheep is similar to grazers (Gordon and Illius, 1988). Goats are also more thorough at initial processing of plant material than are sheep, breaking it down to finer particles prior to swallowing (Domingue *et al.*, 1991).

Anatomically, the alimentary tract of goats is more like that of a browser whereas that of sheep is like that of a grazer (Hofmann, 1983). Goats have a relatively larger rumen volume for their size than sheep (Demment and Van Soest, 1985; Focant *et al.*, 1986; Domingue *et al.*, 1991), and a similar rate of passage of both the liquid and particulate phases of the digesta (Domingue *et al.*, 1991) when fed on the same diet. Although the evidence is equivocal (Devendra, 1978), recent comparative studies show differences in the ability of sheep and goats to digest the same forages (Focant *et al.*, 1986; Domingue *et al.*, 1991). Sheep and goats appear to digest high digestibility forages to a similar extent but goats tend to be more efficient at digesting low digestibility diets (Doyle *et al.*, 1984; Domingue *et al.*, 1991). Goats also appear to be able to digest forages containing secondary compounds more extensively than are sheep (McCabe and Barry, 1988; Howe *et al.*, 1988). This may result from the differential adaptation of the goats' rumen microflora (Giad *et al.*, 1980) and salivary proteins (Provenza and Malecheck, 1984; Robbins *et al.*, 1987; Austin *et al.*, 1989) which nullify the effect of digestion-inhibiting tannins which are present in willow (*Salix* spp.), gorse (*Ulex europaeus*) and other browse species.

The Biology of Grazing Systems

A prerequisite to understanding the foraging strategy of sheep and goats is an understanding of the biology of their grazing ecosystems. The processes within these ecosystems can be described in terms of the flow and partitioning of material along alternative pathways within the system (Figure 7.2; Grant and Maxwell, 1988). This component-based approach requires an adequate description of the variables which mediate plant/animal interactions. The plant resource, for example, can be described in terms of such characteristics as its phenological state, the size and distribution of vegetation communities within the ecosystem, the species composition of each community, the sward canopy structure and its variability, and the digestibility of the species components.

The topics which have attracted greatest research in intensively managed sown pasture systems differ from those relevant to extensive systems on rangelands. In sown pasture, where pasture digestibility is high, pasture quantity is the factor most likely to limit consumption. The manner in which sward characteristics influence consumption rate is the primary focus of research (Hodgson, 1982a). Sward height is a key component affecting intake rate in these systems and its control has proved an effective management tool (Hodgson *et al.*, 1986). In rangeland ecosystems the greater diversity of vegetation allows the animal the opportunity for selection of a diet higher in quality than the average for the environment and overall vegetation quantity is rarely limiting. Factors such as digestibility of different plant species within vegetation communities, their spatial distribution and seasonality of plant growth are likely to be limiting to nutrient intake and result in the uneven distribution of grazing pressure both within and across plant communities (Hunter, 1962; Grant and Maxwell, 1988; Gordon, 1989).

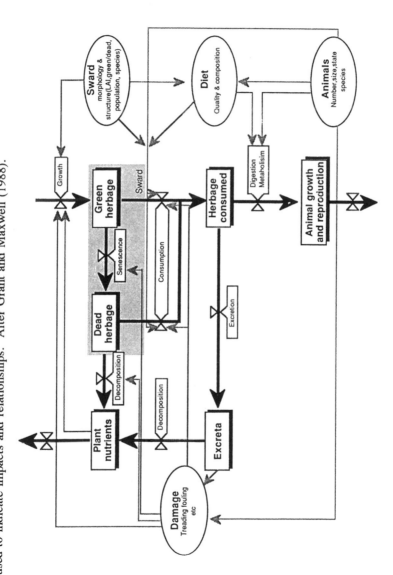

Figure 7.2 Plant and animal inter-relationships in grazing systems. The bold lines and arrows indicate the flow and partitioning of material through the system; the factors which are altered by the grazing processes (growth, consumption, etc.) are circled; faint lines are used to indicate impacts and relationships. After Grant and Maxwell (1988).

Nutritional wisdom

Unlike carnivores which consume a balanced diet in each meal, herbivores feed on plant material which is of low quality and which rarely provides adequate levels of all nutrients if a single foodstuff is consumed (Belovsky, 1978; Hanley and McKendrick, 1983). The suggestion that the primary objective of the foraging herbivore was to consume a balanced diet within the constraint of minimizing the consumption of secondary compounds (Westoby 1974) implies the "nutritional wisdom" of herbivores (Zahorik and Houpt, 1977; Rosenthal and Jansen, 1979; Crawley, 1983). Animals with simple digestive systems such as rats, chickens, turkeys and pigs exhibit dietary choice in order to maintain level of protein or energy intake (Rose and Michie, 1982; Kyriazakis *et al.*, 1990). It had been assumed that herbivores would be unlikely to have the same ability because of the complexity of the diet they consume, the time delay between the consumption of a food item and its physiological consequences and the intervention of the microbial trophic level between the food and the herbivore (host) trophic level (Westoby, 1974).

There is little evidence that herbivores can sense the level of nutrients in the plant material they consume (Arnold, 1981). Over 14 characteristics of plants were required to account for 90% of the variation in preference among plant species eaten by sheep. However, two lines of evidence suggest that herbivores are able to discriminate between foods. Firstly, using the operant technique, Hou *et al.* (1991a, 1991b) have shown that, on pelleted rations, sheep are willing to work hard to maintain the level of protein intake to which they had become accustomed during a previous self-selection regime. Secondly the burgeoning literature on learning as a factor influencing diet selection in herbivores suggests that herbivores may be able to assess the poisonous qualities of the plants they consume (Provenza and Balph 1990; Zahorik *et al.*, 1990). Herbivores can learn to associate the negative effects of ingesting a food plant with the ingestion of that food plant (Zahorik *et al.*, 1990), even if the gastrointestinal consequences occur two hours after ingestion of the poisonous food (Thorhallsdottir *et al.*, 1987). To date, the majority of studies on dietary learning in ruminants have involved taste aversion. However, in order to test whether herbivores have "nutritional wisdom" it must be shown that ruminants are able to learn about beneficial (e.g. protein and energy content) characteristics of their diet.

Foraging Strategy

Optimal foraging

Optimality theory provides a framework within which to investigate the behaviour of herbivores foraging within complex environments. The theory addresses the way costs and benefits of characters such as the animal's morphology, physiology or behavioural processes hypothetically affect fitness. Optimal foraging theory attempts to describe the behaviour of animals while foraging (MacArthur and Pianka, 1966; Schoener, 1971; Charnov, 1976; Pyke *et al.*, 1977; Pyke, 1984; Stephens and Krebs, 1986). The primary assumption underlying optimal foraging theory is that natural selection is likely to favour individuals which maximize their harvest of nutrients or energy (Pyke 1984), while minimizing associated costs such as the time taken for feeding, energy used and exposure to predation. For herbivores not subject to predation, this often means maximizing the rate of energy intake whilst minimizing the amount of energy used (Stephens and Krebs, 1986). Two fundamental components of optimal foraging theory deal with the selection of a diet, from a range of foods, which maximizes intake rate (the prey model), and the length of time an animal should exploit a patch of food as it becomes progressively depleted (the patch model).

Several authors (Belovsky, 1978, 1986; Belovsky and Jordan, 1978; Owen-Smith and Novellie, 1982) have attempted to develop and test optimal foraging models for generalist non-domestic ungulate herbivores, with varying results, although only one study (Bazely, 1990) specifically tested whether domestic species forage in a way predictable by optimal foraging theory. Given the structural complexity of vegetation, the fact that herbivores take up to 40,000 bites in a day and the apparent need for animals to sample the food to gain information about its intake rate (Illius and Gordon, 1990) it is unlikely that grazers can achieve strictly optimal diets. Possibly browsing species feeding on discrete food items of high quality are constrained to a lesser extent (Lundberg *et al.*, 1990).

A foraging herbivore is faced with a range of foods varying in nutrient concentration, availability and handling (prehension and initial processing) times. The optimal diet can be defined by ranking resources in order of profitability (energy yield per unit handling time) and calculating the returns in terms of net yield against handling and search times for strategies incorporating lower and lower ranked resources. The optimal diet includes all food items which cause an increase in the long term average net energy return and excludes those which cause the average net energy intake to fall. As such the inclusion of a food in the diet depends not upon its own availability but the availability of higher ranked foods.

Since the food items of large herbivores do not occur in discrete packages and herbivores spend little time handling food it is commonly believed that patch use models are more appropriate for describing the behaviour of herbivores than prey models (Stephens and Krebs, 1986). Patch use models predict when the foraging herbivore should leave a patch offering a declining intake rate because of reductions in resource availability during the feeding bout (Charnov, 1976). Patch use models predict that, in the majority of cases, the animal should leave before all available food has been consumed. The extent to which this might occur could depend upon the travel time between patches and the average patch depletion rate; where travel time between patches is large, it is predicted that animals will spend longer on a patch. In an experiment using patches of tall grass of potentially higher nutritive value planted into a short background sward of the same species, Bazely (1990) found that sheep foraged in a manner consistent with such a marginal value theorem. Patch residence time was increased and taller grass patches were more intensively grazed than shorter ones when travel times were longer.

There are a number of critiques of the approach of optimal foraging as applied to large herbivores (Crawley, 1983; Provenza and Balph, 1990). The criticisms centre around, firstly, the poor ability of herbivores to sense specific nutritional components of the vegetation on offer (Kreuger *et al.*, 1974; Arnold, 1981) and their consequent inability to determine the rate of intake of specific nutrients. Secondly, the need for herbivores to minimize the ingestion of plant defence compounds. These characteristics should be seen as constraints on herbivore foraging strategies rather than a cause for rejection of the application of optimality theory to the decisions made by herbivores about where and on what to forage.

Hierarchy of scales and decision-rules

Foraging behaviour can be described in terms of a hierarchy of scales, from the food taken in a single bite to patches of selected vegetation to vegetation communities and through landscape to region (Senft *et al.*, 1987). The boundaries between levels of the hierarchy are defined by the perceptual abilities of the animal (Kotliar and Wiens, 1990). The foraging animal must make decisions at all levels of this hierarchy but it is misleading to assume that the goals of decision-making are unique to each scale. We assume that decisions are all based on the need to attain a positive nutrient balance by, for example, maximizing intake rate and minimizing energy expenditure. Thus decisions apparently made at the landscape and vegetation community level resolve into decisions taken about bite and patch selection. A good example of this is the early observation by Hunter (1962) that sheep selected vegetation communities in relation to nutrient availability and that their vegetation community selection varied seasonally.

We are unlikely to be able fully to understand the diet selection and foraging strategy of large herbivores unless we generate and test hypotheses concerning these decision-making mechanisms. In the following two sections of this chapter we will describe firstly research relevant to bite selection and secondly to patch and vegetation community selection and relate this to the premises of optimal foraging theory outlined above.

Bite selection

Sown swards offer a simple model for understanding foraging strategy. Their apparent structural uniformity offers a good opportunity for studying factors affecting intake rate and diet selection.

Short term (instantaneous) intake rate is determined by the rate of biting and the weight of dry matter in each bite; grazing time is an additional factor determining long term daily intake rate (Figure 7.3; Hodgson, 1982b). Some of the underlying behavioural responses of grazing sheep to sward characteristics and how these affect long term intake rate have been elucidated. In general, the rate of intake at pasture increases with the digestibility of the forage (Hodgson, 1977; Birrell, 1989), herbage mass or sward surface height (Hodgson, 1982b; Penning, 1986; Penning *et al.*, 1991). Sheep respond to a low sward height by taking more bites per minute and grazing for longer but on swards below 6 cm surface height these compensatory measures cannot counter the reduced intake per bite; the net effect is an overall fall in daily intake (Figure 7.4; Hodgson and Grant, 1985; Penning, 1986). Therefore, on sown swards, bite size is seen as being the determining factor for intake rate (Hodgson, 1982b). The bite dimensions responsible for differences in bite sizes are the bite depth and area, and the product of these is bite volume. On simple sown swards the bite depth of sheep is positively correlated with the height of the sward surface (Milne *et al.*, 1982; Barthram and Grant, 1984) and on very short swards it may be limited by the presence of the pseudostem layer (Barthram, 1981). Bite area is positively associated with sward surface height (Burlison *et al.*, 1991).

The relationship between intake rate of food in herbivores and food availability is known as the functional response (Noy-Meir, 1975; Crawley, 1983). Intake rate increases with plant biomass or height until an asymptote is reached (Penning, 1986); this asymptote occurs because of the saturation of food processing capacity. During grazing, the animal's jaw movements serve two conflicting functions: gathering herbage in each prehension bite and bolus formation during chewing bites. The balance between these is determined by mass of food taken in each prehension bite, since large bites require more processing while the shorter processing time of small bites permits a rapid rate of prehension bites (Penning, 1986). Clearly, the allocation of time to gathering and chewing food can be considered as part of an optimization process.

Figure 7.3 Components of daily herbage intake.

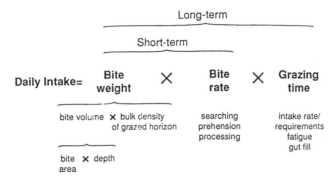

Figure 7.4 The influence of sward height on the ingestive behaviour and herbage intake of ewes under continuous stocking management. After Bircham, 1980.

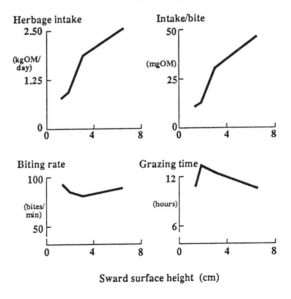

Sward surface height (cm)

Despite the general applicability of the functional response, there is always a large proportion of the total variability in daily intake which remains unexplained by sward height alone (Short, 1985; Milne *et al.*, 1988). Much of this could be explained by variation in instantaneous intake rate with sward height, with animals selecting patches of herbage which are at the top end of the height distribution (see below). Another contributing factor to variation is that the vertical distribution of plant biomass within the sward has a strong effect on the components of instantaneous intake rate. For example, the depth of the lamina layer within the sward canopy and the bulk density within this are significant determinants of bite weight

(Illius and Gordon, 1987; Burlison *et al.*, 1991).

On simple sown swards, the composition of the diet is often considered to be the consequence of a largely unselective grazing habit superimposed on a vertically stratified distribution of plant tissue. However, sheep have been shown to be selective for sward components within the horizon in which they graze. In swards with a low component of clover sheep showed a selectivity for clover within the grazed horizon (Milne *et al.*, 1982); when present in greater abundance, the proportion of clover in the diet was found to be simply related to the proportion in the grazed horizon. Therefore selection for clover is not a static phenomenon and varies with the proportion of clover in the grazed horizon. Where clover is rare there is a preference shown, where it is common no preference is shown. This may reflect the requirement of the animal to sample uncommon components in the sward in order to gain information about such things as intake rate and nutritional contents (Newman *et al.*, 1992). In general, preference for a species within the sward is not static but varies with the proportion of the species in the sward, its vertical and horizontal distribution, and also varies seasonally (Arnold, 1987) and with the animals's prior experience (Newman *et al.*, 1992).

No similar work has been conducted on goats foraging on grass/legume swards although it has been suggested that, as compared to sheep, goats select against legumes in temperate grasslands (Clark *et al.*, 1982), whereas goats consumed a higher proportion of legumes in their diet than did sheep on tropical grasslands (Norton *et al.*, 1990). Norton *et al.* (1990) suggest that these conflicting findings result from the interaction between the feeding style of goats and sheep and the position of legumes in the vertical horizon of the sward in temperate as compared to tropical grassland. Goats browse from the top of the sward whereas sheep tend to graze more deeply (Grant *et al.*, 1987; Norton *et al.*, 1990). Norton *et al.* (1990) state that in temperate grassland legumes tend to grow in the lower strata whereas in tropical grasslands they occur higher in the upper strata. However, the data on distribution of legumes within temperate swards suggests that this may not necessarily be so (Marriot and Grant, 1990).

Within rangelands, preferences between plant species can be more pronounced because of larger differences in morphology and nutritional content of the species available. The intake of the less preferred component of the diet is determined by the availability of the preferred component (Figure 7.5; Grant *et al.*, 1985) and relative species preferences vary seasonally (Grant *et al.*, 1985, 1987).

Research on the relationship between social behaviour and both instantaneous and long-term intake could help improve our understanding of variations in intake between individuals since it has been shown that herbage allowance contributes significantly to variance in intake rate (Hodgson and Milne, 1978). Thouless (1990) found that red deer (*Cervus elaphus*) hinds had a reduced bite rate when they were closer to dominant individuals but their bite rate was unaffected when they were close to subordinates.

Figure 7.5 Relationship between the amount of *Nardus* in the diets of sheep and cattle and the biomass (g DM/m^2) of the preferred live between-tussock grasses. After Grant *et al.* (1985).

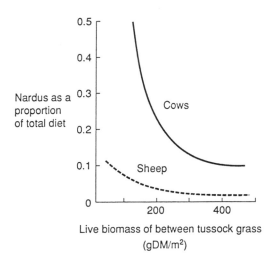

Patch and vegetation community selection

Habitat selection is known to be influenced by many factors such as the forage quantity and quality, the availability of water, the topography, climate and social factors. Ecosystems are composed of a heterogeneous amalgamation of vegetation communities which vary in abundance of preferred food items. In nature, food items tend to be distributed in discrete patches and the foraging animal has to make decisions about which patch to feed on (Weins, 1976; Pyke *et al.*, 1977; Pyke, 1984; Stephens and Krebs, 1986). Therefore, one important foraging option for the herbivore is the selection of patches or vegetation communities on which to feed.

If an animal's net rate of nutrient intake is determined by the abundance and availability of food plants within the vegetation community then we would expect animals to feed selectively on those vegetation communities which maximize their net food or energy intake (Schoener, 1971; Pyke *et al.*, 1977; Pyke, 1984; Stephens and Krebs, 1986). Thus vegetation community selection is analogous to the diet choice at the finer scale described above. Several studies have shown that intake rate increases in areas where forage items are dense or most abundant (Dudzinski and

Arnold, 1973; Trudell and White, 1981; Wickstrom *et al.*, 1984). Indoor trials have also shown that sheep prefer forage which give them a higher instantaneous intake rate (Kenny and Black, 1984), in spite of the fact that this might result in a diet of lower digestibility. In the longer term (over a day or feeding bout), however, herbivores may trade off intake rate for diet digestibility and not prefer the plant species, patch or community which offers the highest instantaneous rate but the highest long term intake rate.

Testing these predictions in the field requires an experimental approach offering animals, in the first instance, relatively simple choices. For example, as has been described previously, Bazely (1990) found that sheep grazed tall grass patches more intensively than shorter ones. Similarly, Arnold (1987) found that sheep consistently consumed more from high biomass patches within a field. Assessing the choices of animals given swards which vary in height, and therefore intake rate, provides a test of the effect of intake rate or correlated variables on patch choice. Clark *et al.* (unpublished, cited in Illius and Gordon, 1990) offered sheep, cattle and goats a range of paired height comparisons while grazing in small plots. The time that the species spent grazing on each sward height patch was measured, as were the bite rate, bite size and intake rate. Taller swards were generally preferred by all species, with sheep showing the least sensitivity to sward height and intake rate differences, with goats being the most sensitive. Goats were found to have a different grazing style from sheep or cattle, taking shallower bites from the sward surface. During the experiment the animals frequently moved between patches of differing sward height but never stayed long on the shorter patch.

Rangelands offer animals more complex mixtures of plant communities from which to choose. In Scotland the intake, diet composition and foraging behaviour of sheep and cattle have been described for many rangeland plant communities (Grant *et al.* 1985; Hodgson *et al.*, 1991) and descriptive research has shown that one of the primary determinants of community choice is likely to be the intake rate on the community (Hunter, 1962; Gordon, 1989). As described above, on grassland communities, sward height is a major determinant of the intake rate an animal can achieve on a given sward or patch choice. Recent research provides evidence in support of the hypothesis that the relative nutrient density and intake rate affect the choices of foraging sheep, even when it is the nutrient density of or intake rate on the normally less preferred community that has changed (Gordon and Beattie, 1990).

These results give some indication of the factors affecting vegetation community choice in a herbivore but much more research is required on the decision-making processes of foraging sheep and goats, in both the short- and long-term, before we can fully understand their foraging strategy.

Modelling Rangeland Systems

The complexity of foraging behaviour and of the interactions between components of rangeland ecosystems limits the ability of an empirically derived approach to provide an understanding of the dynamics of these ecosystems. An approach towards overcoming these problems is the use of simulation modelling (Christian, 1981; Rice *et al.*, 1983). By changing the inputs into these models the role of a range of interrelationships within the system can be tested. Simulation models fall into two overlapping categories: research and management models.

A number of animal-based research models have been developed describing various aspects of foraging tactics (Illius and Gordon, 1987; Demment and Greenwood, 1988). These vary in their complexity. The model of Demment and Greenwood (1988) integrates ingestive behaviours (bite size, movement rate) and food processing behaviours (mastication and rumination requirements) with rumen function to predict the combination of behaviours which produces the highest energy digestion per unit time. These animal-based models have a limited capacity to inform us about foraging strategy as they contain few sward variables. Ungar and Noy-Meir (1988) produced a model which incorporates herbage availability and sward structure but it has only a limited capability to incorporate animal variables such as those described by Demment and Greenwood (1988).

For management purposes there are a number of models available which assess the impact of the foraging herbivore on the dynamics of plant components within ecosystems, be they simple sown swards or more complex mosaics of indigenous communities. However, few of these incorporate foraging rules for the herbivore at anything more than the simple level of selection for plant species. Therefore environmental and ecosystem heterogeneity is poorly described by these models. An approach using Artificial Intelligence rule-based behavioural responses and Geographic Information System data formats is likely to give a more realistic description of the impact of herbivores on their environment (Saarenmaa *et al.*, 1988; Folse *et al.*, 1989).

As an example, within the upland ecosystem of the Scottish hills there are a least eight main vegetation communities which result from variation in soil, microclimate and aspect. These communities form mosaics which occur at both the micro- and macro-scale. They also differ in the quantity and nutritive value of the herbage on offer both within a season and between seasons. It would take a vast research effort to understand such an ecosystem using simply an experimental approach. Computer simulation models are particularly useful at describing and analysing such a complex biological system. Consequently, a model is currently being developed at the Macaulay Land Use Research Institute to predict the level of offtake by herbivores from a range of these vegetation communities on a monthly basis (Sibbald *et al.*, 1988; Armstrong, 1990). Features judged to be most important to incorporate into the model as it is being developed are:

- the inclusion of all common large vertebrate grazing species,
- the prediction of offtake from each vegetation type based on the foraging behaviour of each species,
- the incorporation of specific vegetation types,
- estimates of vegetation production on a site specific basis,
- a seasonal component to vegetation biomass production and offtake,
- the prediction of not only quantity but also quality of food available from each vegetation type,
- the impact of the grazing regime on animal productivity and performance.

A generalized model of foraging behaviour which can be applied to grazers has been incorporated as a sub-model and a vegetation sub-model has been designed so that the two are compatible. The foraging sub-model requires height, diet digestibility as well as plant component biomass to be predicted from the vegetation-submodel. Information has been collated on production, both seasonal and annual, of upland vegetation types under different conditions and on height /biomass relationships in the development of the vegetation sub-model. The final output of the model allows the user to monitor both vegetation production and offtake through time. Outputs include monthly vegetation utilization rates, offtake and digestibilities for each vegetation type for each species of grazer. It is thus able to give quantitative predictions of different grazing regimes for combinations of different vegetation types in a range of sites. Decisions on the optimal management strategy remain with the user and will depend upon management objectives.

Field Techniques

Several developments have been added to the various techniques available to quantify the variables relevant pertinent to the decision-making processes in the foraging animal (Leaver, 1982): intake rate (both short- and long-term), diet composition, foraging behaviour (e.g. bite rate, feeding time, mastication rate, movement rate), location of the animal.

Intake rate

Long-term

Many methods have been developed for measuring long-term (daily) intake rate (Leaver, 1982; Mayes, 1989). Many of the pasture- and animal-based methods of intake estimation are of limited value except on simple swards (Meijs *et al.*, 1982; Coates *et al.*, 1987). The more reliable methods for estimating the intake of animals feeding on varied communities rely on the estimation of faecal output and diet digestibility for individual animals. These techniques involve the use of external markers such as chromic oxide to estimate faecal output and *in vitro* estimates of diet digestibility (Kotb and Luckey, 1972). If the component is totally indigestible, the digestibility can be determined using the concentration of the component in the diet and faeces. Recently, Mayes *et al.* (1986) have shown that n-alkanes found in the cuticles of plants can be used as both internal and external markers to estimate diet digestibility and intake in sheep, goats, cattle and red deer.

Short-term

Changes in body weight before and after grazing, measured using very accurate balances, can be used to estimate short-term intake rate (a few hours) (Penning and Hooper, 1985). It is necessary to correct weight changes for faecal, urinary, respiratory and perspiratory weight losses. Using the fact that there is a strong correlation between the number of boluses swallowed by an animal while feeding and its intake of herbage (Forwood and Hulse, 1989), a number of methods have been developed to determine the number of swallows and the rate of swallowing of the foraging animal (Stuth *et al.*, 1981; Forwood *et al.*, 1985).

Diet composition

A range of techniques have been developed for assessing the diet composition of the free-ranging animal (McInnis *et al.*, 1983). To date there is no effective, reliable technique for estimating diet composition of animals foraging across a mosaic of vegetation communities.

The use of oesophageal fistulation has been relatively successful in determining the diets of domestic animals (Vavra *et al.*, 1978); however, it has not been extensively used in wild herbivores. The problems with this technique are 1) surgery (Rice, 1970); 2) incomplete collection because of the short time period and the small area over which the samples are collected (Le Sperance *et al.*, 1974; Coates *et al.*, 1987) and 3) the grazing behaviour of fistulated animals may differ from that of intact animals

(Engels and Malan, 1973; but see Forbes and Beattie, 1987). Overall the results from oesophageal fistula extrusa provide the most accurate method for determining diet composition as long as the fistulated animals are accustomed to the vegetation community before the sample is gathered (McInnis *et al.*, 1983).

Foraging behaviour

The primary foraging behaviour variables which researchers are interested in are:

- movement rate
- bite rate
- prehension
- harvesting
- mastication rate
- feeding time

A number of mechanical and electronic devices have been developed to measure one or more of the variables (Penning, 1983; Matsui and Okubo, 1989, 1990). The majority of these devices are carried on the animal and require the animal to be handled regularly in order to retrieve the data. A more flexible system involves the data being transmitted to a remote receiver and will allow instantaneous data capture in relation to such things as social interactions and vegetation community choice within a mosaic (Janeau and Lécrivain, 1982). An alternative approach is to use a manually operated computer based program to capture foraging data (Owen-Smith, 1979; Demment and Greenwood, 1987). This requires a lot of manpower, relies on the animals being easily tractable and the presence of a human observer can alter the behaviour of even tame animals (Owen-Smith and Cooper, 1987).

An approach which provides a useful control over the sward variables affecting foraging strategy are sward boards, mini-swards and grazing cage techniques (Kenny and Black, 1984; Illius and Gordon, 1990; Burlison *et al.*, 1991). These small-scale trials have several advantages over conventional, large-scale grazing trials. They allow observers to work close to the grazing animal and make detailed measurements of sward characteristics and bite depth, bite weight, bite rate and diet selection. Direct estimates of true bite area and volume can also be obtained, allowing the complete set of bite variables to be related to the characteristics of the vegetation grazed.

Location

If we are to gain an understanding of the foraging strategy of herbivores across mosaics of vegetation communities it is important to gather information on the spatial distribution of individuals. At its most simple, this type of information can be gathered by direct observation (Attwood and Hunter, 1957). However, this is time consuming and requires specialized night-vision equipment if information on nocturnal distribution is to be gathered (Boag *et al.*, 1990). Systems using video cameras and computer aided image analysis or radio-telemetry (Kenward, 1987) are likely to provide the required information more efficiently. These systems are still in their infancy.

Conclusions

To date much of the research on diet selection and foraging behaviour in herbivores has been limited to sown swards or simple descriptions of the diet selected. In the former case there is limited heterogeneity both in terms of species composition on both the horizontal and vertical axes, in the latter case little predictive value can be gained from the research effort. If we are to better understand the foraging strategy of herbivores in a range of circumstances we must define the mechanisms and rules herbivores use when foraging on swards, mosaics of communities and ecosystems varying in complexity. Only through this understanding will we be able to predict the behaviour of these herbivores in a range of circumstances.

Acknowledgements

We thank Prof. T.J. Maxwell and Drs J.A. Milne and I.A. Wright for commenting on previous drafts of the manuscript. C. Haggarty kindly drew the figures.

References

Armstrong, H.M. (1990) Modelling the effects of vertebrate herbivore populations on heather moorland vegetation. In: Whitby, M. and Grant, S. (eds.),
Modelling Heather Management. A Workshop Report. Department of Agric-
ultural Economics and Food Marketing, University of Newcastle upon Tyne,
pp. 56-67.
Arnold, G.W. (1981) *Grazing Behaviour.* In: Morley, F.H.W. (ed.), *Grazing
Animals.* Elsevier Scientific Publishing Company, Amsterdam, pp. 79-104.
Arnold, G.W. (1987) Influence of the biomass, botanical composition and sward
height of annual pastures on foraging behaviour by sheep. *Journal of Applied
Ecology* 24, 759-772.

Attwood, P.R. and Hunter, R.F. (1957) A method for studying the preferential grazing of hill sheep. *Journal of Animal Behaviour* 5, 149-152.

Austin, P.J., Suchar, L.A., Robbins, C.T. and Hagerman, A.E. (1989) Tannin-binding proteins in saliva of deer and their absence in saliva of sheep and cattle. *Journal of Chemical Ecology* 15, 1335-1347.

Barthram, G.T. (1981) Sward structure and the depth of the grazed horizon. *Grass and Forage Science* 36, 130-131.

Barthram, G.T. and Grant, S.A. (1984) Defoliation of ryegrass-dominated swards by sheep. *Grass and Forage Science* 39, 211-219.

Bazely, D.R. (1990) Rules and cues used by sheep foraging in monocultures. In: Hughes, R.N. (ed.), *Behavioural Mechanisms of Food Selection.* Springer-Verlag, New York, pp. 343-368.

Belovsky, G.E. (1978) Diet optimization in a generalist herbivore: The moose. *Theoretical Population Biology* 14, 105-134.

Belovsky, G.E. (1986) Optimal foraging and community structure: implications for a guild of generalist grassland herbivores. *Oecologia, Berlin* 70, 35-52.

Belovsky, G.E. and Jordan, P.A. (1978) The time energy budget of moose. *Theoretical Population Biology* 14, 76-104.

Bircham, J.S. (1980) *Herbage Growth and Utilisation Under Continuous Management.* Ph.D. Thesis, University of Edinburgh.

Birrell, H.A. (1989) The influence of pasture and animal factors on the consumption of pasture by grazing sheep. *Australian Journal of Agricultural Research* 40, 1261-1275.

Boag, B., McFarlane Smith, W.H. and Griffiths, D.W. (1990) Observations on the grazing of double low oilseed rape and other crops by roe deer. *Applied Animal Behaviour Science* 28, 213-220.

Browning, B.M. and Monson, G. (1980) Food. In: Monson, G. and Sumner, L. (ed.), *The Desert Bighorn: Its Life History, Ecology, and Management.* University of Arizona Press, Tucson, pp. 80-99.

Bullock, D.J. (1985) Annual diets of hill sheep and feral goats in Southern Scotland. *Journal of Applied Ecology* 22, 423-433.

Burlison, A.J., Hodgson, J. and Illius, A.W. (1991) Sward canopy structure and the bite dimensions and bite weight of grazing sheep. *Grass and Forage Science* 46, 29-38.

Charnov, E.L. (1976) Optimal foraging : the marginal value theorem. *Theoretical Population Biology* 9, 129-136.

Christian, K.R. (1981) Simulation of grazing systems. In: Morley, F.H.W. (ed.), *Grazing Animals.* Elsevier Scientific Publishing Company, Amsterdam, pp. 361-377.

Clark, D.A., Lambert, M.G., Rolston, M.P. and Dymock, N. (1982) Diet selection by goats and sheep on hill country. *Proceedings of the New Zealand Society of Animal Production* 42, 155-157.

Coates, D.B., Schachenmann, P. and Jones, R.J. (1987) Reliability of extrusa samples collected from others fistulated at the oesophagus to estimate the diet of resident animals in grazing experiments. *Australian Journal of Experimental Agriculture* 27, 739-745.

Crawley, M.J. (1983) *Herbivory: The Dynamics of Plant-Animal Interactions.* Blackwell Scientific Publications, Oxford.

Demment, M.W. and Greenwood, G.B. (1987) The use of a portable computer for real-time recording of observations of grazing behaviour in the field. *Journal of Range Management* 40, 284-285.

Demment, M.W. and Greenwood, G.B. (1988) Forage ingestion: effects of sward characteristics and body size. *Journal of Animal Science* 66, 2380-2392.

Demment, M.W. and Van Soest, P.J. (1985) A nutritional explanation for body-size patterns of ruminant and non-ruminant herbivores. *American*

Naturalist 125, 641-675.
Devendra, C. (1978) The digestive efficiency of goats. *World Review of Animal Production* 14, 9-22.
Domingue, D.M.F., Dellow, D.W., Wilson, P.R. and Barry, T.N. (1991) Comparative digestion in deer, goats and sheep. *New Zealand Journal of Agricultural Research* 34, 45-53.
Doyle, P.T., Egand, J.K. and Thalen, A.J. (1984) Intake, digestion and nitrogen and sulphur retention in Angora goats and Merino sheep fed herbage diets. *Australian Journal of Experimental Agriculture and Animal Husbandry* 24, 165-169.
Dudzinski, M.L. and Arnold, G.W. (1973) Comparison of diets of sheep and cattle grazing together on sown pastures on the Southern Tablelands of New South Wales by principal component analysis. *Australian Journal of Agricultural Research* 24, 899-912.
Engels, E.A.N. and Malan, A. (1973) Sampling of pastures in nutritive evaluation studies. *Agronanimalia* 5, 89-94.
Fitzhugh, H.A., Hodgson, H.J., Scoville, O.J., Nguyen, T.D. and Byerly, T.C. (1978) *The Role of Ruminants in Support of Man*. Winrock International Livestock Research and Training Centre, Morrilton, USA.
Focant, M., Vanbelle, M. and Goldfroid, S. (1986) Comparative feeding behaviour and rumen physiology in sheep and goats. *World Review of Animal Production* 22, 89-95.
Folse, L.J., Packard, J.M. and Grant, W.E. (1989) AI modelling of animal movements in a heterogenous habitat. *Ecological Modelling* 46, 57-72.
Forbes, T.D.A. and Beattie, M.M. (1987) Comparative studies of ingestive behaviour and diet composition in oesophageal-fistulated and non-fistulated cows and sheep. *Grass and Forage Science* 42, 79-84.
Forwood, J.R. and Hulse, M.M. (1989) Electronic measurement of grazing time and intake in free roaming livestock. In: Jarrige, R. (ed.), *Proceedings of the 16th International Grassland Congress*. Nice, France, pp. 799-800.
Forwood, J.R., Hulse, M.M. and Ortbals, J.L. (1985) Electronic detection of bolus swallowing to measure forage intake of grazing livestock. *Agronomy Journal* 77, 969-972.
Giad, E.A., El-Bedawy, T.M. and Mehrez, A.Z. (1980) Fiber digestibility by goats and sheep. *Journal of Dairy Science* 63, 1701-1706.
Gordon, I.J. (1989) Vegetation community selection by ungulate on the Isle of Rhum. II. Vegetation Community Selection. *Journal of Applied Ecology* 26, 53-64.
Gordon, I.J. and Beattie, M.M. (1990) Develop and test foraging strategy theories for ruminants grazing mixed indigenous hill communities. *Macaulay Land Use Research Institute Annual Report* 1989/90, pp. 53.
Gordon, I.J. and Illius, A.W. (1988) Incisor arcade structure and diet selection in ruminants. *Functional Ecology* 2, 15-22.
Grant, S.A. and Maxwell, T.J. (1988) Hill vegetation and grazing animals : the biology and definition of management options. In: Usher, M.B. and Thompson, D.B.A. (eds.), *Ecological Change in the Uplands*. Special Publication No. 7 of the British Ecological Society, Blackwell Scientific Publications, Oxford, pp. 201-214.
Grant, S.A., Suckling, D.E., Smith, H.K., Torvell, L., Forbes, T.D.A. and Hodgson, J. (1985) Comparative studies of diet selection by sheep and cattle grazing individual hill plant communities as influenced by season of the year. 1. The indigenous grasslands. *Journal of Ecology* 73, 987-1004.
Grant, S.A., Torvell, L., Smith, H.K., Suckling, D.E., Forbes, T.D.A. and Hodgson, J. (1987) Comparative studies of diet selection by sheep and cattle: blanket bog and heather moor. *Journal of Ecology* 75, 947-960.
Hanley, T.A. (1982) The nutritional basis of food selection by ungulates. *Journal*

of Range Management 35, 146-151.
Hanley, T.A. and McKendrick, J.D. (1983) *Seasonal Changes in Chemical Composition and Nutritive Value of Native Forages in Spruce-Hemlock Forest, Southeastern Alaska.* US Forest Service Northwest Forest Range Experiment Station, Portland, Oregon. Research Paper PNW-312.
Hodgson, J. (1977) Factors limiting herbage intake by the grazing animal. *Proceedings of the International Meeting from Temperate Grassland,* Dublin, pp. 70-75.
Hodgson, J. (1982a) Ingestive behaviour. In: Leaver, J.D. (ed.), *Herbage Intake Handbook.* The British Grassland Society, Hurley, pp. 113-138.
Hodgson, J. (1982b) Influence of sward characteristics on diet selection and herbage intake by the grazing animal. In: Hacker, J.B. (ed.), *Nutritional Limits to Animal Production from Pastures.* Proceedings of an International Symposium held at St. Lucia, Queensland, Australia, August 24-28, 1981.
Hodgson, J. and Grant, S.A. (1985) The grazing ecology of hill and upland swards. In: Maxwell, T.J. and Gunn, R.G. (eds.), *Hill and Upland Livestock Production.* British Society for Animal Production Occasional Publication No. 10, pp. 77-84.
Hodgson, J. and Milne, J.A. (1978) The influence of weight of herbage per unit area and per animal upon the grazing behaviour of sheep. *7th General Meeting of the European Grassland Federation,* pp. 31-38.
Hodgson, J., Mackie, C.K. and Parker, J.W.G. (1986) Sward surface height for efficient grazing. *Grass Farmer* 24, 5-10.
Hodgson, J., Forbes, T.D.A., Armstrong, R.H., Beattie, M.M. and Hunter, E.A. (1991) Comparative studies of the ingestive behaviour and herbage intake of sheep and cattle grazing indigenous hill plant communities. *Journal of Applied Ecology* 28, 205-227.
Hofmann, R.R. (1983) Adaptive changes of gastric and intestinal morphology in response to different fibre content in ruminant diets. *Royal Society of New Zealand Bulletin* 20, 51-58.
Hofmann, R.R. (1989) Evolutionary steps of ecophysiological adaptation and diversification of ruminants: a comparative view of their digestive system. *Oecologia, Berlin* 78, 443-457.
Hou, Z.X., Emmans, G.C., Anderson, D., Illius, A.W. and Oldham, J.D. (1991a) The effect of different pairs of feeds as a choice on food selection by sheep. *Proceedings of the Nutrition Society* (in press).
Hou, Z.X., Lawrence, A.B., Illius, A.W., Anderson, D. and Oldham, J.D. (1991b) Operant studies on food selection in sheep. *Proceedings of the Nutrition Society* (in press)
Howe, J.C., Barry, T.N. and Popay, A.I. (1988) Voluntary intake and digestion of gorse (*Ulex europaeus*) by goats and sheep. *Journal of agricultural Science,* Cambridge 111, 107-114.
Hunter, R.F. (1962) Hill sheep and their pasture: a study of sheep-grazing in South East Scotland. *Journal of Ecology* 50, 651-680.
Illius, A.W. and Gordon, I.J. (1987) The allometry of food intake in grazing ruminants. *Journal of Animal Ecology* 56, 989-999.
Illius, A.W. and Gordon, I.J. (1990) Constraints on diet selection and foraging behaviour in mammalian herbivores. In: Hughes, R.N. (ed.), *Behavioural Mechanisms of Food Selection.* Springer-Verlag, New York, pp. 369-394.
Janeau, G. and Lécrivain, E. (1982) Use of biotelemetry in feeding behaviour: application to ewes living outdoors. *7th International Symposium on Biotelemetry,* Stanford, California. pp. 124-127.
Kenny, P.A. and Black, J.L. (1984) Factors affecting diet selection by sheep. I. Potential intake rate and acceptability of feed. *Australian Journal of Agricultural Research* 35, 551-563.
Kenward, R. (1987) *Wildlife Radio Tagging. Equipment, Field Techniques and*

Data Analysis. Academic Press, London, New York.

Kobt, A.R. and Luckey, T.D. (1972) Markers in nutrition. *Nutrition Abstracts and Reviews* 42, 813-845.

Kotliar, N.B. and Wiens, J.A. (1990) Multiple scales of patchiness and patch structure: a hierarchical framework for the study of heterogeneity. *Oikos* 59, 253-260.

Kyriazakis, I., Emmans, G.C. and Whittemore, C.T. (1990) Diet selection in pigs: choices made by growing pigs given foods of different protein concentrations. *Animal Production* 51, 189-199.

Kreuger, W.C., Laycock, W.A. and Price, D.A. (1974) Relationships of taste, smell, sight and touch to forage selection. *Journal of Range Management* 27, 258-262.

Leaver, J.D. (1982) *Herbage Intake Handbook.* The British Grassland Society, Hurley.

Le Sperance, A.L., Clanton, D.C., Nelson, A.B. and Theurer, C.B. (1974) *Factors Affecting the Apparent Chemical Composition of Fistula Samples.* West Region Coordinating Committee Publication No. 8. Nevada Agricultural Research Station, Reno.

Lu, C.D. (1988) Grazing behaviour and diet selection of goats. *Small Ruminant Research* 1, 205-216.

Lundberg, P., Astrom, M. and Danell, K. (1990) An experimental test of frequency-dependent food selection: winter browsing by moose. *Holarctic Ecology* 13, 177-182.

MacArthur, R. and Pianka, E. (1966) On optimal use of a patchy environment. *American Naturalist* 100, 603-609.

MacDonald, D. (1984) *The Encyclopaedia of Mammals: 2.* Gauld Publishing, London.

Malechek, J.C. and Narjisse, H. (1990) Behavioural ecology of sheep and goats: implications to sustain production on pastures and rangelands. Presentation to Commission on Sheep and Goat Production, 41st Annual Meeting, *European Association of Animal Production,* 8-12 July, Toulouse, France.

Marriott, C.A. and Grant, S.A. (1990) Plant species balance on sown swards in low input and extensive grazing systems. *Macaulay Land Use Research Institute Annual Report* 1989-90, pp. 24-30.

Mason, I.L. (1984) Goat. In: Mason, I.L. (ed.), *Evolution of domesticated animals.* Longman, London and New York, pp. 85-99.

Matsui, K. and Okubo, T. (1989) A 24-hour automatic recording of grazing and ruminating behaviour in cattle on pasture by a data logger method. *Japanese Journal of Zootechnical Sciences* 60, 940-945.

Matsui, K. and Okubo, T. (1990) Automatic recording over a 24-hour period for biting rate during grazing, and chewing rate and number of boluses during rumination of cow on pasture. *Japanese Journal of Zootechnical Sciences* 61, 493-500.

Mayes, R.W. (1989) The quantification of dietary intake, digestion and metabolism in farm livestock and its relevance to the study of radionuclide uptake. *The Science of the Total Environment* 85, 21-51.

Mayes, R.W., Lamb, C.S. and Colgrove, P.M. (1986) The use of dosed and herbage n-alkanes as markers for the determination of herbage intake. *Journal of agricultural Science,* Cambridge 107, 161-170.

McCabe, S.M. and Barry, T.N. (1988) Nutritive value of willow for sheep, goats and deer. *Journal of agricultural Science,* Cambridge 111, 1-9.

McInnis, M.L., Vavra, M. and Krueger, W.C. (1983) A comparison of four methods used to determine the diets of large herbivores. *Journal of Range Management* 36, 302-306.

Meijs, J.A.C., Walter, R.J.K. and Keen, A. (1982) Sward methods. In: Leaver, J.D. (ed.), *Herbage Intake Handbook.* British Grassland Society, Hurley,

Maidenhead, Berkshire, pp. 11-36.

Meyer, J.H., Lofgreen, G.P. and Hull, J.H. (1957) Selective grazing by sheep and cattle. *Journal of Animal Science* 16, 766-772.

Milne, J.A., Colgrove, P.M., Kerr, W.G. and Elston, D.A. (1988) Herbage intake of ewes in the autumn as influenced by sward herbage type and amount of supplementation and fatness of the ewe. In: *British Grassland Society Research Meeting No. 1 Session VI*, paper 3.

Milne, J.A., Hodgson, J., Thomson, R., Souter, W.G. and Barthram, G.T. (1982) The diet ingested by sheep grazing swards differing in white clover and perennial ryegrass content. *Grass and Forage Science* 37, 209-218.

Newman, J.A., Parsons, A.J. and Harvey, A. (1992) Not all sheep prefer clover: Diet selection revisited. *Journal of agricultural Science*, Cambridge (in press).

Norton, B.W., Kennedy, P.J. and Hales, J.W. (1990) Grazing management studies with Australian cashmere goats. 3. Effect of season on the selection of diets by cattle, sheep and goats from two tropical grass-legume pastures. *Australian Journal of Experimental Agriculture* 30, 783-788.

Noy-Meir, I. (19 75) Stability of grazing systems: an application of predator-prey graphs. *Journal of Ecology* 63, 459-481.

Owen-Smith, N. (1979) Assessing the foraging efficiency of a browsing ungulate, the kudu. *South African Journal of Wildlife Research* 9, 102-110.

Owen-Smith, N. and Cooper, S.M. (1987) Foraging strategies of browsing ungulates: Comparisons between goats and African wild ungulates. In: O.P. Santana, A.G. da Silva and W.C. Foote (eds.), *Proceedings of the IVth International Conference on Goats*, Departmento de Difusao de Tecnologia, Brazilia, pp. 957-969.

Owen-Smith, N. and Novellie, P. (1982) What should a clever ungulate eat? *American Naturalist* 119, 151-178.

Penning, P.D. (1983) A technique to record automatically some aspects of grazing and ruminating behaviour of sheep. *Grass and Forage Science* 38, 89-96.

Penning, P.D. (1986) Some effects of sward conditions on grazing behaviour and intake by sheep. In: Gudmundsson, O. (ed.), *Grazing Research at Northern Latitudes*. Plenum Press, London, in association with NATO Scientific Affairs Division, pp. 219-226.

Penning, P.D. and Hooper, G.E. (1985) An evaluation of the use of short term weight changes in grazing sheep for estimating herbage intake. *Grass and Forage Science* 40, 79-84.

Penning, P.D., Parsons, A.J., Orr, R.J. and Treacher, T.T. (1991) Intake and behaviour responses by sheep to changes in sward characteristics under continuous stocking. *Grass and Forage Science* 46, 15-28.

Provenza, F.D. and Balph, D.F. (1990) An assessment of five explanations for diet selection by ruminants facing five foraging challenges. In: Hughes, R.N. (ed.), *Behavioural Mechanisms of Food Selection*. Springer-Verlag, New York, pp. 423-460.

Provenza, F.D. and Malechek, J.C. (1984) Diet selection by domestic goats in relation to blackrush twig chemistry. *Journal of Applied Ecology* 21, 831-841.

Pyke, G.H. (1984) Optimal foraging theory : a critical review. *Annual Review of Ecology and Systematics* 15, 523-575.

Pyke, G.H., Pulliam, H.R. and Charnov, E.L. (1977) Optimal foraging: a selective review of theory and tests. *Quarterly Review of Biology* 52, 137-153.

Rice, R.W. (1970) Stomach contents analysis : a comparison of rumen vs. oesophageal techniques. In: *Range and Wildlife Habitat Evaluation - A Research Symposium*. USDA Forest Service Miscellaneous Publication 1147, pp. 127-132.

Rice, R.W., MacNeil, M.D., Jenkins, T.G. and Koong, L.J. (1983) Simulation of the herbage/animal interface of grazing lands. In: Lauenroth, W.K., Stogerboe, G.K. and Flug, M. (eds.), *Analysis of Ecological Systems: State-of-the-Art in Ecological Modelling, 5.* Elsevier, Amsterdam, pp. 475-488.

Robbins, C.T., Hanley, T.A., Hagerman, A.E., Hjeljord, O., Baker, D.L., Schwartz, C.C. and Mautz, W.W. (1987) Role of tannins in defending plants against ruminants: reduction in protein availability. *Ecology* 68, 98-107.

Roberts, T.J. (1977) *The Mammals of Pakistan.* Ernest Benn Ltd., London and Tonbridge.

Rose, S.P. and Michie, W. (1982) The food intakes and growth of choice-fed turkeys offered balancer mixtures of different compositions. *British Poultry Science* 23, 547-554.

Rosenthal, G.A. and Janzen, D.H. (1979) *Herbivores. Their Interactions with Secondary Metabolites.* Academic Press, London.

Ryder, M.L. (1984) Sheep. In: Mason, I.L. (ed.), *Evolution of Domesticated Animals.* Longman, London and New York, pp. 63-84.

Saarenmaa, H., Stone, N.D., Folse, L.J., Packard, J.M., Grant, W.E., Makela, M.E. and Coulson, R.N. (1988) An artificial intelligence modelling approach to simulating animal/habitat interactions. *Ecological Modelling* 44, 125-141.

Schoener, T.W. (1971) Theory of feeding strategies. *Annual Review of Ecology and Systematics* 2, 369-404.

Senft, R.L., Coughenour, M.B., Bailey, D.W., Rittenhouse, L.R., Sala, O.E. and Swift, D.M. (1987) Large herbivore foraging and ecological herbivores. *Bioscience* 37, 789-799.

Short, J.C. (1985) The functional response of kangaroos, sheep and rabbits in an arid grazing system. *Journal of Applied Ecology* 22, 435-447.

Sibbald, A.R., Grant, S.A., Milne, J.A. and Maxwell, T.J. (1988) Heather moorland management - a model. In: Bell, M. and Bunce, R.G.M. (eds.), *Agriculture and Conservation in the Hills and Uplands.* Institute of Terrestrial Ecology, Grange over Sands, pp. 107-108.

Skogland, T. (1980) Comparative summer feeding strategies of arctic and alpine reindeer. *Journal of Animal Ecology* 49, 81-89.

Skogland, T. (1984) Wild reindeer foraging-niche organisation. *Holarctic Ecology* 7, 345-379.

Stephens, D.W. and Krebs, J.R. (1986) *Foraging Theory.* Princeton University Press, New Jersey.

Stuth, J.W., Kanouse, K.J., Hunter, J.F. and Pearson, H.A. (1981) Multiple electrode impedance plethysmography system for monitoring grazing dynamics. *Biomedical Science Instrumentation* 17, 121-124.

Taylor, St. C.S., Murray, J.I. and Illius, A.W. (1987) Relative growth of incisor arcade breadth and eating rate in cattle and sheep. *Animal Production* 45, 453-458.

Thorhallsdottir, A.G., Provenza, F.D. and Balph, D.F. (1987) Food aversion learning in lambs with or without a marker: discrimination, novelty and persistence. *Applied Animal Behaviour Science* 18, 327-340.

Thouless, C.R. (1990) Feeding competition between grazing red deer hinds. *Animal Behaviour* 40, 105-111.

Trudell, J. and White, R.G.. (1981) The effect of forage structure and availability on food intake, biting rate, bite size and daily eating time in reindeer. *Journal of Applied Ecology* 18, 63-81.

Ungar, E.D. and Noy-Meir, I. (1988) Herbage intake in relation to availability and sward structure: Grazing processes and optimal foraging. *Journal of Applied Ecology* 25, 1045-1062.

Van Dyne, G.M., Brockington, N.R., Szocs, Z., Duek, J. and Ribic, C.A.

(1980) Large animal subsystem. In: Breymeyer, A.I. and Van Dyne, G.M.(eds.), *Grasslands, Systems Analysis and Man.* International Biome Programme 19. Cambridge University Press, Cambridge, pp. 269-528.

Vavra, M., Rice, R.W. and Hansen, R.M. (1978) A comparison of oesophageal fistula and fecal material to determine steer diets. *Journal of Range Management* 31, 11-13.

Weins, J.A. (1976) Population responses to patchy environments. *Annual Review of Ecology and Systematics* 7, 81-120.

Westoby, M. (1974) An analysis of diet selection by large generalist herbivores. *American Naturalist* 108, 290-302.

Wickstrom, M.L., Robbins, C.T., Hanley, T.A., Spalinger, D.E. and Parish, S.M. (1984) Food intake and foraging energetics of elk and mule deer. *Journal of Wildlife Management* 48, 1285-1301.

Zahorik, D.M. and Houpt, K.A. (1977) The concepts of nutritional wisdom: applicability of laboratory learning models to large herbivores. In: Barker, L.M., Best, M.R. and Domjam, M. (eds.), *Learning Mechanisms in Food Selection.* Baylor University Press, Waco, Texas, pp. 45-67.

Zahorik, D.M., Houpt, K.A. and Swartzman-Andert, J. (1990) Taste-aversion learning in three species of ruminants. *Applied Animal Behaviour Science* 26, 27-39.

Chapter 8

Parasites, Immunity and Anthelmintic Resistance

A.R. Sykes, R.G. McFarlane and A.S. Familton

Animal and Veterinary Sciences Group, Lincoln University, Canterbury, New Zealand

Introduction

Small ruminant pastoral production systems are extremely susceptible to the effect of a wide range of nematode endoparasites. Indeed so extensive is the list of endemic internal parasites that it is a tribute to man's ingenuity that, during the past 40 years, the technology has been developed to enable at least temporary control of the threat they pose. The development of chemicals with greater than 95% effectiveness against the parasite burdens, when administered to the host at safe levels, has not only allowed control of parasite populations but also, by their removal, allowed assessment of the true impact of the worm burden for host productivity. These latter studies have demonstrated much greater than anticipated nutritional penalties to the host, even in circumstances of modest exposure to infection (Coop *et al.*, 1982; McAnulty *et al.*, 1982; Brunsdon and Vlassof, 1986).

One objective in the development of health care in man and animals is to move from chemotherapy to enhancement of the host's own ability to resist infection through controlled exposure or vaccination. This has proved particularly difficult in relation to nematode parasite infections with very limited exceptions, for example against the lungworm *Dictyocaulus viviparus*, where an attenuated vaccine has proved practically efficacious. We still have a very poor understanding of the host response to parasites which limits our ability to enhance "self protection".

The need to change this situation is becoming imperative. As would be expected, the target parasite populations have shown evidence of genetic variation in resistance to anthelmintic and of selection towards resistant genotypes. In some geographic regions the utility of certain chemical families has become exhausted. This trend and the increasing consumer concern about agricultural practices, have caused us to examine more closely the epidemiology of parasites in traditional production systems, and to develop management systems in which the animal can be shielded from large infections, with minimal use of anthelmintics.

It would be impossible in this chapter to cover these issues for all parasites. We will, therefore, concentrate on those nematode parasites which inhabit the gastrointestinal tract and have been shown to be of economic concern in sheep.

The Effect of Endoparasites on the Host

Endoparasites have a predilection for a particular site in the alimentary tract (Table 8.1). The precise damage to the host caused by a particular parasite therefore depends on its site of predilection and the intimacy of its contact with host tissues. Thus, while infection in the abomasum, and subsequent damage to the secretory cells of that organ, will result in reduced acid secretion and therefore elevated pH, the extent of that change varies with the particular nematode at that site. *Ostertagia* (*Teladorsagia*) *circumcincta* for example, causes profound changes in pH (Bang *et al.*, 1990) which may have implications for solubility of and absorption of nutrients in subsequent sections of the tract whereas other abomasal dwellers, for example *Trichostrongylus axei*, have much less intimate contact with tissues and less effect on pH. In the duodenum and proximal small intestine *T. colubriformis* and *T. vitrinus* cause extensive villous atrophy in the mucosa (Coop *et al.*, 1979). These changes have, in the past, been considered likely to result in reduction in absorption of protein. However, a consequence of recent detailed nutritional physiological studies has been the recognition of the integrated function of the alimentary tract. Distal regions of the tract have been shown to compensate by not only absorbing nutrients not absorbed at the site of infection but, in the case of protein, resorbing the considerable additional endogenous secretions induced by damage caused by the parasite and shown in increased plasma leakage, cell sloughing and mucus production (Symons and Jones, 1983; Poppi *et al.*, 1986).

Digestion of food *per se*, with regard to energy- and protein-yielding constituents is not now considered the major impact of infection. Of much greater significance is the marked reduction in feed intake which is universally observed, the extent of which depends on the severity of larval challenge (Symons, 1985). This reduction in intake is probably the major factor in reduction in efficiency of use of metabolizable energy. In addition, particularly in infection of the intestine, the repair of gastrointestinal tissue and increased mucus and plasma protein secretions into the tract as part of the host immune response, even though these are subsequently absorbed, does increase body protein synthesis. This also results in reduced efficiency of use of metabolizable energy for productive processes (Figure 8.1) and in an induced protein deficiency (Bown *et al.*, 1991).

Table 8.1 Some important nematode endoparasites of sheep and their location in the host.

Site	Species
Abomasum	*Haemonchus contortus* *Ostertagia (Teladorsagia) circumcincta* *Trichostrongylus axei*
Duodenum and proximal small intestine	*Nematodirus battus* *Nematodirus spathiger* *Trichostrongylus colubriformis* *Trichostrongylus vitrinus* *Cooperia curticei* *Strongyloides papillosus* *Capillaria longipes* *Bunostomum trigonocephalum*
Caecum, colon and distal ileum	*Oesophagostomum venulosum* *Oesophagostomum columbianum* *Chabertia ovina* *Trichuris ovis*

Figure 8.1 Relationship between metabolizable energy intake (MEI) and energy retention (ER) in uninfected and infected sheep. Reduction in ER due to infection may occur as a result of reduction in efficiency of use of ME (A) - typical of mild intestinal infections with nematodes such as *T. colubriformis* - or due predominantly to reduction in food intake (B) - typical of mild infection with *O. circumcincta*. With increasing level of infection at either site or in mixed infection reduction in feed intake increases in importance. Adapted from Sykes *et al.* (1988).

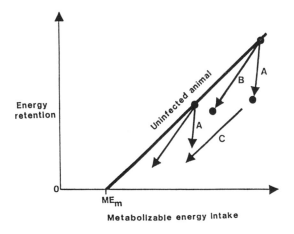

Changes in the gastrointestinal tract have been shown to interfere with mineral nutrient absorption in the case of P (Wilson and Field, 1983), Cu (Bang *et al.*, 1990) and possibly Ca (Sykes *et al.*, 1988).

Host Immunity

Infection with gastrointestinal nematodes and exposure to their numerous antigens generates a complex immune response in the host, both against ingested larvae and the resident adults (Miller, 1984; Wakelin, 1986). A non-specific inflammatory response follows primary exposure but as most host - parasite relationships involve continual cycles of infection, regulation of parasites is largely manifested through acquired immune mechanisms.

Following primary infection, blood borne mediators including Interleukin (IL) -1, IL-6 and tumour necrosis factor may be released from local tissue in the alimentary tract, and induce the release of acute phase proteins from the liver which may stimulate lymphocyte activity (Finkelman *et al.*, 1991).

The development of specific acquired immunity requires parasite antigen recognition. This process occurs principally in the peripheral lymphoid system. Generally, little is known of the parasite antigens that stimulate protective host responses although the characterization of antigens from *T. colubriformis* (O'Donnell *et al.* 1989; Rothwell and Sangster, 1991) and *O. ostertagi* (Mansour *et al.*, 1990) has recently been described. These are undoubtedly presented by macrophages at cell surfaces in association with major histocompatibility complex (MHC) proteins.

One consequence is the production of systemic humoral antibodies though a precise role for these is equivocal (Adams *et al.*, 1980; Smith *et al.*, 1983; Miller, 1984). Researchers have more consistently demonstrated adoptive immunity between animals using transfers of lymphocytes (Smith *et al.*, 1984). It is thought that local antibody is important in preventing pathology within the alimentary tract. An enhanced production of IgA follows secondary challenge of lambs with *O. circumcincta* (Duncan *et al.*, 1978; Smith *et al.* 1983) and these antibodies may complex with and remove some parasitic components as well as preventing adhesion to the gut mucosa. Local hypersensitivity reactions are associated with the immune expulsion of gastrointestinal nematodes (Stewart, 1953). The production of IgE is frequently elevated and is dependent on the presence of functional thymus-derived lymphocytes (Ahmad *et al.*, 1990). Parasite antigens may cause cross linking of IgE bound to mucosal mast cells which leads to release of vasoactive amines that increase vascular permeability, thereby increasing the transfer of serum antibodies into the lumen of the gut.

The precise function of thymus-derived (T) lymphocytes in regulating the immune response is only now being revealed, but these cells are undoubtedly involved in both antibody production and hypersensitivity reactions, in association with proteins of the MHC complex. There are two classes of T-cell which are involved in immune function, T helper and T suppressor cells. The former has a major role in the immuno-inflammatory responses against gastrointestinal parasites through the secretion of cytokines. These may include IL-4, IL-5 and IL-10 or IL-2 and g interferon, depending on the subsets of T helper cells stimulated (Mosmann *et al.*, 1986). These cytokines in turn may regulate antibody (IgE, IgG) production (Romagnani, 1990) and the development of an eosinophilia (Finkelman *et al.*, 1991).

An immediate response to infection is shown by an increase in density of mucosal mast cells (MMC) in the gastric mucosa possibly under cytokine control (Finkelman *et al.*, 1991). The number of globule leucocytes (a likely modified mast-cell) in the mucosa has been associated with increased resistance to nematodes in ruminants (Douch *et al.*, 1986), although whether this reflects increased immune function is debatable (Mayrhofer, 1984). These cells may well be involved in the production of leukotriene - like material which has been shown in recent in-vitro studies to depress the mobility of nematode larvae (Douch *et al.*, 1983). Other inflammatory mediators such as histamine and 5-hydroxytryptamine, produced in the alimentary tract mucosa, have long been considered to be active against nematodes (Stewart, 1953; Rothwell and Dineen, 1972). Their precise role is unclear, however, since they do not appear necessary for worm expulsion in all species (Befus *et al.*, 1979).

A feature of the ruminant host response to gastrointestinal nematode infection is the between-animal variation in response. This undoubtedly has a significant genetic component (Wakelin, 1986), which has been associated with particular MHC proteins expressed on lymphocytes (Outteridge *et al.*, 1988). Development of this immunity has both age and experience components. Naive lambs under 5 months of age mount a less effective local immune response than older lambs and additionally have fewer T helper cells in the gut mucosa (Gorrell *et al.*, 1988). The ability to mount an immune response has improved markedly by 12 months of age, though our recent studies have shown that the full resistance shown by mature sheep may not be achieved until 2 years of age (Familton *et al.*, 1991). There is now good evidence that the development of resistance is influenced by protein intake or status (Bown *et al.*, 1991), though whether this operates through the immune response has yet to be established. Ewes in the *peripartum* period have a diminished capacity to respond to parasites (O'Sullivan and Donald, 1970) which may be related to the production of immunosuppressive hormones during late pregnancy and lactation.

Resistance to Anthelmintics

Modern anthelmintics have their genesis with the introduction of phenathiazine in the 1940s. Subsequently, only three main classes of anthelmintic have been developed - the benzimidazoles, levamisole and the milbemycins though many sub-group derivatives have been produced and marketed (Figure 8.2). These operate by interfering either with the nervous system or oxidative metabolism of the nematode. All anthelmintics are relatively rapidly broken down within the body though the more complex the chemical structure the more slowly this occurs (Figure 8.3). Nematode genotypes capable of resistance to anthelmintics are probably present in any population of parasites at a very low frequency (Le Jambre, 1978). Those worms which can survive the effect of benzimidazole drugs may have the ability to reduce their energy demands, to use other energy producing pathways or may simply have poorer uptake of anthelmintic (Taylor and Hunt, 1989). Resistant genes have been considered to exist only in the heterozygous state and in the absence of anthelmintic remain at a low frequency.

The situation, however, is not clearly understood. For example, it has recently been suggested that benzimidazole resistance in *Caenorhabditis elegans* and *T. colubriformis* may be conferred by deletion of benzimidazole-susceptible B-tubulin genes (Driscoll *et al.*, 1989; Lenane and Le Jambre, 1991). Nevertheless, a new anthelmintic when used at a dose rate above LD 100 (100% lethal dose) would be expected to remove all worms and nematode selection for resistance not to occur (Anderson, 1990). However, use of the drug at this level rarely occurs.

Factors affecting the development of resistance

It is recognized that the rate of development of resistance is dependent on several factors. Firstly, the timing of use of anthelmintic. The proportion of the parasite population which is free-living (on pasture) and that which is within the host when anthelmintic is administered is crucial. If the larval population on pasture is large, eggs from resistant worms in the host will be diluted in the susceptible population not exposed to anthelmintic and resistance will be relatively slow to develop. Conversely, when the population on pasture is small, larvae from the 'resistant' worms are likely to comprise a larger proportion of the population. Dosing sheep during the periparturient period to control *H. contortus* infection in some temperate areas can, because of this parasite's susceptibility to cold and the presence, therefore, of a very small free-living population of larvae on pasture, contribute to the development of anthelmintic resistance in that worm

Figure 8.2 The chemical structure of some of the major chemical groups of anthelmintic -a) Levamisole, b) Morantel, c) and d) the benzimidazoles - Thiobendazole and Oxfendazole, respectively and e) the Avermectins.

a

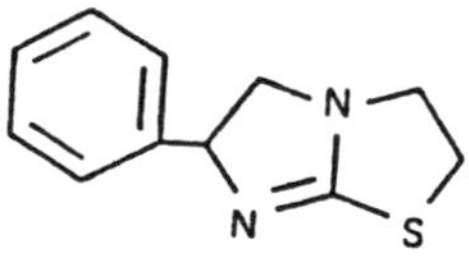

b

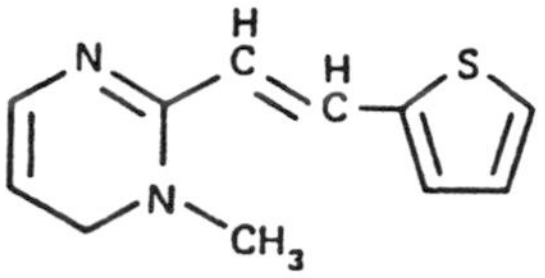

c

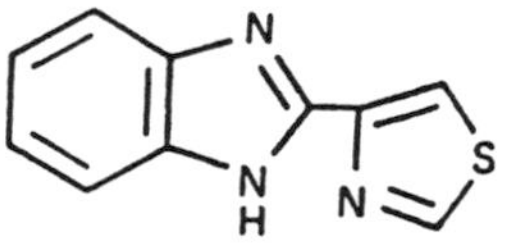

d

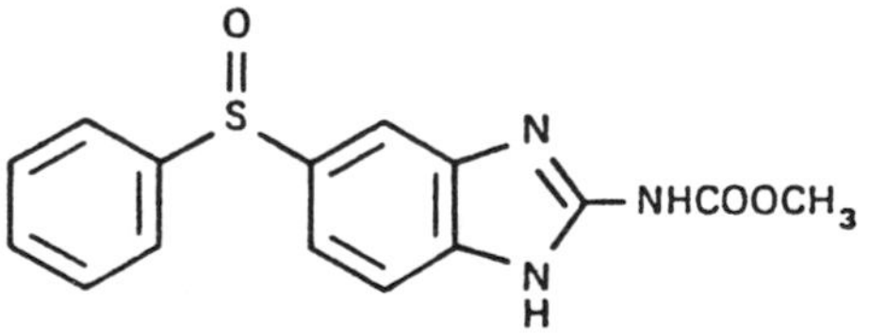

e

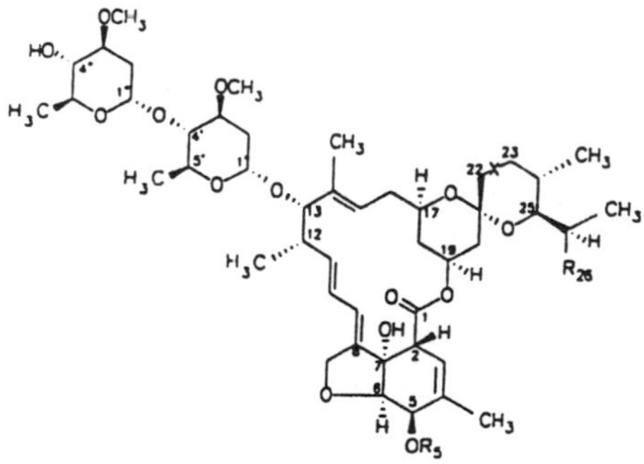

Figure 8.3 The typical time course of realized plasma concentration of several chemical groups of anthelmintic following administration - adapted from Pritchard (1978) and Campbell (1989).

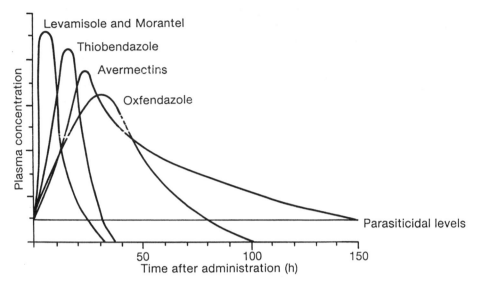

species (Taylor and Hunt, 1989). In contrast, the cold tolerance of larval populations of *O. circumcincta* means that significant numbers of over-wintered larvae are likely to be present on pasture and the development of resistance, following anthelmintic use at this time, would be less rapid. Treatment of stock with anthelmintic in summer dry regions may similarly enhance development of resistance, because larval populations outside the host and the faecal mass are diminished by high temperature and dessication. These factors may help explain the particularly rapid development of resistance to several anthelmintics in certain climatic regions.

Drenching animals at a frequency interval which approaches the prepatent period of the parasite in the host (approximately 21 days) increases the number of resistant parasites both in the host and on the pasture. It has also been suggested that regimes in which sheep are dosed and moved to minimally infested (safe) pasture enhances selection for resistance since larval populations on pasture become, increasingly, survivors of the drenching programme (Taylor and Hunt, 1988). However, where this management practice has been followed the development of resistance does not appear to have accelerated.

Once anthelmintic resistance is present in a nematode population it may persist for several years even in the absence of continuous selection pressure by the specific anthelmintic (Boorgsteede and Duyn, 1989).

The development of resistance is aided by poor quality control in the administration of anthelmintic. The effectiveness of an anthelmintic depends on the height of and area achieved under the blood anthelmintic concentration-time relationship (Figure 8.3). Underdosing with anthelmintic, lack of attention to the setting of drench guns and underestimation of the live weight of animals, all reduce effective blood concentration and therefore exposure of parasites to the anthelmintic, circumstances which favour survival of resistant nematode genotypes. An Australian survey has shown that only 27% of farmers estimated live weight to within 20% of the correct value and that 30% of the farmers used a dosage which differed by more than 10% from that recommended (Besier and Hopkins, 1988). Lack of awareness of differences in anthelmintic dose rates for effective treatment of different species of parasite undoubtedly contributes to this problem. Moreover, it is known that manufacturer recommendations may not necessarily completely remove all worms, e.g. *Nematodirus battus* (Gibson and Parfitt, 1971). To combat the build up of resistance a change of drench class family at strategic intervals has been advocated (Waller *et al.* 1989; Pomroy, 1990). This should coincide with the end of a 3-4 week interval drenching programme or occur at a time when the change to safe pasture is most important, such as weaning.

Detection and avoidance of resistance

Detection of anthelmintic resistance is achieved by the faecal egg count reduction (FECR) test and other *in vitro* assays, such as the egg hatch assay, tubulin binding assay and a larval development assay. All show similar efficacy in assessing benzimidazole resistance (Johansen and Waller, 1989), although interpretation of the FECR test is dependent on allowing sufficient time to elapse between administration of anthelmintic and faecal collection (Scott *et al.*, 1991).

Annual checks on anthelmintic efficiency using the FECR test are important to confirm that resistance is not present. The best insurance to avoid bringing resistant nematode genotypes onto a property is prior treatment with an anthelmintic to which drug resistance is least likely to have developed, such as the recently developed milbemycin products. The use of combinations of benzimidazole and levamisole drugs has been advocated in the strategic alteration of different classes of anthelmintic (McKenna, 1990). Ruminal slow release anthelmintic devices (Alzieu *et al.*, 1990) are effective against incoming 3rd stage larvae at a very much lower drug concentration than against adult worm populations and may lead to a less rapid development of resistance (Anderson, 1985). In fact such devices, delivering oxfendazole, have produced good control against certain benzimidazole-resistant strains of parasite (Martin, 1988).

More precise targeting of anthelmintic and therefore reduction in use will become increasingly important in slowing down development of resistance. Recognition that 65-70% of nematode larvae on pasture may be derived from the ewe rather than from the lamb, as shown in a recent study in New Zealand (Familton *et al.*, 1991) may well direct anthelmintic use to that category of animal and reduce the need for very freqent use of anthelmintic in lambs to protect them from a large nematode larval population on pasture.

The escalating development of anthelmintic resistance and the difficulty of unravelling the complexity of the host immune response mean that the major production and economic losses caused by nematode endoparasites will persist in the foreseeable future. Medium term control will require managemental strategies which incorporate an ecological approach. Immunological technology may be the key for control in the long term.

References

Adams, D.B., Merritt, G.C. and Cripps, A.W. (1980) Intestinal lymph and the local antibody and immunoglobin response to infection by *Trichostrongylus colubriformis* in sheep. *Australian Journal of Experimental Biology and Medical Science* 58, 167-177.

Ahmad, A., Wang, C.H. and Bell, R.G. (1990) Rapid expulsion of *Tricheaella spiralis* mediated by 0X8-0X22-Th cells and purified IgE. *Proceedings VII International Congress for Parasitology* p. 75.

Alzieu, J.P., Bichet, H., Dorchies, P., and Pothier, F. (1990) Results of four years of clinical trials with an albendazole slow-release device in sheep. 2. Effects on fertility, lambing rate and productivity of ewes. *Revue de Médecine Vétérinaire* 141, 199-204.

Anderson, N. (1985) The controlled release of anthelmintics for helminth control in ruminants. In: Anderson N. and Waller, P.J. (eds.) *Resistance in Nematodes to Anthelmintic Drugs*. CSIRO Division of Animal Health and Australian Wool Corporation, Glebe, NSW.

Anderson, N. (1990) Developments in the control of nematode infections. *Proceedings of the New Zealand Society of Animal Production* 50, 215-227.

Bang, K.S., Familton, A.S. and Sykes, A.R. (1990) Effect of ostertagiasis on copper status in sheep. A study involving use of copper oxide wire particles. *Research in Veterinary Science* 49, 306-314.

Befus, A.D., Johnston, N. and Bienenstock, J. (1979) *Nippostrongylus brasiliensis*: Mast cell and histamine levels in tissues of infected and normal rats. *Experimental Parasitology* 48, 1-8.

Besier, R.B. and Hopkins, D.L. (1988) Anthelmintic dose selection by farmers. *Australian Veterinary Journal* 65, 193-194.

Boorgsteede, F.H.M. and Duyn S.P.J. (1989) Lack of reversion of a benzimidazole resistant strain of *Haemonchus contortus* after six years of levamisole usage. *Research in Veterinary Science* 46, 270-272.

Bown, M.D., Poppi, D.P. and Sykes, A.R. (1991) The effect of post-ruminal infusion of protein or energy on the patto physiology of *Trichostrongylus colubriformis* infection and body composition in lambs. *Australian Journal of Agricultural Research* 42, 253-267.

Brunsdon, R.V. and Vlassof, A. (1986) Longer term benefits seen for post lamb ewe drenching. *New Zealand Journal of Agriculture* 51, 36-37.

Campbell, W.C. (1989) In: *Ivermectin and abamectin*. Springer-Verlag, New York.

Coop, R.L., Angus, K.W. and Sykes, A.R. (1979) Chronic infection with *Trichostrongylus vitrinis* on sheep. Pathological changes in the small intestine. *Research in Veterinary Science* 26, 363-371.

Coop, R.L., Sykes, A.R. and Angus, K.W. (1982) The effect of three levels of intake of *Ostertagia circumcincta* larvae on growth rate, food intake and body composition of growing lambs. *Journal of agricultural Science,* Cambridge, 98, 247-255.

Douch, P.G.C., Harrison, G.B.L., Buchanan, L.L. and Greer, K.S. (1983) *In vitro* bio-assay of sheep gastrointestinal mucus for nematode paralysing activity mediated by a substance with some properties characteristic of SRS-A. *International Journal of Parasitology* 13, 207-212.

Douch, P.G.C., Harrison, G.B., Elliot, D.C., Buchanan, L.L. and Greer, K.S. (1986) Relationship of gastrointestinal histology and mucus antiparasitic activity with the development of resistance to trichostrongyle infections in sheep. *Veterinary Parasitology* 20, 315-331.

Driscoll, M., Dean, E., Reilly, E., Bergholz, E., Chalfie, M. (1989) Genetic and molecular analysis of a *Caenorhabditis elegans* b-tubulin that conveys benzimidazole sensitivity. *Journal of Cell Biology* 109, 2993-3003.

Duncan, J.L., Smith, W.D. and Dargie, J.D. (1978) Possible relationship of levels of mucosal IgA and serum IgG to immune responsiveness of lambs to *Haemonchus contortus. Veterinary Parasitology* 4, 21-27.

Familton, A.S., McAnulty, R.W. and Sykes, A.R. (1991) The annual contribution of adult sheep to the trichostrongylid pasture larval population under New Zealand conditions (unpublished data).

Finkelman, F.D., Pearce, E.J., Urban, J.F. and Sher, A. (1991) Regulation and biological function of helminth-induced cytokine responses. *Immunology Today* 12, 62-66.

Gibson, T.E. and Parfitt, J.W. (1971) The evaluation of three anthelmintics against *Nematodirus battus* in lambs by means of the improved control test. *British Veterinary Journal* 127, 28-31.

Gorrell, M.D., Willis, G., Brandon, M.R. and Lascelles, A.K. (1988) Lymphocyte phenotypes in the intestinal mucosa of sheep infected with *Trichostrongylus colubriformis. Clinical and Experimental Immunology* 72, 274-279.

Johansen, M.J. and Waller, P.J. (1989) Comparison of three *in vitro* techniques to estimate benzimidazole resistance in *Haemonchus contortus* of sheep. *Veterinary Parasitology* 34, 213-221.

Le Jambre, L.F. (1978) *The epidemiology and control of gastrointestinal parasites of sheep in Australia.* CSIRO Division of Animal Health, Melbourne, p. 109.

Lenane, I.J. and Le Jambre, L.F. (1991) Molecular evolution of benzimidazole resistance in a *Trichostrongylus colubriformis* ben - 1 homologue. *Proceedings Queenstown Molecular Biology Meeting.*

Martin, P.J. (1988) Anthelmintic resistance. In: Howell M.J. (ed.) *Sheep Health and Production.* Proceedings No. 110. Postgraduate Foundation of Veterinary Science, Sydney University. pp 347-367.

Mayrhofer, G. (1984) Physiology of the Intestinal Immune System. In: Newby, T.J. and Stokes, C.R., (eds). *Local Immune Responses of the Gut.* CRC Press, Boca Raton, Florida.

McAnulty, R.W., Clark, V.R. and Sykes, A.R. (1982) The effect of use of clean pasture and anthelmintic frequency on growth rate of hoggets. *Proceedings of the New Zealand Society of Animal Production* 42, 187-188.

McKenna, P.B. (1990) The use of benzimidazole - levamisole mixtures for the control and prevention of anthelmintic resistance in sheep nematodes: an assessment of their likely effects. *New Zealand Veterinary Journal* 38, 45-49.

Mansour, M.M., Dixon, J.B., Clarkson, M.J., Carter, S.D., Rowan, T.G. and Hammet, N.C. (1990) Bovine immune recognition of *Ostertagia ostertagi* larvae antigen. *Veterinary Immunology and Immunopathology* 24, 361-371.

Miller, H.R.P. (1984) The protective mucosal response against gastrointestinal nematodes in ruminants and laboratory animals. *Veterinary Immunology and Immunopathology* 6, 167-259.

Mosmann, T.R., Cherwinski, H., Bond, M.W., Giedlin, M.A., Coffman, R.L. (1986) Two types of murine helper T-cell clones. I. Definition according to profiles of lymphokine activities and secreted proteins. *Proceedings National Academy of Science (USA)* 83, 5654-5658.

O'Donnell, I.J., Dineen, J.K., Wagland, B.M., Letho, S., Dopheide, T.A.A., Grant, W.N. and Ward, C.W. (1989) Characterisation of the major immunogen in the excretory - secretory products of exsheathed third-stage larvae of *Trichostrongylus colubriformis*. *International Journal for Parasitology* 19, 793-802.

O'Sullivan, B.M. and Donald, A.D. (1970) A field study of nematode parasite populations in the lactating ewe. *Parasitology* 61, 301-315.

Outteridge, P.M., Windon, R.G. and Dineen, J.K. (1988) An association between a lymphocyte antigen in sheep and responsiveness to vaccination against the parasite *Trichostrongylus colubriformis*. *International Journal of Parasitology* 15, 121-127.

Pomroy, W.E. (1990) Strategies to combat anthelmintic resistance. *Proceedings of the Sheep and Beef Cattle Society of the New Zealand Veterinary Association* 20, 21-26.

Poppi, D.P., MacRae, J.C., Brewer, A.C. and Coop, R.L. (1986) Nitrogen transactions in the digestive tract of lambs exposed to the intestinal parasite, *Trichostrongylus colubriformis*. *British Journal of Nutrition* 55, 593-602.

Pritchard, R.K. (1978) Anthelmintics. In: *The therapeutic jungle*. Proceedings No. 39, Postgraduate Committee in Veterinary Science University of Sydney pp. 421-463.

Romagnani, S. (1990) Regulation and deregulation of human IgE synthesis. *Immunology Today* 11, 316-321.

Rothwell, T.L.W. and Dineen, J.K. (1972) Cellular reactions in guinea pigs following primary and challenge infection with *Trichostrongylus colubriformis* with special reference to the roles played by eosinophils and basophils in rejection of the parasite. *Immunology* 22, 733-745.

Rothwell, T.L.W. and Sangster, N.C. (1991) Localisation of protective antigens in *Trichostrongylus colubriformis*. *International Journal for Parasitology* 21, 115-117.

Scott, E.W., Baxter, P. and Armour, J. (1991) Fecundity of anthelmintic resistant adult *Haemonchus contortus* after exposure to ivermectin or benzimidazoles *in vivo*. *Research in Veterinary Science* 5, 247-249.

Smith, W.D., Jackson, F., Jackson, E. and Williams, J. (1983) Local immunity and *Ostertagia circumcincta*: Changes in the lymph node of immune sheep after a challenge infection. *Journal of Comparative Pathology* 93, 479-488.

Smith, W.D., Jackson, F., Jackson, E., Williams, J., Willadsen, S.M and Fehilly, G.B. (1984) Resistance to *Haemonchus contortus* transferred between genetically histocompatible sheep by immune lymphocytes. *Research in Veterinary Science* 37, 199-204.

Stewart, D.F. (1953) Studies in resistance of sheep to infections with *Haemonchus contortus* and *Trichostrongylus* spp. and on the immunological reactions of sheep exposed to infestation. *Australian Journal of Agricultural Research* 4, 100-117.

Sykes, A.R., Poppi, D.P. and Elliot, D.C. (1988) Effect of concurrent infection with *Ostertagia circumcincta* or *Trichostrongylus colubriformis* on the performance of growing lambs offered fresh herbages. *Journal of agricultural Science*, Cambridge, 110, 531-541.

Symons, L.E.A. (1985) Anorexia: occurrence, pathophysiology and possible causes in parasitic infections. *Advances in Parasitology* 24, 103-133.

Symons, L.E.A. and Jones, W.O. (1983) Intestinal protein synthesis in guinea pigs infected with *Trichostrongylus colubriformis*. *International Journal for Parasitology*, 1309-1312.

Taylor, M.A. and Hunt, K.R. (1988) Field observations on the control of ovine parasitic gastroenteritis in south-east England. *Veterinary Record* 123, 241-245.

Taylor, M.A. and Hunt, K.R (1989) Anthelmintic drug resistance in the United Kingdom. *Veterinary Record* 125, 143.

Wakelin, D. (1986) The role of the immune response in helminth population regulation. *Proceedings of the 6th International Congress of Parasitology* pp. 549-557.

Waller, P.J., Donald, A.D., Dobson, R.J., Lacey, E., Hennessy, D.R. Allerton, G.R., and Prichard, R.K. (1989) Changes in anthelmintic resistance status of *Haemonchus contortus* and *Trichostrongylus colubriformis* exposed to different anthelmintic selection pressures in grazing sheep. *International Journal of Parasitology* 19, 99-110.

Wilson, W.D. and Field, A.C. (1983) Absorption and secretion of calcium and phosphorus in the alimentary tract of lambs infected with daily doses of *Trichostrongylus colubriformis* or *Ostertagia circumcincta* larvae. *Journal of Comparative Pathology* 93, 61-71.

Chapter 9

Selection for Lean Meat Production in Sheep

G. Simm

The Scottish Agricultural College Edinburgh, West Mains Road, Edinburgh EH9 3JG, UK

Introduction

The increasing preference of consumers for leaner meat over the past few decades is well documented (e.g. Kempster, 1983; Woodward and Wheelock, 1990). This preference was originally based on considerations of taste and reduced waste. More recently it has been reinforced by reports linking high consumption of saturated fats in many Western countries with an increased incidence of several chronic degenerative diseases (see reviews in Wood and Fisher, 1990). However, the validity of recommendations to reduce saturated fat intake is still questioned (Blaxter, 1991).

Sinclair and O'Dea (1990) give a fascinating account of fat in the human diet through history. They conclude that the modern Western diet is out of step with diets typical of the hunter-gatherer lifestyle which has dominated human history and set an "evolutionary precedent". These authors have studied the diet of Aborigines in north western Australia who temporarily reverted to a traditional lifestyle. Aborigines represent one of the few hunter-gatherer societies to survive into the 20th Century. Despite the high proportion of food of animal origin in the traditional diets of the Aborigines (principally kangaroo meat and fish) the overall fat content of the diet was low. This was due to the particularly low fat content of the muscle meat eaten, and the relatively small fat depots which are present in wild animals for most of the year.

In general the proportion of fat in the carcasses of wild animals, and the proportion of lipid in their muscles are both much lower than in domesticated species (see Sinclair and O'Dea (1990) for results of a number of studies). Old drawings, and more recent objective analyses, show that breeds which have undergone little or no selection for growth and meat production are also leaner than "improved" breeds (e.g. Soay *v.* the Southdown and Oxford Down breeds; McClelland *et al.*, 1976). But from the 18th Century until the earlier part of this century, selection, management and feeding practices in Britain were geared to producing animals which we would now regard as obese. For example, George Culley (1801) in his book "Observations on Livestock" refers to his 3 year old wether of the Dishley (Leicester) breed "which measured seven inches and one eighth (about 18 cm) of solid fat on the ribs, cut straight through without any slope". This breed was developed by Robert Bakewell, the pioneer of live-stock improvement, and is the forerunner of many modern sheep breeds (Hall and Clutton-Brock, 1989).

As a consequence of this selection history, together with often ambiguous market signals and traditional husbandry methods, the proportion of fat in sheep carcasses is now higher than that desired by consumers. For example, Kempster *et al.* (1986) estimated that the average lamb carcass in Britain in 1984 had 20 percentage units of fat in excess of consumer demand. As a result, fat is trimmed from carcasses or joints before sale and further trimming often takes place before cooking and during eating. Although it may not be most biologically or economically efficient to precisely match consumer demands in the live animal, excess fat production is clearly a source of inefficiency in sheep production in many countries. In this chapter, I aim to examine the contribution which selection can make to aligning the carcass composition of lambs more closely with consumer requirements.

Genetic Improvement in Context

Options to reduce fatness

There are a number of non-genetic means for reducing the fatness of carcasses in the short term, including slaughtering lambs at lighter weights, ceasing castration or altering the quality and quantity of feed offered (e.g. Bass *et al.*, 1990). The composition of carcasses may also be manipulated by pharmacological means, although these are becoming less acceptable to consumers and have been prohibited in several countries. Several of these short-term measures have disadvantages - for example, reducing the slaughter weight of animals may lead to lower output per unit of land or capital, and altering feed quality and quantity may be impractical in extensive production systems. Thus, genetic improvement, either alone or in combination with some of the measures outlined, is an attractive option for effecting longer-term, permanent improvement in carcass composition.

Between and within-breed selection

Selection may take place either between breeds or between animals within breeds. When there are clearly "better" breeds for a given set of character-istics, selection between breeds can lead to rapid improvement. Selection between breeds is often based on the results of designed breed comparisons. Taylor (1980) describes a conceptual framework which is valuable in designing such comparisons, in predicting the relative performance of different breeds and in matching breeds to production systems. Taylor's approach is based on the observation that, when breeds are compared at similar degrees of maturity in live weight, most of the differences in composition disappear (McClelland *et al.*, 1976). Comparisons on this basis are also valuable in highlighting breeds which depart from this general relationship. For example, as already mentioned, the Soay has less fat and more bone than expected (McClelland *et al.*, 1976) and the Texel has less fat and more lean than expected (Wolf *et al.*, 1980). Breed comparisons involving direct measurements of carcass composition are expensive, but they are required relatively infrequently, for example when new breeds are imported or after periods of effective within-breed selection.

Differences in mature size may well explain much of the within-breed variation in carcass composition at the same weight. However, it will often be more practical to base selection on immature live weights and measurements of carcass composition than to prolong selection decisions until mature size can be measured. Moreover, there is evidence of genetic variation in rates of maturing in fatness and in fatness at maturity in laboratory animals and sheep (Roberts, 1979; Butterfield *et al.*, 1983; Thompson *et al.*, 1985, 1987).

Direct measurement of carcass composition poses problems in within-breed selection. It requires either (i) collection and freezing of semen from candidates for selection, prior to slaughter and evaluation, or (ii) slaughter and evaluation of relatives of the candidates for selection. The first option will be unattractive to breeders if the animals are potentially of high value (although this may change if advances in embryo technology allow the production of several genetically identical animals at low cost). The second option is complicated by the fact that breeding flocks are frequently geographically dispersed and not "genetically linked", which makes it harder to pre-select candidates for progeny testing. For these reasons there has been much interest in methods of estimating carcass composition in live animals. These are discussed in more detail later.

Selection Objectives and Criteria for Within-Breed Selection

One of the most important steps in any breeding programme is to define the objective or goal for selection and the criteria or measurements on which selection will be based. The selection objective will depend on the role of the breed. In specialized meat breeds, the objective is usually restricted to improving growth and carcass characteristics. However, in multi-purpose breeds, wool characteristics, reproduction, milk production, and other maternal traits may be equally or more important than growth and composition, which complicates definition of the selection goal. Discussion here is confined to specialized meat breeds, although much of it will apply to other breeds where growth and composition are important parts of a wider overall objective.

In specialized meat breeds, the selection objective is usually to alter carcass tissue weights or proportions at a given age, live weight or carcass weight. Selection for this objective is usually based on measurements of live weight, together with *in vivo* estimates of carcass composition of candidates for selection or direct carcass measurements on the progeny or other relatives of candidates for selection. The heritabilities of some of these traits are shown in Table 9.1, and estimates of the correlations amongst these traits are given by the authors cited there. In general, the traits of interest are moderately highly heritable.

Table 9.1 Average estimates of heritabilities from three reviews.

	Wolf and Smith 1983		Simm *et al.* 1987b[1]		Young 1989[1]
Basis of evaluation (constant age or weight)	Weight	Age	Age	Weight	Age
Trait					
Live weight (post 100 d)			0.24		0.28
Carcass weight	(0.02)[2]	0.44		0.22	0.16
Carcass lean or protein %	0.41	(0.51)			
Carcass lean weight			0.27	(0.39)	
Carcass fat/ether extract %	0.46	(0.50)			
Carcass fat weight			0.29	(0.44)	
Carcass bone %	0.20	0.18			
Carcass bone weight		(0.43)		(0.28)	
Carcass eye-muscle area	0.29	0.34			(0.53)
Carcass eye-muscle depth			0.24[3]	(0.62)	0.38
Carcass backfat depth	0.37	0.28		0.25	
Ultrasonic backfat depth			0.23		

[1]Where possible estimates which appear in previous columns have been excluded.
[2]Figures in parentheses are single estimates.
[3]Includes some muscle areas

In theory, economic selection indices, as proposed by Hazel (1943), weight different measurements optimally to maximize response in overall economic merit. However, derivation of an economic index requires (i) estimates of phenotypic and genetic variances and covariances for goal traits and measurements and (ii) estimates of the relative economic values of traits in the selection goal. Each of these requirements presents difficulties. Firstly, estimating genetic parameters is time consuming and expensive. Also, published estimates are rarely complete, or fully compatible with any new breeding scheme (for example, breeds, management systems, ages or weights at measurement differ widely). Secondly, deriving relative economic values is a complex task. Moav (1973) showed that relative economic values varied, depending on whether they were calculated from a national viewpoint, the producer's viewpoint, or that of a new investor.

Carcass "conformation" or overall shape provides a good example of this dilemma. Carcasses of good conformation produce a higher financial return for producers and meat wholesalers in many countries. However, there are several studies which show that, at least within breeds, conformation is a poor indicator of carcass lean content and proportion of lean in the higher-priced joints, in carcasses of equal weight and subcutaneous fat cover (e.g. Kempster *et al.*, 1981; Wolf *et al.*, 1981). Thus, good conformation would not appear to be of particular value from the consumers' or national viewpoint. This result may well be specific to the particular methods of assessing conformation used, the level of fatness of the animal's involved and possibly also the breeds concerned; other studies show that conformation score, after adjusting for carcass weight and fatness, may be a useful predictor of muscle weight (Bass *et al.*, 1984). Also, more objective measures of muscularity have been proposed which may prove more useful predictors of tissue weights and distribution (e.g. Young, 1990; Purchas *et al.*, 1991)

There has been much debate on the most appropriate method of deriving relative economic values and methods have been proposed to unify the different approaches (e.g. Brascamp *et al.*, 1985; Smith *et al.*, 1986). Partly because of these difficulties, together with incongruous payment schemes which make it difficult to calculate relative economic values for carcass components, and the scarcity of genetic parameters, there have been few attempts to derive economic selection indices in meat sheep (e.g. Ponzoni and Walkley, 1981; Rae, 1984; Simm *et al.*, 1987b).

Fowler *et al.* (1976) examined the use of lean tissue growth rate and lean tissue food conversion as selection objectives for pigs, which avoid calculation of economic weights and the need for genetic parameters. Purchas *et al.* (1985) proposed a lean growth index for sheep which is essentially equivalent to the lean tissue growth rate of Fowler *et al.* Although attractive at first sight, there are limitations to selection on these "biological indices" (Simm *et al.*, 1987a; Bennett, 1990). Where there is a large imbalance in the variation in component traits of a biological index (for example in ruminants, weight or growth rate is usually much more variable than estimated lean content) then the most variable component will dominate the index. Although biological indices avoid the need to calculate economic weights their component traits do have *implied* economic weights. These are uncontrolled and possibly widely different from the true economic weights.

There are a number of approaches which combine the rigour and control of an economic selection index with the simplicity of a biological index. For example, restricted selection indices or desired gains indices (Brascamp, 1984; Rae, 1984) allow more control over the outcome of selection than a biological index. Simm and Dingwall (1989) examined expected responses in carcass lean weight and carcass fat weight from various combinations of relative economic values to assist in the final choice of appropriate values. These approaches, or the strict economic approach, are likely to give more reliable selection responses even after allowing for possible errors in genetic parameters or relative economic values.

Techniques for *In Vivo* Assessment

As mentioned earlier, *in vivo* measurements of carcass composition are being used increasingly as selection criteria in sheep breeding. There are a number of reviews of techniques for *in vivo* assessment in sheep and other livestock (e.g. Alliston, 1983; Lister, 1984; Simm, 1987; Allen, 1990). I do not intend to review techniques in detail here but only to outline those which appear to be of most use at present or are likely to be of most value in the future.

Live weight

Live weight is obviously an important selection criterion in its own right in specialized meat sheep, whether the aim is to increase it or maintain it at a fairly constant level. It is also a valuable predictor of composition. As animals grow, the weight and proportion of fat in the body and carcass generally increase, and the proportion of lean tissue decreases. Thus, as mentioned earlier, when animals of the same breed and sex are compared at a similar age, much of the variation in carcass composition can be explained by variation in live weight. For this reason *in vivo* techniques are often compared on the basis of the additional variation in carcass composition which they account for, after that accounted for by live weight.

Ultrasonic measurements

Ultrasonic measurements are probably the most widely used *in vivo* measurements in farm animals. Two types of ultrasonic measurement have been used in *in vivo* assessment: (i) pulse-echo techniques and (ii) transmission techniques. Pulse-echo techniques are used to map tissue boundaries in one (usually "A-mode") or two ("B-mode") dimensions (see Figure 9.1 and Simm (1987) for more details). Transmission techniques rely on the relationship between the reciprocal of the speed of transmission of ultrasound and the fat content of the tissue through which the sound travels (see Miles *et al.,* 1984).

There are several published evaluations of the precision of fat or muscle depths or areas from pulse-echo ultrasonic machines in predicting carcass composition (see Simm, 1987). These reports vary quite widely, both in the carcass characteristics considered and the precision achieved. In some studies, ultrasonic measurements have added little to the precision achieved by using measurements of live weight alone (e.g. Cuthbertson *et al.*, 1984: residual standard deviation (r.s.d.) in lean % 1.4 after fitting live weight (LW), 1.3-1.5 after fitting LW and ultrasonic measurements) whilst in others, larger improvements have been achieved (e.g. Purchas and Beach, 1981: r.s.d. in fat % 3.9 after fitting LW alone, 2.6-2.9 after adding ultrasonic fat depth measured by different operators; Simm, 1987: r.s.d. in lean % 2.9 after fitting LW, 2.4 after adding ultrasonic measurements).

Figure 9.1 A real-time "B" mode (Vetscan) ultrasonic scan at the 3rd lumbar vertebra of a Suffolk ram.

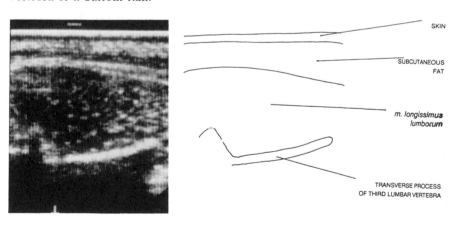

Some of the variation in precision achieved in different studies is probably a result of differences in the breed of sheep used, the type of machine used, the site of measurement and operator experience. The mean and variation of tissue depths are probably also important since the ability

Figure 9.2 An X-ray CT image at the 3rd lumbar vertebra of a Suffolk ram (lying on its back in a scanning cradle)

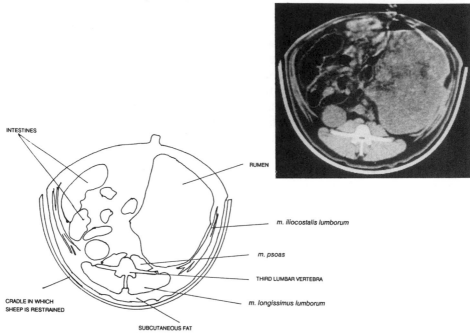

of ultrasonic machines to discriminate between small differences in tissue depths and the error in interpretation of scans are likely to be proportionally less important as tissue depths, or variation in them, increase. Generally, sheep have a lower proportion of fat in the subcutaneous depot than pigs (Kempster *et al.*, 1986) and have lower absolute fat depths than either pigs or most breeds of cattle. This may explain the lower level of precision often achieved by ultrasonic measurements in sheep, and the apparent improvement in precision when the sheep used are fatter or show greater variation in tissue proportions.

Although the level of precision achieved with ultrasonic scanning measurements in sheep is usually only moderate, the technique has advantages in ease and speed of operation, mobility and cost, with the potential for measuring large numbers of animals.

Advanced imaging techniques

Over the past few years there has been considerable interest in the application of more advanced imaging techniques, developed primarily for use in human medicine. Perhaps the two strongest contenders for application in animal science are X-ray Computed Tomography (CT) and Magnetic Resonance Imaging (MRI). X-ray CT is based on the fact that the rate of attenuation of X-rays differs from one tissue to another, depending on tissue density. X-ray CT involves rotating an X-ray source around the body of the subject, firing pulses of radiation at discrete intervals and measuring the amount of radiation received after transmission through the subject. The resulting matrix of attenuation values, or CT numbers, is valuable in its own right but for most medical applications this information is processed to provide a 2-dimensional image (see Figure 9.2).

Although there have been few studies applying CT to sheep, the results to date are encouraging. Sehested (1984) reported that 89-92 per cent of the variation in carcass tissue weights could be explained by the frequency of CT numbers in different ranges from a single scan and the animal's live weight. In a study involving an older CT machine, with access only to images and not distributions of CT numbers, Simm (1987) reported that 60 per cent and 73 per cent of the variation in lean percentage and lean weight was explained by CT tissue depths and areas and live weight. Equivalent R^2 results for ultrasonic machines were 37-44 per cent for lean percentage and 58-64 per cent for lean weight.

MRI is based on nuclear magnetic resonance, a property of atomic nuclei that have an odd number of protons or neutrons, the most common of which in body tissue is the hydrogen atom. The properties of these nuclei also differ, depending on the "chemical background" of the tissue, such as the state of hydration and fat content. Consequently, MRI has potentially greater power to discriminate between tissues than X-ray CT. The density, distribution and properties of protons can be examined by measuring the electromagnetic signals which they induce in a coil surrounding the subject in a magnet, in response to changes in the magnetic field characteristics.

Although the application of these techniques in animal science is fairly new (Groeneveld *et al.*, 1984; Sehested, 1984; Foster *et al.*, 1989), they do have potentially large advantages in precision over ultrasonic scanning techniques. Additionally they offer new possibilities such as *in vivo* estimation of muscle volumes and tissue distribution (Knopp, 1985) for which ultrasonic techniques are less suitable. The main disadvantages are cost and throughput, although the cost-effectiveness of these techniques may well be favourable at a national level, particularly if used in two-stage selection, following wider screening on, for example, ultrasonic measurements.

Precision of *in vivo* measurements and response to selection

Simm and Dingwall (1989) examined the expected responses from index selection on different *in vivo* measurements with a range of relative economic values (REVs) for carcass lean and fat weights. Figure 9.3 shows their results for an index with REVs of $+3$ for lean weight and -1 for fat weight. These results indicate that including ultrasonic measurements of fat

and muscle depth in an index is expected to lead to similar responses in carcass lean weight to those achieved from selection on live weight alone but with progressive reductions in the correlated increase in fat weight.
There are few estimates of genetic parameters for measurements with the advanced imaging techniques. These authors therefore examined expected responses to selection, assuming that carcass composition could be predicted perfectly from *in vivo* measurement, to show the potential for improving responses with improved *in vivo* techniques. The figure shows that, with REVs of +3 and -1, the annual rate of response in lean weight could be up to 1.35 times as great as that possible with ultrasonic measurements, whilst simultaneously decreasing fat weight.

Figure 9.3 Expected annual responses in carcass lean and fat weights following selection on live weight (LW) or on indices including LW, ultrasonic fat and muscle depths (UFD, UMD) or perfect *in vivo* measurements (relative economic values +3 lean, -1 fat). (After Simm and Dingwall, 1989).

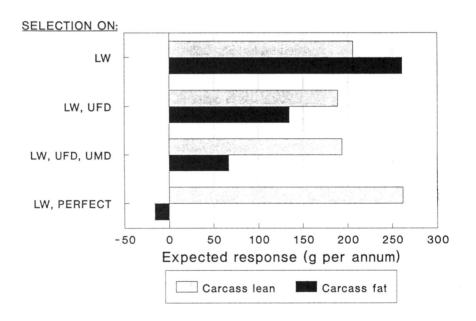

Physiological Predictors

A number of workers have examined physiological predictors of growth and carcass composition and this area is reviewed by Blair *et al.* (1990). Frequently potential predictors are identified by screening for physiological differences between genetically divergent lines. For example, Peterson and Purchas (1989) have reported higher liver catalase activity in sheep from a Massey University high backfat selection line than in those from a low backfat line. Higher plasma urea concentrations have also been reported in these high backfat line sheep (Carter *et al.*, 1989). Similar results for urea concentrations have been reported for other selected lines by Cameron (1991). Suttie *et al.* (1989) have shown that growth hormone (GH) pulsatility differs between lean and fat line sheep, and that GH levels were lower in the fatter sheep within each line. Sheep selection lines (high, low and control) have recently been established at Massey University for plasma levels of insulin-like growth factor (Blair *et al.*, 1990).

Another study on unselected sheep has shown an association between blood concentration of glutathione and meat yield (Fiebrand *et al.*, 1989).

Blair *et al.* (1990) suggest that, although selection solely for a physiological trait has yet to yield startling results, the inclusion of appropriate traits in selection indices may permit faster rates of gain than those currently possible.

Methods of Testing and Evaluation

The majority of improvement programmes in meat sheep are based on on-farm recording of performance (e.g. Croston *et al.*, 1980; Atkins *et al.*, 1986; Nicoll *et al.*, 1986; Johnson *et al.*, 1989; Meat and Livestock Commission, 1989). However, central performance tests also operate in France (Bouix, 1988), Canada (Trus and Wilton, 1989), the USA (Waldron *et al.*, 1986), Finland (Puntila *et al.*, 1990), Denmark (Jensen, 1990) and Northern Ireland (and on a smaller scale in other parts of the UK), and in a few other European countries (Croston *et al.*, 1980). In France (Bouix, 1988), Norway (Olesen and Klemetsdal, 1990), Iceland (Thorgeirsson and Thorsteinsson, 1989), and to a lesser extent in some other European countries (Croston *et al.*, 1980), progeny testing for carcass composition operates in conjunction with recording performance on-farm or in a central station. In recent Icelandic progeny tests, possible evidence has emerged of a major gene affecting carcass composition (Thorgeirsson and Thorsteinsson, 1989).

The evaluation of results is most commonly done by contemporary comparisons or indices of various types. However, there is growing interest in applying more sophisticated methods of evaluation. Conventionally, estimates of the genetic merit of animals (estimated breeding values, EBVs) are derived in two steps. Firstly, records of performance are adjusted for "fixed effects" such as age at measurement, birth rank and dam age. Secondly, BVs are estimated from the adjusted records of performance and presented for individual traits or for some weighted combination of these. This two-stage process often assumes that the average genetic merit of animals in different contemporary groups is equal, whereas in practice there is often confounding of genetic merit with contemporary group. Consequently, animals cannot be fairly compared across contemporary groups within flocks, or across flocks and years. Ram circles, which were developed in Norway (Steine, 1982), group breeding schemes, which have been established in several countries (Barton and Smith, 1982) and central performance tests are all tactical methods of surmounting this problem.

Since the 1950s, considerable effort has been devoted to statistical methods of surmounting the problem outlined, initially in the evaluation of progeny-tested dairy sires (Henderson, 1973) and more recently in other species. The result is a procedure known as Best Linear Unbiased Prediction (BLUP) which is now the method of choice for estimating BVs in animal breeding programmes.

The most important advantage of BLUP is that it estimates fixed effects and BVs simultaneously, resulting in more accurate estimates of each of these. In order to do this, BLUP requires "genetic links" across contemporary groups. In practice, to fulfil this requirement, all relationships between animals for which BVs are being estimated are included in the statistical model. This leads to additional benefits which include more accurate EBVs, since these are based on records of performance of relatives as well as the individual animal's performance. BLUP methods are now widely used in dairy cattle evaluation worldwide. The development of more sophisticated BLUP models and computing strategies, together with the wider availability of powerful computers are leading to a rapid uptake of BLUP procedures in other species. BLUP methods have already been introduced, or are planned, in either national or individual sheep breeding programmes in Canada, Australia, New Zealand, Norway, Finland, Sweden, France and the UK (Johnson *et al.*, 1989; Trus and Wilton, 1989; Kurowska, 1990; Oleson and Klemetsdal, 1990; Poivey *et al.*, 1990; Puntila *et al.*, 1990).

Providing that adequate genetic links exist across flocks and across years, EBVs of animals estimated by BLUP can be compared across flocks and years. This has several important benefits. Firstly, the number of candidate animals for selection which can be directly compared is greatly increased. This has direct benefits on the rate of response which can be achieved. Secondly, if genetic parameters have been estimated in the population under selection (Thompson, 1986) the genetic trend in performance can be charted year-by-year, by comparing the average EBVs in the breed or flock in successive years. This provides a valuable check on genetic progress, both for breeders themselves and for their customers.

In most dairy cattle populations, genetic links between herds arise naturally because of the widespread use of artificial insemination (AI). However, in most meat sheep breeds the use of AI is very low and genetic links need to be created deliberately in most cases. Over the past decade, Sire Referencing Schemes have been established in a number of sheep breeds in Australia, New Zealand (e.g. Lewer, 1987; Anon, 1989) and, more recently, in the UK. These schemes involve the use of "reference" sires in common across co-operating flocks. This creates ideal linkage between flocks for BLUP evaluations and for selection across flocks

(Kinghorn and Shepherd, 1990). Additionally, Sire Referencing Schemes probably have advantages over some other co-operative breeding schemes in that they may require less legal and financial commitment. Also, providing sufficient reference sires are used, they probably have a lower risk of genotype x environment interaction affecting the success of the breeding programme than schemes where sires are evaluated in a single nucleus flock or in a central performance test station (Notter and Hohen-boken, 1990).

Results of Selection in Experiments and Industry

During the past decade or so, experimental lines selected for various carcass composition traits have been established in a number of locations, particularly New Zealand and the UK. Table 9.2 summarizes the responses achieved in these experiments. Most of the New Zealand selection lines have been selection for divergent live weight-adjusted ultra-sonic backfat depth. More recently selection lines have been established in the UK and New Zealand, where selection is on an economic index. In most cases substantial responses have been achieved.

In those reports where conversion of results is possible, approximate annual rates of change in overall index score or in weight-adjusted backfat depth usually range from 2.1 to 2.4% per annum but are as high as 4.4% in one case.

Where carcass composition has been measured directly (e.g. Bennett *et al.*, 1988; Lord *et al.*, 1988; Kadim *et al.*, 1989), selection for reduced ultrasonically measured backfat depth has resulted in reduced carcass subcutaneous fat depth. This has been accompanied usually by reduced proportions or weights of fat in the subcutaneous and other depots, as expected from the fairly strong genetic correlations between fat proportions in different depots (Wolf *et al.*, 1981) and by increased proportions or weights of lean tissue.

There are few reports of progress in industry schemes involving selection for improved carcass composition. Steine (1982) reported annual rates of progress in some Norwegian breeds of about 1.25% of the mean in aggregate breeding value for growth rate, carcass quality and fleece weight, with these traits assessed on progeny. Lower rates of progress (0.4 to 0.5% per annum) have been achieved recently as a result of revised weightings for traits in the index (Olesen and Steine, 1988).

The use of ultrasonic machines in on-farm performance testing is a fairly recent development in several countries. For example in Britain, following pilot trials by the Scottish Agricultural College and the Meat and Livestock Commission (MLC) in 1986 and 1987, the MLC offered ultrasonic scanning as an option in its Sheepbreeder recording scheme. Measurements of live weight, fat depth and muscle depth are combined into a lean growth

Table 9.2 Summary of responses to selection for carcass composition in experimental flocks.

Author	Country	Breed	Selection method and criterion	Duration of selection	Response measured in[1]	in units[2]	Response as % of mean	approximate annual %
Bennett *et al.* 1988	NZ	Suffolk + Southdown	Divergent selection on weight adjusted ultrasonic backfat depth	-	Crossbred lambs from Romney and Coopworth ewes at fixed age (L-H)	-0.5 mm carcass backfat** -0.1 kg carcass wt.	14.7% 0.7%	-
Lord *et al.* 1988	NZ	Coopworth	Divergent selection on weight adjusted ultrasonic backfat depth	-	Crossbred lambs from Romney ewes at fixed time (L-H)	-1.1 mm carcass backfat** +1.2 kg carcass wt.	40.0%	-
Kadim *et al.* 1989	NZ	Southdown	Divergent selection on weight adjusted ultrasonic backfat depth	-	Selected purebred rams at constant carcass wt (L-H)	-3.5 mm carcass backfat*** +0.71 kg carcass fat	49.6% 16.5%	-
Beatson 1989	NZ	Dorset Down	Economic index[4] - measurements of LW and ultrasonic fat depth. Selection and control lines.	2 years	Purebred lines at 7 months of age (S-C)	Rams: +2.3 kg LW -0.44 mm US fat -0.2 mm US muscle Ewes: +2.1 kg LW -0.1 mm US fat -0.4 mm US muscle	4.2% 9.0% 0.7% 4.5% 2.9% 1.5%	2.1% 4.5% 0.4% 2.3% 1.5% 0.8%
Bishop 1991 (unpublished)	UK	Scottish Blackface	Divergent selection on index of weight and ultrasonic backfat depth	3 years	Purebred lines at 5 months of age (L-H)	+1.6 kg LW -0.6 mm US fat +1.0 mm US muscle	3.7% 26.2% 4.4%	0.6% 4.4% 0.7%
Cameron 1991	UK	Texel/Oxford synthetic	Divergent selection on index of weight and ultrasonic backfat depth	3 years	Purebred lines at 5 months of age (L-H)	+1.0 kg LW -0.8 mm US fat +0.9 mm US muscle	2.1% 14.4% 3.5%	0.4% 2.4% 0.6%
Simm *et al.* 1991 (unpublished)	UK	Suffolk	Desired gains index[5] - measurements of LW, ultrasonic fat and muscle depths. Selection and control lines	5 years	Purebred lines at 5 months of age (S-C)	Rams: +5.1 kg LW*** -0.7 mm US fat** +1.7 mm US muscle** +47 index pts Ewes: +3.6 kg LW*** -0.5 mm US fat** +1.9 mm US muscle** +47 index pts	7.7% 8.9% 5.6% 11.1% 6.2% 5% 6.3% 10.3%	1.5% 1.8% 1.1% 2.2% 1.2% 1.0% 1.3% 2.1%

[1] Response measured as performance of low minus high lines (L-H) or selection minus control (S-C)

[2] * = P<0.05. ** = P<0.01. *** = P<0.001; LW = live weight; US = ultrasonic

[3] Assumes symmetry of response if divergent selection is practised, and no genotype x environment interaction, if response is measured in a different "environment" from that in which selection is practised

[4] See Simm *et al.* (1987b) for details

[5] See Simm and Dingwall (1989) and Simm *et al.* (1990) for details

index based on that described by Simm and Dingwall (1989). By 1990, around 360 pedigree flocks were using this service, with around 15,000 rams scanned and indexed in that year (D. Croston, personal communication).

As yet there appear to be no estimates of genetic trend in industry flocks in countries using ultrasonic measurements on-farm. Probably one of the major reasons is that many large-scale recording schemes produce EBVs within-flock. The wider use of BLUP procedures will allow estimation of genetic trend, provided that genetic links exist between flocks and years. For example, de Klerk and Heydenrych (1990) have recently employed BLUP procedures to estimate genetic trend since 1948 in industry flocks of the multi-purpose South African Dohne Merino breed. They showed that small but positive changes in weight at several ages (0.05-0.06 kg per annum) had taken place over a 37-year period.

In several countries there is a perceived antagonism between selection for reduced fatness and maintaining or improving the eating quality of lamb. Wood (1990) has reviewed the evidence on the relationship between carcass fatness and meat quality. He concludes that there are both direct and indirect effects of fatness on quality. In cattle, sheep and pigs there are small but positive effects of higher intramuscular fat concentrations on taste, juiciness and sometimes flavour. These appear to be more important when carcass fatness is very low, and Wood (1990) considers that the effects are not large enough to counteract the strong demands for leaner meat. However, there may be opportunities to prevent or limit reductions in intramuscular fat, whilst reducing depot fat, by employing some of the more advanced techniques for *in vivo* assessment mentioned earlier. Higher carcass fatness can have an indirect effect on eating quality by ameliorating the negative effects of some post-slaughter treatments, such as rapid chilling. However, there are direct methods of increasing quality in these cases.

Opportunities to Accelerate Genetic Improvement

In the longer term, molecular genetics techniques may assist in the improvement of lean meat production in sheep but there are a number of opportunities to accelerate genetic improvement in carcass composition in the short to medium term. Some of these have been highlighted in previous sections. For example, the use of more accurate methods of *in vivo* measurement could have a large effect on selection responses. More accurate methods are usually more costly, but their strategic use is probably still highly cost-effective at a national level. This subject, together with optimal design of breeding programmes to integrate different measurement and selection methods (e.g. ultrasonics, CT, progeny testing, two-stage selection) justifies further investigation.

The inability to compare animals across flocks is a major constraint on genetic improvement in many countries. This is particularly so in countries where pedigree flock size is small. As described earlier, BLUP procedures enable across-flock evaluations provided that there are genetic links between flocks. The improvements in AI techniques for sheep, described by Wallace in this book, will be helpful in creating genetic links. Initially, improved AI techniques will probably encourage wider participation in sire referencing schemes. In the longer term, if sheep AI becomes as widely used as AI in dairy cattle, the need for the deliberate creation of links will diminish, and larger scale or national across-flock BLUP evaluations will be feasible. Apart from these benefits the wider use of AI could have direct benefits on genetic progress by permitting much higher selection intensities amongst males. Also, more rapid dissemination of improvement to commercial flocks would be possible.

Improved methods of superovulation, embryo recovery and transfer also have important implications for genetic improvement programmes. Smith (1986) examined the use of multiple ovulation and embryo transfer (MOET) in breeding schemes for growth and carcass composition in sheep. He estimated that genetic progress in schemes using MOET could be up to double that possible in conventional schemes. However, Smith assumed a very large nucleus flock (1000 ewes). Steane *et al.* (1988) showed that in smaller flocks, and compared at a constant rate of inbreeding, responses in MOET schemes were likely to be about 1.3 times higher than those in conventional schemes. Their results also illustrate the fact that extra responses from more accurate *in vivo* estimation, and the use of MOET are additive. For example, the use of a perfect *in vivo* technique together with MOET was expected to increase economic response by about a factor of 3 compared to selection on live weight alone, with conventional breeding. Toro and Silio (1990) have shown further advantages from combining BLUP procedures for estimating BVs with MOET and perfect *in vivo* estimation of carcass composition.

Early studies on MOET in livestock breeding schemes ignored variation in family size and the reduction in genetic variance in selected populations. Recent simulation studies for beef cattle have shown that this leads to overestimates of absolute rates of response in both conventional and MOET schemes, although the relative advantage of MOET schemes is only slightly reduced (Wray and Simm, 1990). More seriously, rates of inbreeding can be underestimated by factors of 1.5 to 2.5. However, even after accounting for these problems, MOET schemes are still expected to give substantial additional responses to those from conventional schemes, at acceptable levels of inbreeding (Toro and Silio, 1990; Wray and Simm, 1990). New approaches to optimize response while restricting rates of in-breeding may allow more of the potential extra response from MOET schemes to be realized (Toro and Perez-Enciso, 1990).

Conclusions

For a variety of reasons consumer demand in many countries is for increasingly lean meat. Lamb is disadvantaged by this preference because of a relatively high fat content but also, because of the lower palatability of lamb fat, the relatively high waste from traditional lamb cuts and, in some countries, the relatively high price of lamb compared to some other meats. Carcasses often contain fat levels considerably in excess of consumer demand. This may be due partly to the widespread use of traditional husbandry practices but, in many countries, pricing schemes for carcasses do not, or have only recently begun to, discriminate effectively between carcasses of different fatness.

Genetic improvement is a permanent and cumulative method of improving carcass composition which should be highly cost-effective. In the short to medium term, selection between breeds will be most effective if there are differences in economically important characteristics. In the medium and long term, selection within the most appropriate breeds will be valuable. Direct comparison of the carcass composition of different breeds or crosses poses no particular problems. However, within-breed selection for improved carcass composition is more complicated and requires either direct measurements on relatives of the candidates for selection, or *in vivo* measurements on the candidates themselves. A wide variety of techniques for *in vivo* assessment have been used in research. Of these, ultrasonic measurements of tissue depths or areas are now becoming quite widely used in industry improvement programmes in several countries. Whilst ultrasonic measurements can be taken on large numbers of geographically dispersed animals at a relatively low cost, there is considerable scope for increasing responses to selection from the use of more precise methods of *in vivo* assessment. Advanced imaging techniques developed for use in human medicine, such as X-ray CT or MRI, appear to offer the most immediate opportunities in this respect. Although these machines are very costly, their strategic use in sheep breeding programmes, for example in two-stage selection or in large nucleus flocks, is likely to be cost-effective at a national level.

One of the greatest constraints on genetic improvement in many countries is the inability to compare breeding values of sheep across flocks. Estimating breeding values across flocks requires: (i) genetic links between these flocks and (ii) appropriate statistical procedures. Improved techniques for AI in sheep are helping to fulfil the first of these requirements. In some countries, Sire Referencing Schemes have been established in order to create links between flocks and exploit the opportunities for faster progress which these links offer. The second requirement is being met by the wider application of BLUP procedures for estimating breeding values. These procedures are more accurate in their own right, as well as allowing comparisons across flocks and estimation of genetic trend in a given breed or population.

Individually, these developments in *in vivo* techniques, reproductive technology and statistical procedures for estimating breeding values give much greater prospects for genetic improvement of lean meat production over the next few decades than has been possible in the past. However, perhaps the greatest challenge for those involved in research, extension and sheep breeding is to design and implement breeding programmes which combine several or all of these techniques to allow even greater progress.

Acknowledgements

I am grateful to my colleagues Gerry Emmans, Harry McClelland, Sue Murphy, John Oldham and Naomi Wray for their valuable comments on this text and to Steven Bishop for access to his unpublished results. I am also grateful to the Scottish Office Agriculture and Fisheries Department, the Meat and Livestock Commission and the Suffolk Sheep Society for their support of the SAC Suffolk Project reported on above.

References

Allen, P. (1990) New approaches to measuring body composition in live meat animals. In: Wood, J.D. and Fisher, A.V. (eds.), *Reducing Fat in Meat Animals*. Elsevier, London, pp. 201-254.

Alliston, J.C. (1983) Evaluation of carcass quality in the live animal. In: Haresign, W. (ed.), *Sheep Production*. Butterworths, London, pp. 75-95.

Anon (1989) *Sire Reference Schemes and Across-Flock Genetic Evaluation*. Proceedings of a one-day seminar, Massey University, New Zealand.

Atkins, K.D., McGuirk, B.J. and Thompson, R. (1986) Intra-flock genetic improvement programmes in sheep and goats. *Proceedings of the 3rd World Congress on Genetics Applied to Livestock Production*, Vol. IX, pp. 605-618.

Barton, R.A. and Smith, W.C. (eds.) (1982) *Proceedings of the World Congress on Sheep and Beef Cattle Breeding*. Dunmore Press, Palmerston North, New Zealand.

Bass, J.J., Butler-Hogg, B.W. and Kirton, A.H. (1990) Practical methods of controlling fatness in farm animals. In: Wood, J.D. and Fisher, A.V. (eds.), *Reducing Fat in Meat Animals*. Elsevier, London, pp. 145-200.

Bass, J.J., Carter, W.D., Woods, E.G. and Moore, R.W. (1984) Evaluation of the ability of two carcass conformation systems to predict carcass composition of sheep. *Journal of agricultural Science*, Cambridge 103, 421-427.

Beatson, P.R. (1989) Responses in Dorset Down sheep selected for lean tissue growth. In: Purchas, R.W. (ed.), *Proceedings of the 12th Workshop on Overfatness and Lean Meat Production from Sheep*. New Zealand Society of Animal Production, Hamilton, New Zealand, p. 48.

Bennett, G.L. (1990) Selection for growth and carcass composition in sheep. *Proceedings of the 4th World Congress on Genetics Applied to Livestock Production*, Vol. XV, pp. 27-36.

Bennett, G.L., Meyer, H.H. and Kirton, A.H. (1988) Effects of selection for divergent ultrasonic fat depth in rams on progeny fatness. *Animal Production*

47, 379-386.

Blair, H.T., McCutcheon, S.N. and Mackenzie, D.D.S. (1990) Components of the somatotropic axis as predictors of genetic merit for growth. *Proceedings of the 4th World Congress on Genetics Applied to Livestock Production*, Vol. XVI, pp. 246-255.

Blaxter, K.L. (1991) Current and future problems. *British Society of Animal Production, Winter Meeting, Sir John Hammond Memorial Lecture.*

Bouix, J. (1988) Genetic determination of carcass quality in sheep. *Proceedings 3rd World Congress on Sheep and Beef Cattle Breeding*, Vol. I, pp. 397-413

Brascamp, E.W. (1984) Selection indices with constraints. *Animal Breeding Abstracts* 52, 645-654.

Brascamp, E.W., Smith, C. and Guy, D.R. (1985) Derivation of economic weights from profit equations. *Animal Production* 40, 175-179.

Butterfield, R.M., Griffiths, D.A., Thompson, J.M., Zamora, J. and James, A.M. (1983) Changes in body composition relative to weight and maturity in large and small strains of Australian Merino rams. 1. Muscle, bone and fat. *Animal Production* 36, 29-37.

Cameron, N.D. (1991) Physiological responses to selection for carcass lean content in a terminal sire breed of sheep. *British Society of Animal Production, Winter Meeting*, Paper No. 25.

Carter, M.L., McCutcheon, S.N. and Purchas, R.W. (1989) Plasma metabolite and hormone concentrations as predictors of genetic merit for lean meat production in sheep: effects of metabolic challenges and fasting. *New Zealand Journal of Agricultural Research* 32, 343-353

Croston, D., Danell, O., Elsen, J.M., Flamant, J.C., Hanrahan, J.P., Jakubec, V., Nitter, G. and Trodahl, S. (1980) A review of sheep recording and evaluation of breeding animals in European countries: a group report. *Livestock Production Science* 7, 373-392.

Culley, G. (1801) *Observations on live stock; containing hints for choosing and improving the best breeds of the most useful kinds of domestic animals*. G.G. and J Robinson, London, Third edition.

Cuthbertson, A., Croston, D. and Jones, D.W. (1984) *In vivo* estimation of lamb carcass composition and lean tissue growth rate. In: Lister, D (ed.), *In Vivo Measurement of Body Composition in Meat Animals*. Elsevier, London, pp. 163-166.

Fiebrand, G., Dzapo, V., Wassmuth, R. and Freudenreich, P. (1989) Association of blood concentration of glutathione with meat and fat gain and fitness in various sheep breeds and crosses. *Journal of Animal Breeding and Genetics* 106, 59-66

Foster, M.A., Fowler, P.A., Cameron, G., Fuller, M. and Knight, C.H. (1989) NMR imaging studies of live animals. In: Kallweit, E., Henning, M. and Groeneveld. E. (eds.), *Application of NMR Techniques on the Body Composition of Live Animals*. Elsevier, London, pp. 107-120.

Fowler, V.R., Bichard, M. and Pease, A. (1976) Objectives in pig breeding. *Animal Production* 23, 365-387.

Groeneveld, E., Kallweit, E., Henning, M. and Pfau, A. (1984) Evaluation of body composition of live animals by X-ray and nuclear magnetic resonance computed tomography. In: Lister, D. (ed.) *In Vivo Measurement of Body Composition in Meat Animals*. Elsevier, London, pp. 84-88.

Hall, S.J.G. and Clutton-Brock, J. (1989) *Two hundred years of British farm livestock*. British Museum (Natural History), London.

Hazel, L.N. (1943) The genetic basis for constructing selection indexes. *Genetics* 28, 476-490.

Henderson, C.R. (1973) Sire evaluation and genetic trends. *Proceedings of Animal Breeding and Genetics Symposium in Honor of Dr J L Lush*. American Society of Animal Sciences and American Dairy Science Association,

Champaign, Illinois, pp. 10-40.

Jensen, N.E. (1990) Performance tests of ram lambs 1990. *Beretning fra Statens Husdyrbrugsforsog* No. 684.

Johnson, D.L., Rae, A.L. and Clarke, J.N. (1989) Technical aspects of the Animalplan system. *Proceedings of the New Zealand Society of Animal Production* 49, 197-202.

Kadim, I.T., Purchas, R.W., Rae, A.L. and Barton, R.A. (1989) Carcass characteristics of Southdown rams from high and low backfat selection lines. *New Zealand Journal of Agricultural Research* 32, 181-191.

Kempster, A.J. (1983) Carcass quality and its measurement in sheep. In: Haresign, W. (ed.) *Sheep Production*. Butterworths, London, pp. 59-74.

Kempster, A.J., Croston, D. and Jones, D.W. (1981) Value of conformation as an indicator of sheep carcass composition within and between breeds. *Animal Production* 33, 39-49.

Kempster, A.J., Cook, G.L. and Grantley-Smith, M. (1986) National estimates of the body composition of British cattle, sheep and pigs with special reference to trends in fatness. A review. *Meat Science* 17, 107-138.

Kinghorn, B.P. and Shepherd, R.K. (1990) The impact of across-flock genetic evaluation on sheep breeding structures. *Proceedings of the 4th World Congress on Genetics Applied to Livestock Production*, Vol. XV, pp. 7-16.

de Klerk, H.C. and Heydenrych, H.J. (1990) BLUP-analysis of genetic trends in the Dohne Merino. *Proceedings of the 4th World Congress on Genetics Applied to Livestock Production*, Vol. XV, pp. 77-80.

Knopp, T.C. (1985) Quantitative analysis of computed tomographic images. *MSc thesis*, University of Otago, New Zealand.

Kurowska, Z. (1990) Selection strategies in Swedish sheep breeding. *41st Annual Meeting of the European Association for Animal Production*, Abstracts, Vol. 1, p. 102.

Lewer, R.P. (1987) Progress in establishing an Australian Merino sire referencing scheme. In: McGuirk, B.J. (ed.), *Merino Improvement Programmes in Australia*. Australian Wool Corporation, Melbourne, pp. 413-420.

Lister, D. (ed.) (1984) *In Vivo Measurement of Body Composition in Meat Animals*. Elsevier, London.

Lord, E.A., Fennessy, P.F. and Littlejohn, R.P. (1988) Comparison of genotype and nutritional effects on body and carcass characteristics of lambs. *New Zealand Journal of Agricultural Research* 31, 13-19.

McClelland, T.H., Bonaiti, B. and Taylor, St. C.S. (1976) Breed differences in body composition of equally mature sheep. *Animal Production* 23, 281-293.

Meat and Livestock Commission (1989) *Sheep Yearbook 1989*. Meat and Livestock Commission, Milton Keynes, UK.

Miles, C.A., Fursey, G.A.J. and York, R.W.R. (1984) New equipment for measuring the speed of ultrasound and its application in the estimation of body composition of farm livestock. In: Lister, D. (ed.), *In Vivo Measurement of Body Composition in Meat Animals*. Elsevier, London, pp. 93-105.

Moav, R. (1973) Economic evaluation of genetic differences. In: Moav, R. (ed.) *Agricultural Genetics. Selected Topics*, John Wiley and Sons, New York, pp. 319-352.

Nicoll, G.B., Bodin, L. and Jonmundson, J.V. (1986) Evaluation of interflock genetic improvement programs for sheep and goats. *Proceedings of the 3rd World Congress on Genetics Applied to Livestock Production*, Vol. IX, pp. 619-636.

Notter, D.R. and Hohenboken, W.D. (1990) Industry breeding structures for effecting and evaluating genetic improvement. *Proceedings of the 4th World Congress on Genetics Applied to Livestock Production*, Vol. XV, pp. 347-356.

Olesen, I. and Klemetsdal, G. (1990) Multitrait reduced animal model (RAM) in Norwegian sheep breeding. *41st Annual Meeting of the European Association of Animal Production*, Abstracts, Vol. 1, p. 100.

Olesen, I. and Steine, T. (1988) Genetic change and utilization of artificial insemination in Norwegian sheep breeding. *Proceedings of the 3rd World Congress on Sheep and Beef Cattle Breeding*, Vol. 2, pp. 512-514.

Peterson, S.W. and Purchas, R.W. (1989) Liver catalase in Southdown sheep selected for high and low backfat depth. *Proceedings of the New Zealand Society of Animal Production* 49, 143-146.

Poivey, J.P., Cournut, J., Jullien, E., Bibe, B., Perret, G., Elsen, J.M., Berny, F., Bouix, J. and Bodin, L. (1990) Estimation of meat sheep breeding values in French on farm performance recording system. *41st Annual Meeting of the European Association for Animal Production*, Abstracts, Vol. 1, p. 96.

Ponzoni, R.W. and Walkley, J.R.W. (1981) Objectives and selection criteria for Dorset sheep in Australia. *Livestock Production Science* 8, 331-338.

Puntila, M-L., Maki-Tanila, A. and Nylander, A. (1990) Genetic evaluation for a Finnsheep nucleus flock. *41st Annual Meeting of the European Association for Animal Production*, Abstracts, Vol. 1, p. 104.

Purchas, R.W. and Beach, A.D. (1981) Between-operator repeatability of fat depth measurements made on live sheep and lambs with an ultrasonic probe. *New Zealand Journal of Experimental Agriculture* 9, 213-220.

Purchas, R.W., Bennett, G.L. and Dodd, C.J. (1985) The calculation of a simple lean-growth index for young sheep. *Proceedings of the New Zealand Society of Animal Production*, 45, 73-76.

Purchas, R.W., Davies, A.S. and Abdullah, A.Y. (1991) An objective measure of muscularity: Changes with animal growth and differences between genetic lines of Southdown sheep. *Meat Science* 30, 81-94.

Rae, A.L. (1984) Development of selection programs for increasing lean meat production in sheep. *Proceedings of the 4th Conference of the Australian Association of Animal Breeding and Genetics*, pp. 3-7.

Roberts, R.C. (1979) Side effects of selection for growth in laboratory animals. *Livestock Production Science* 6, 93-104.

Sehested, E. (1984) Computerized tomography of sheep. In: Lister, D. (ed.) *In Vivo Measurement of Body Composition in Meat Animals*. Elsevier, London, pp. 67-74.

Simm, G. (1987) Carcass evaluation in sheep breeding programmes. In: Marai, I.F.M. and Owen, J.B. (eds.) *New Techniques in Sheep Production*. Butterworths, London, pp. 125-144.

Simm, G. and Dingwall, W.S. (1989) Selection indices for lean meat production in sheep. *Livestock Production Science* 21, 223-233.

Simm, G., Dingwall, W.S., Murphy, S.V. and FitzSimons, J. (1990) Selection for improved carcass composition in Suffolk sheep. *Proceedings of the 4th World Congress on Genetics Applied to Livestock Production*, Vol. XV, pp. 100-103.

Simm, G., Smith, C. and Thompson, R. (1987a) The use of product traits such as lean growth rate as selection criteria in animal breeding. *Animal Production* 45, 307-316.

Simm, G., Young, M.J. and Beatson, P.R. (1987b) An economic selection index for lean meat production in New Zealand sheep.. *Animal Production* 45, 465-475.

Sinclair, A.J. and O'Dea, K. (1990) Fats in human diets through history: Is the Western diet out of step? In: Wood, J.D. and Fisher, A.V. (eds.) *Reducing Fat in Meat Animals*. Elsevier, London, pp. 1-47.

Smith, C. (1986) Use of embryo transfer in genetic improvement of sheep. *Animal Production* 42, 81-88.

Smith, C., James, J.W. and Brascamp, E.W. (1986) On the derivation of economic weights in livestock improvement. *Animal Production* 43, 545-551.

Steane, D.E., Simm, G. and Guy, D.R. (1988) The use of embryo transfer in terminal sire sheep breeding schemes. *Proceeding of the 3rd World Congress on Sheep and Beef Cattle Breeding*, Vol. 1, pp. 211-213.

Steine, T.A. (1982) Fifteen years' experience with a co-operative sheep breeding scheme in Norway. *Proceedings of the World Congress on Sheep and Beef Cattle Breeding*, Vol. II, pp. 145-148.

Suttie, J.M., Veenvliet, B.A., Littlejohn, R.P., Corson, I.D., Fennessy, P.F. and Gluckman, P.D. (1989) Growth hormone pulsatility in relation to carcass composition in ram lambs of two genotypes. In: Purchas, R.W. (ed.), *Proceedings of the 12th Workshop on Overfatness and Lean Meat Production from Sheep*. New Zealand Society of Animal Production, Hamilton, New Zealand, p. 25.

Taylor, St. C.S. (1980) Genetic size-scaling rules in animal growth. *Animal Production* 30, 161-165

Thompson, J.M., Butterfield, R.M. and Perry, D. (1985) Food intake, growth and body composition in Australian Merino sheep selected for high and low weaning weight. 2. Chemical and dissectible body composition. *Animal Production* 40, 71-84.

Thompson, J.M., Butterfield, R.M. and Perry, D. (1987) Food intake, growth and body composition in Australian Merino sheep selected for high and low weaning weight. 4. Partitioning of dissected and chemical fat in the body. *Animal Production* 45, 49-60.

Thompson, R. (1986) Estimation of realized heritability in a selected population using mixed model methods. *Genetique, Selection, Evolution* 18, 475-484.

Thorgeirsson, S. and Thorsteinsson, S.S. (1989) Growth, development and carcass characteristics. In: Dyrmundsson, O.R. and Thorgeirsson, S. (eds.) *Reproduction, Growth and Nutrition in Sheep. Dr. Halldor Palsson Memorial Publication.* Agricultural Research Institute and Agricultural Society, Reykjavik, Iceland, pp. 169-204.

Toro, M.A. and Perez-Enciso, M. (1990) Optimization of selection response under restricted inbreeding. *Genetique, Selection, Evolution* 22, 93-107.

Toro, M.A. and Silio, L. (1990) Selection for lean meat production in sheep: A simulation study. *Proceedings of the 4th World Congress on Genetics Applied to Livestock Production*, Vol. XV, pp. 96-99.

Trus, D. and Wilton, J.W. (1989) *Genetic Evaluation of Sheep in Ontario. Final Report.* Centre for Genetic Improvement of Livestock, University of Guelph, Canada.

Waldron, D.F., Thomas, D.L., Stookey, J.M. and Fernando, R.L. (1986) Relationship between growth of Suffolk rams on central performance test and growth of their progeny. *Proceedings of the 3rd World Congress on Genetics Applied to Livestock Production*, Vol. IX, pp. 639-644.

Wolf, B.T. and Smith, C. (1983) Selection for carcass quality. In: Haresign, W. (ed.), *Sheep Production*. Butterworths, London, pp. 493-514.

Wolf, B.T., Smith, C., King, J.W.B. and Nicholson, D. (1981) Genetic parameters of growth and carcass composition in crossbred lambs. *Animal Production* 32, 1-7.

Wolf, B.T., Smith, C. and Sales, D.I. (1980) Growth and carcass composition in the crossbred progeny of six terminal sire breeds of sheep. *Animal Production* 31, 307-313.

Wood, J.D. (1990) Consequences for meat quality of reducing carcass fatness. In: Wood, J.D. and Fisher, A.V. (eds.) *Reducing Fat in Meat Animals*. Elsevier, London, pp. 344-397

Wood, J.D. and Fisher, A.V. (eds.) (1990) *Reducing Fat in Meat Animals*. Elsevier, London.

Woodward, J. and Wheelock, V. (1990) Consumer attitudes to fat in meat. In: Wood, J.D. and Fisher, A.V. (eds.) *Reducing Fat in Meat Animals*. Elsevier,

London, pp. 66-100.

Wray, N.R. and Simm, G. (1990) The use of embryo transfer to accelerate genetic impovement in beef cattle. *Proceedings of the 4th World Congress on Genetics Applied to Livestock Production*, Vol. XV, pp. 315-318.

Young, M.J. (1989) Responses to selection for leanness in Suffolk sheep. *MSc thesis* University of Edinburgh.

Young, M.J. (1990) Developmental changes in muscularity and selection to alter body composition. *Proceedings of the 8th Conference of the Australian Association of Animal Breeding and Genetics* pp. 553-554.

Chapter 10

Growth Hormone Manipulation and Growth Promotants in Sheep

J.D. Murray and A.M. Oberbauer

Department of Animal Science, University of California, Davis, CA 95616-8521, USA

Introduction

Control over the efficiency of lean muscle growth and the optimal partitioning of nutrient intake into muscle growth are major objectives of the meat industry (Kempster, 1989). Historically, improvement in these traits has been realized via selective breeding for improved meat breeds and by nutrient supplementation.

Selective breeding can be used to affect the growth characteristics of sheep (Wolf *et al.*, 1981). Weaning weight and live weight have a moderate to high heritability in sheep and are positively correlated with live weight at other ages (Davis and Kinghorn, 1986). When sampled at the same age body weight and backfat thickness in sheep have moderate to high heritabilities (Wolf *et al.*, 1981) indicating that alterations in these traits may still be achieved via selective breeding (Simm *et al.*, 1990). Recent advances in the application of genetic engineering techniques now make it possible to reliably transfer new genetic material into sheep that may also affect the ability to realize improvements in the efficiency of lean muscle production via selection. However, the usefulness of this approach will depend both on the realization of an increase in the biological efficiency of lean muscle production following selection as well as on the consequences of selection on correlated traits such as reproduction and wool production. The area of genetic selection for growth in sheep has recently been reviewed by Thompson (1990) and readers are referred to that paper for further information.

The growth of an animal is a complex interaction of genetic, environmental, nutritional, and hormonal influences; influences that are all inter-related at some level. Focusing on man's ability to intervene in the regulation of growth, we will concern ourselves with the recent advances in biotechnology that can accelerate genetic improvement and the administration of exogenous compounds that can enhance the endogenous growth potential of sheep.

Manipulation of Growth Through Exogenous Compounds

The growth hormone axis

Regulation of tissue growth and metabolism is dependent upon many inputs, including genetic, environmental, and nutritional. But the primary, direct input regulating tissue growth is the interplay of hormones and growth factors. In this regard growth hormone (GH) assumes a critical role. The growth promotant effects of GH in pituitary extract were first described in the early 1930s when it was observed that treatment of rats with a crude preparation of GH significantly enhanced body weight gain (Evans and Simpson, 1931; Lee and Schaffer, 1934). More importantly, these workers found that the composition of the gain was altered; increased protein deposition and diminished lipid deposition characterized GH treated rats. This suggested that compositional gain in the growing animal could be manipulated by the administration of GH.

As mentioned above, GH is a principle regulator of growth at the tissue (muscle, bone/cartilage, and adipose) level; the reader is referred to several excellent reviews for tissue specific actions of GH (Florini, 1987; Isaksson *et al.*, 1987; Boyd and Bauman, 1989; Vernon and Flint, 1989). How GH exerts its effects at the tissue level has been the subject of a great deal of research. Salmon and Daughaday (1957) identified a factor produced in the liver believed to mediate the endocrine actions of GH. This factor, initially termed somatomedin to signify its dependence upon GH and mediator role, is now referred to as insulin-like growth factor (IGF). Originally, all GH actions were attributed to the production of IGF by the liver in response to GH stimulation.

However, in the past decade many nonhepatic tissues have been identified as producers of IGF (D'Ercole *et al.*, 1984), some of which synthesize IGF in a GH dependent manner (Murphy *et al.*, 1987; Turner *et al.*, 1988; Gaskins *et al.*, 1990; Nilsson *et al.*, 1990). These observations have necessitated a revision of how GH exerts its physiological effects. In the modified scenario, the endocrine effects of GH would be mediated by hepatic IGF, while the local effects of GH would be due, in part, to the target tissue synthesis and secretion of IGF which would act in a paracrine or autocrine manner. Alternatively, some cell types may respond directly to GH independent of IGF. Finally, the action of GH has also been described as a "dual effector" (Green *et al.*, 1985). Growth hormone in this strategy directly induces, without intermediates, preadipocytes and C3H10T1/2 cells to differentiate with the differentiated cell type then responding to IGF by proliferating.

The identification of the central role played by growth hormone in carcass composition has resulted in attempts to manipulate the growth of animals by altering the concentration of growth hormone in circulation.

Strategies for altering GH levels have included the direct administration of exogenous GH, genetic selection and genetic engineering.

Effects of injected GH

The enhanced protein deposition observed in rats in response to GH injections (Lee and Schaffer, 1934) led to studies of the effect of GH in animals of economic importance. One of the first studies that successfully demonstrated improvement of a production trait by GH was one in which Brumby (1959) injected GH in growing calves and observed increased growth rates. Machlin (1972), by giving daily injections of pituitary-derived GH to growing pigs, demonstrated a striking increase in muscling due to the exogenous administration of GH. However, the limited availability and the relative high cost associated with pituitary-purified GH rendered widespread applicability of exogenous GH administration impractical. With the report by Bauman *et al.* (1982) that recombinantly derived GH was as effective as pituitary-purified GH in the stimulation of milk production in dairy cows, the feasibility of extending the use of injected recombinant GH to other animal industries became possible.

In comparing extant literature pertaining to exogenous GH administration to sheep, one must be cognizant of the many variables that exist in the studies, such as diet, age at slaughter, length of treatment, age at which study was initiated, and the breed of the animals. However, the general trend in response to GH administration in sheep is that carcass composition is dramatically improved. This response to GH in lambs is realized irrespective of route and mode of GH administration (Johnsson *et al.*, 1987).

In other species it has been shown that the optimal dose for GH administration is dependent upon the response criteria being measured. For example, in swine a GH dose of 100 µg/kg body weight maximizes daily gain while 200 µg/kg is the dose that maximizes feed conversion. A few studies have specifically looked at dose-response effects in sheep and the optimal dose with regard to daily gain (increased 12-19%) was between 100 and 200 µg/kg body weight (Johnsson *et al.*, 1987; Zainur *et al.*, 1989; Sinnett-Smith *et al.*, 1989). An early report from Wagner and Veenhuizen (1978) indicated an increased rate of growth, improved feed conversion efficiency, and an alteration in body composition in lambs treated with GH, but in a larger trial only feed conversion efficiency was improved which may have been due to the significantly decreased fat deposition (Muir *et al.*, 1983).

Some of the variability in response to GH administration may be due to the use of inappropriate dosages for the response criterion under consideration. A study reported by Johnsson *et al.* (1985) using 100 µg/kg pituitary bovine GH, detected a significant increase in daily gains (63 g/d), final carcass weight, and lean tissue deposition (2.2 kg) with a reduction of fat deposition (1.0 kg), but in a subsequent report using sheep of similar genotypes to their previous study, no anabolic response to biosynthetic GH was found (Johnsson *et al.*, 1987). However, a dose-dependent linear decrease in dissectable and visceral fat was observed, with the lack of a GH effect being attributed to deficiencies in the diet formulation rather than the recombinantly produced GH as the same group had previously observed an increase in daily gain with biosynthetic GH (Pullar *et al.*, 1986).

Recent investigations have reported increased daily gains (Koppel *et al.*, 1988; Zainur *et al.*, 1989; Beermann *et al.*, 1990; Pell *et al.* 1990) and enhanced feed conversion efficiency (Sinnett-Smith *et al.*, 1989; Zainur *et al.*, 1989; Beermann *et al.*, 1990; Pell *et al.*, 1990) in sheep treated with GH, while others report no effect on daily gain (Rosemberg *et al.*, 1989; Sinnett-Smith *et al.*, 1989) or feed efficiency (Rosemberg *et al.*, 1989). However, in all studies of GH treated lambs, there was substantial improvement of body composition. For example, Beermann *et al.* (1990) report a 22% decrease in lipid accretion rate and a 36% increase in protein accretion rate in the carcasses of lambs treated with 160 μg/kg ovine GH for 7 weeks. The source of protein in the sheep diet also played a role in the magnitude of response to GH, with diets formulated with fishmeal resulting in even greater gains, feed efficiencies, and hind leg muscle weights (Beermann *et al.*, 1990).

These results demonstrate the efficacy of GH administration in improving the efficiency of growth and in improving the composition of the carcass to one of lower fat content. However, modification of carcass composition raises the issue of consumer acceptance. Sinnett-Smith *et al.* (1989) concluded that meat quality indicators were unaffected by GH and meat produced from GH injected lambs would be of acceptable quality. This is also true of GH treated pork (Novakofski *et al.*, 1988).

The exogenous administration of recombinant ovine GH favourably alters the chemical composition of the sheep carcass and in many instances, improves the efficiency of growth in the animal. But the quantity of GH needed to achieve these improvements is problematic as well as the requirement for daily injections. A more efficient method to achieve the positive effects of GH may be to manipulate the endogenous hormonal environment.

The compounds which regulate anterior pituitary secretion of GH are regulated by stimulatory and inhibitory neurohormones released from the hypothalamus: growth hormone releasing factor (GRF) and somatostatin (also known as somatotropin releasing inhibitory factor, SRIF), respectively. It was speculated that small quantities of injected GRF would be sufficient to produce an endogenous surge of GH with the net effect on production being equivalent to directly injecting exogenous GH. Hodate *et al.* (1985) tested this hypothesis by injecting 0.25 μg/kg body weight GRF in rams and ewes of varying ages. Within 5 to 10 minutes post-injection, circulating GH levels reached maximal values. The amplitude of the GH peaks was dependent upon age and sex of the sheep with young rams having the greatest amplitude. This elevation in GH by GRF translates into increased growth rate, increased feed conversion efficiency, lowered fat content of the carcass and enhanced rate of protein deposition in the growing lamb (Beermann *et al.*, 1990). A decrease in carcass lipid was also reported by Pastoureau *et al.* (1989) in lambs receiving GRF from birth until slaughter at 45 or 90 days of age.

GRF administered to pregnant ewes, resulted in increased birthweight of lambs from treated ewes. The lambs from GRF treated ewes continued to show sustained effects of the GRF with elevated growth rates relative to lambs born from placebo treated control ewes (Kann and Martinet, 1989).

Immunization against components of the growth hormone axis

An alternative approach to directly stimulating GH is to inactivate the inhibitory regulation of GH; that is, neutralize the effects or activity of somatostatin. The approach has been to reduce the activity of somatostatin by immunizing animals with somatostatin conjugated to a globulin protein. Early reports from Spencer and Williamson (1981) showed a dramatic effect of immunoneutralization of somatostatin on growth rate: an increase of 76% in St. Kilda lamb growth rate. Even though significant somatostatin antibody titres were obtained GH levels were not affected, however IGF was elevated. Subsequent results from this group have demonstrated increased growth rate (Spencer, 1984; Spencer *et al.*, 1985) and feed conversion efficiency (Spencer, 1984) in response to immunization to somatostatin, although the proportion of carcass fat, muscle, and bone were unaltered. These results would suggest that the mechanism of action of somatostatin antibodies is not through GH directly.

Despite these initial encouraging results, the overall effects of immunoneutralization of somatostatin have been much less striking and are countered by numerous reports characterizing the lack of efficacy of such an approach. Bass *et al.* (1987) and Sun *et al.* (1990) independently reported enhanced growth rate in sheep actively immunized against somatostatin while Hoskinson *et al.* (1988) and Trout and Schanbacher (1990) failed to detect any growth advantage from somatostatin immunization in sheep or beef, respectively. Furthermore, lambs born to ewes actively immunized against somatostatin during the final gestational trimester did not show any growth advantage over lambs born from control ewes (Van Kessel *et al.*, 1990). Thus, the merit of such immunization strategies remains, to date, equivocal.

A second mode of GH axis manipulation based upon immunization that has been successful in a laboratory setting is the potentiation of endogenous growth hormone by active immunization stimulating region specific GH antibody production (Pell *et al.*, 1991). Though no published reports exist as to the efficacy of such an approach in livestock species, the work with rats has proved promising (James and Cottingham, 1989).

Compounds impinging on GH axis

Other hormones and synthetic compounds may be used to alter the growth rate and body composition of the growing animal. Some of these factors act directly on the tissues themselves or act in concert with a modification of circulating GH. The steroids oestradiol and testosterone have both been reported to elevate circulating GH and IGF-I concentrations in calves (Coxam *et al.*, 1990). Manipulation of day length in rams caused an increase in circulating prolactin and also elevated GH with the net result being increased body weight gain (Barenton *et al.*, 1988). However, attempts to reproduce a prolactin effect on GH concentrations and growth rate have been unsuccessful (Spoon and Hallford, 1989).

In vitro cultures of ram lamb anterior pituitary cells treated with thyrotropin-releasing hormone (TRH) respond with elevated GH secretion (Blanchard *et al.*, 1987), though *in vivo* dose-response studies administering TRH to growing lambs fail to demonstrate an increase in either circulating GH concentrations or growth rate (Wrutniak *et al.*, 1987). Administration of triiodothyronine directly did not affect circulating GH concentrations, nor are carcass and growth characteristics improved (Hinrichs and Hallford, 1987; Rosemberg *et al.*, 1989). Finally, the beta-agonist, cimaterol (to be discussed below), has been shown to elevate circulating GH concentrations over two-fold in the lamb (Beermann *et al.*, 1987) with a concomitant depression in circulating IGF-I. In contrast, Coxam *et al.* (1990) and Claeys *et al.* (1989) were unable to detect an effect of beta-agonists on either circulating GH or IGF-I concentrations in calves and lambs, respectively.

Steroids in growth

Many studies have verified the anabolic actions of steroid hormones on the growth of livestock species. In general, studies of sheep have shown that daily gain, feed conversion efficiency, and carcass lean are increased in response to oestradiol and/or trenbolone acetate (reviewed in Roche and Quirke, 1986; Beermann, 1989).

At present, the trend is away from the application of anabolic steroids due to lack of consumer acceptance. This has prompted heightened interest in less controversial means of enhancing animal growth. Recent research in the use of steroids has focused on the possibility of altering post-natal growth of an animal by the prenatal exposure of the animal to steroids. DeHann *et al.* (1990) implanted pregnant ewes with trenbolone acetate at day 40 to 60 of gestation and, although subsequent post-natal growth rate and carcass merit were not improved, feed conversion efficiency was. However, steroid implants maintained until parturition increased the occurrence of dystocia and depressed milk production in the ewe. Clearly, additional research must be done to determine the efficacy of this approach.

Beta-agonists

The β-adrenergic agonists are a group of pharmacologic analogs of the catecholamines epinephrine and norepinephrine; they are so named because of their interaction with the β-receptors (for a review of adrenergic hormones and neurotransmitters the reader is referred to Mersmann, 1989). These compounds were designed to elicit changes in body composition and either preferentially bind to the β_2-receptors or to the atypical β-receptors of brown adipose (Timmerman, 1987). Though clenbuterol was the first agonist to be described, the second generation agonists, such as ractopamine and cimaterol, apparently are more effective in stimulating lipolysis and decreasing lipogenesis. In general, the β-agonists act as "repartitioning agents" that markedly favour lean deposition in all major livestock species when included in the diet. However, there is great variability in the degree of responsiveness to the β-agonists depending on the species, for example, poultry respond less dramatically than mammals (Hanrahan *et al.*, 1986). Cattle and sheep exhibit the greatest response to these compounds (Mersmann, 1989). In contrast to the anabolic sex steroids, β-agonists do not exhibit sexual dimorphism in the degree of response.

Initially, the majority of data centered on the efficacy of the use of β-agonists to modify feed conversion efficiency, growth rate, and carcass composition (Baker *et al.*, 1984; Thornton *et al.*, 1985; Beermann *et al.*, 1986). These results were supplemented by functionality studies of how the agonists were achieving the increased lean tissue and decreased fat content in the carcass. Kim *et al.* (1987) reported significant increases in weight gain, feed conversion efficiency, dressing percentage, and cross sectional areas of economically important meat cuts. This enhancement was accomplished without change in fiber type numbers. However, the diameter of fast twitch fibers (type II) was increased (Hamby *et al.*, 1986; Kim *et al.*, 1987). DNA concentration in muscles isolated from cimaterol treated lambs was decreased suggesting that the enhanced protein accretion was independent of nuclei added from satellite cells.

Recent studies of β-agonist induced muscle hypertrophy suggest that fractional protein synthesis rates in lambs are elevated following β-agonist treatment (Claeys *et al.*, 1989), although earlier studies with a variety of species (Reeds *et al.*, 1986; Bohorov *et al.*, 1987; Williams *et al.*, 1987) failed to detect alterations in protein synthesis and attributed the muscle hypertrophy to diminished rates of protein degradation. The amino acid pool also plays a role in the action of β-agonist induced muscle hypertrophy because fishmeal used as a portion of dietary protein improved all measures of muscle development (Beermann *et al.*, 1986). The effect of β-agonists on adipose *in vivo* is lipolytic (Beermann *et al.*, 1987) as determined by elevated free fatty acid concentrations as well as inhibiting lipogenesis (O'Connor *et al.*, 1992) as evaluated by plasma glycerol and non-esterified fatty acids.

Although compositional alterations of the β-agonist treated lamb are impressive and the efficacy of the technique is undisputed, the effects on carcass quality are not as positive. Cimaterol treated lambs that have experienced pre-slaughter stress tend to have an elevated pH in leg muscle which may be due to lower muscle glycogen stores in treated lambs (Hanrahan *et al.*, 1987). The β-agonists also increase the force required to shear samples of the longissimus dorsi in both lambs and pigs (Jones *et al.*, 1985; Hamby *et al.*, 1986; Hanrahan *et al.*, 1987).

The effects on toughness appear to be independent of variation in the subcutaneous fat depots. Breeds of sheep that have inherently different fat deposition patterns, when treated with β-agonists, exhibited improved carcass composition patterns relative to their contemporaries, but still maintained differences in the degree of fat deposition relative to other breeds. However, all β-agonist treated animals had demonstrably elevated shear force values when compared to controls (Hanrahan *et al.*, 1987).

Manipulation of Growth by Genetic Engineering

Transgenic animals

Genetic engineering, or the transfer of foreign DNA into an animal's genome, was first successfully carried out in mice (Gordon *et al.*, 1980). The initial report was rapidly followed by a number of papers demonstrating that foreign DNA, or transgenes, can be permanently incorporated into an animal's genome. Once incorporated, transgenes are stable and can be transmitted to an animal's progeny in the course of normal breeding (for a review see Gordon, 1989).

Palmiter *et al.* (1982, 1983) established that the alteration of the pattern of expression of a gene, or the expression of a foreign gene, could result in a marked change in a quantitative trait like body growth. These workers produced mice transgenic with a rat growth hormone (rGH) gene under the control of the mouse metallothionein I (MT) gene promoter, thus directing the production of GH in the liver rather than the anterior pituitary. The transgenic mice expressing the transgene grew 1.5 to 2 times larger than their non-transgenic littermates.

Table 10.1 Growth promoting transgenes in sheep

Transgene	Lambs transgenic	Transgenic lambs expressing No.	%
mMT-hGH[a]	1		-
mMT-bGH[a]	2	2	100
mMT-hGRF[a]	9	1/7	14
mTF-bGH[a]	11	3	27
mAL-hGRF[a]	4	2	50
Mt1a-oGH5[b]	5	0	0
Mt1a-oGH9[b]	4	4	100

[a] Rexroad *et al.*, 1990. mMT=mouse metallothionein 1; mTF=mouse transferrin; mAL=mouse albumin; hGH=human growth hormone; bGH=bovine growth hormone; hGRF=human GH releasing factor.
[b] Murray *et al.*, 1989; Nancarrow *et al.*, 1991. Mt1a=ovine metallothionein 1a; oGH=ovine growth hormone; Mt1a-oGH5 transgene incorporated SV40 viral enhancer sequences, while the Mt1a-oGH9 construct did not.

The first transgenic sheep was reported in 1985 by Hammer *et al.* This single animal demonstrated the possibility of transferring foreign genes, in this case a mouse metallothionein human growth hormone (mMThGH) fusion gene, into sheep, albeit at a very low efficiency (see review of Rexroad and Pursel, 1988).

The consequences of growth promoting transgene expression in sheep

Thirty-six lambs have been produced by microinjecting fusion genes encoding either growth hormone or growth hormone releasing factor (Table 10.1), with 12 expressing the transgene. Nine of these transgenic sheep carried transgenes that expressed high levels of growth hormone, while the remaining 3 lambs expressed growth hormone releasing factor. Many of the 24 non-expressing sheep have transmitted the transgene to their progeny, which have also failed to express the transgene (Murray *et al.*, 1989; Rexroad *et al.*, 1989, 1990; Nancarrow *et al.*, 1991).

The transgenic lambs expressing either GH or GRF grew at approximately the same rate as non-transgenic littermates, even though circulating plasma GH levels in these animals ranged from 3 to >1000 times higher than found in control animals (Rexroad *et al.*, 1989; Murray *et al.*, 1989; Nancarrow *et al*, 1991). The expressing transgenic animals also had elevated levels of circulating IGF-I and insulin (Rexroad *et al.*, 1990; Nancarrow *et al.*, 1991). Plasma levels of prolactin and the thyroid hormones were depressed in expressing females, but were not different from controls in a single expressing transgenic male. Lower levels of thyroid hormone are normally associated with a reduced basal metabolic rate (BMR) but, in the case of two of the GH expressing females, BMR was increased by 30% and metabolic heat production by 20-50% (Nancarrow *et al.*, 1991). Expressing transgenic sheep do not appear to have increased feed efficiency (Rexroad *et al.*, 1989), but they do have significantly reduced amounts of body fat (Ward *et al.*, 1990; Nancarrow *et al.*, 1991).

The secretion of other hormones, FSH and LH, from the anterior pituitary gland was normal in both males and females, as were the serum levels of the sex steroids. However, both expressing the ram and the ewes appear to have delayed sexual development when compared to controls (Nancarrow *et al.*, 1991). This may be similar to the situation observed in transgenic pigs expressing growth hormone, where females remain anoestrous and males lack libido, but produce viable spermatozoa (Pursel *et al.*, 1990).

The continuous high levels of circulating growth hormone observed in the expressing transgenic sheep resulted in severe health problems (Nancarrow *et al.*, 1991). None of the 12 expressing transgenic sheep has attained puberty; all died before 1 year of age. The cause of death has varied, but there is clear data that the over expression of GH adversely affects liver, kidney and cardiac function (Nancarrow *et al.*, 1991). Plasma levels of insulin and glucose are also abnormal indicating an inability to maintain serum sugar levels resulting in a diabetic condition (Rexroad *et al.*, 1990).

Future Directions

Research

The growth of an animal is a process that occurs over time and is influenced by many factors, both intrinsic, such as the genetic constitution of the individual, and external, such as nutritional status. In addition, there are varying levels of interactions between all of these factors with both immediate and delayed consequences. Although the central role of growth hormone in promoting growth has been known for 60 years (Evans and Simpson, 1931), the nature of many of the interactions is still poorly known. Additionally, the reasons for the varying responses between species when either transgenic for growth hormone or given exogenous growth hormone are not understood.

For example, transgenic GH pigs are more feed efficient and leaner, but do not grow larger than their non-transgenic littermate controls (Pursel *et al.*, 1989). Transgenic GH sheep do not grow larger than non-transgenic controls, but they are leaner (Nancarrow *et al.*, 1991) though not more feed efficient (Rexroad *et al.*, 1989). Finally, transgenic GH mice grow significantly larger than controls and are more feed efficient, but show much smaller changes in fat content (Pomp *et al.*, 1992).

These few GH transgenic studies cited above exemplify and highlight the great diversity of response to elevated circulating GH that exists among the species used to date. As yet, these types of transgenic studies have not considered breed differences or the potential effects of different genetic backgrounds. Thus, though some growth mechanisms are known, considerable basic research is needed to enumerate and elucidate the various factors that control growth of an animal along the entire developmental continuum. Furthermore, in order for genetic engineering and selection to be used most effectively, the genetic interactions controlling normal growth need to be fully understood and characterized.

The same holds true for exogenous administration of GH or other hormones that impinge on the GH axis, including both immunoneutralization and active immunization against hormones. The varying responses in feed efficiency and growth rate observed in sheep (Hoskinson *et al.*, 1988; Rosemberg *et al.*, 1989; Sinnett-Smith *et al.*, 1989; Zainur *et al.*, 1989; Beermann *et al.*, 1990; Pell *et al.*, 1990, 1991) illustrate the need for a better understanding of the basic biological mechanisms underlying the action of the various components of the GH axis. Only when the mechanisms are fully understood, can they then be exploited most efficiently to define optimal nutritional regimens, dosages, breeds, etc.

The identification of the major genes influencing growth in animals and a basic understanding of the patterns and consequences of the expression of these genes in an appropriate tissue and developmental context is essential from a practical standpoint of genetic engineering. Detailed knowledge of the number, location and expression patterns of the important genes affecting growth would also aid in the selection of sheep for increased growth, feed efficiency or carcass composition. Furthermore, new controllable and tissue specific promoter elements need to be identified for use with growth promoting genes in order to avoid the health problems seen to date in GH transgenic sheep and pigs.

Application to industry

Clearly, the growth patterns of an animal can be manipulated by the exogenous administration of either natural or synthetic compounds

(Beermann, 1989). The current consumer preference for a carcass of high lean content with minimal fat can be supplied with the battery of compounds available for use today. However, consumer perception of the safety and acceptability of meat produced using such means needs to be addressed. For example, β-agonist treated sheep yield desirable carcasses in terms of composition, but further research into the acceptability and palatability of these carcasses is required. For example, can the increased toughness due to β-agonist treatment be minimized with alternative slaughter practices? Indeed, is tenderness an issue with consumers or are less expensive meat cuts more important?

As previously mentioned researchers and producers need to consider the public perception of growth promotant treatments being proposed for use in animal agriculture. Compound residues present in meat derived from animals treated with steroids or β-agonists or the perceived possibility of residues from the use of GH, pose potential problems for consumers of the product as well as being viewed as unsafe materials in the food system.

Immunoneutralization has an advantage with respect to consumer acceptability that is absent in other forms of manipulation. Active immunization requires only one to two injections given with normal health preventative immunizations and may then be perceived as being routine herd health procedures rather than detrimental to the animal's well being. However, the technologies relying on immunoneutralization or immunopotentiation approaches have not been proven beyond the laboratory setting and thus, considerable further research is required to define and gain an understanding of the range of consequences in treated animals under varying conditions as well as establishing the efficacy of these techniques.

Research over the last decade or so has shown that the growth and body composition of sheep can be manipulated by a wide variety of strategies, including genetic (selective breeding and transgenics) and non-genetic (exogenous GH, β-agonists) technologies. However, in all these cases there is still much to learn concerning the biological mechanisms underlying the changes as well as determining the practical utility of these technologies.

References

Baker, P.K., Dalrymple, R.H., Ingle, D.L. and Ricks, C.A. (1984) Use of a β-adrenergic agonist to alter muscle and fat deposition in lambs. *Journal of Animal Science* 59, 1256-1261.

Barenton, B., Ravault, J.P., Chabanet, C., Daveau, A., Pelletier, J. and Ortavant, R. (1988) Photoperiodic control of growth hormone secretion and body weight in rams. *Domestic Animal Endocrinology* 5, 247-255.

Bass, J.J., Gluckman, P.D., Fairclough, R.J., Peterson, A.J., Davis, S.R. and Carter, W.D. (1987) Effect of nutrition and immunization against somatostatin on growth and insulin-like growth factors in sheep. *Journal of Endocrinology* 112, 27-31.

Bauman, D.E., DeGeeter, M.J., Peel, C.J., Lanza, G.M., Gorewit, R.C. and Hammond, R.W. (1982) Effect of recombinantly derived bovine growth hormone (bGH) on lactational performance of high yielding dairy cows. *Journal of Dairy Science* 65 (Suppl. 1), 121.

Beermann, D.H. (1989) Status of current strategies for growth regulation. In: Campion, D.R., Hausman, G.J. and Martin, R.J. (Eds.) *Animal Growth Regulation*. Plenum Press, New York, pp.377-400

Beermann, D.H., Butler, W.R., Hogue, D.E., Fishell, V.K., Dalrymple, R.H., Ricks, C.A. and Scanes, C.G. (1987) Cimaterol-induced muscle hypertrophy and altered endocrine status in lambs. *Journal of Animal Science* 65, 1514-1524.

Beermann, D.H., Hogue, D.E., Fishell, V.K., Dalrymple, R.H. and Ricks, C.A. (1986) Effects of cimaterol and fishmeal on performance, carcass characteristics, and skeletal muscle growth in lambs. *Journal of Animal Science* 62, 370-380.

Beermann, D.H., Hogue, D.E., Fishell, V.K., Aronica, S., Dickson, H.W. and Schricker, B.R. (1990) Exogenous human growth hormone-releasing factor and ovine somatotropin improve growth performance and composition of gain in lambs. *Journal of Animal Science* 68, 4122-4133.

Blanchard, M., Goodyer, C.G., Charier, J., Dulor, J.P. and Barenton, B. (1987) Effect of hypothalamic hormones (GRF, TRH, somatostatin) and insulin-like growth factor I on growth hormone secretion from prepubertal male lamb pituitary cultures. *Reproduction, Nutrition, Development* 27, 471-480.

Bohorov, O.P., Buttery, P.J., Correia, J.H.E.S. and Soar, J.B. (1987) The effect of the β-2-adrenergic agonist clenbuterol or implantation with estradiol plus trenbolone acetate on protein metabolism in wether lambs. *British Journal of Nutrition* 57, 99-107.

Boyd R.D. and Bauman D.E. (1989) Mechanisms of action for somatotropin in growth. In: Campion, D.R., Hausman G.J. and Martin R.J. (eds.), *Animal Growth Regulation*. Plenum, New York, pp. 257-293.

Brumby P.J. (1959) The influence of growth hormone on growth in young cattle. *New Zealand Journal of Agricultural Research* 2, 683-689.

Claeys, M.C., Mulvaner, D.R., McCarthy, F.D., Gore, M.T., Marple, D.N. and Sartin, J.L. (1989) Skeletal muscle protein synthesis and growth hormone secretion in young lambs treated with clenbuterol. *Journal of Animal Science* 67, 2245-2254.

Coxam, V., Davicco, M.J., Durand, D., Bauchart, D., Opmeer, F. and Barlet, J.P. (1990) Steroid hormone may modulate hepatic somatomedin C production in newborn calves. *Biology of the Neonate* 58, 16-23.

Davis, G.P. and Kinghorn, B.P. (1986) Genetic and phenotypic parameters for growth, wool production and reproduction traits in a line of Merino sheep. *Proceedings 3rd World Congress on Genetics Applied to Livestock Production*, Lincoln II, 145-150.

DeHaan, K.C., Berger, L.L., Kesler, D.J., McKeith, F.K. and Thomas, D.L. (1990) Effect of prenatal trenbolone acetate treatment on lamb performance and carcass characteristics. *Journal of Animal Science* 68, 3041-3045.

D'Ercole, A.J., Stiles, A.D. and Underwood, L.E. (1984) Tissue concentrations of somatomedin C: Further evidence for multiple sites of synthesis and paracrine and autocrine mechanisms of action. *Proceedings of the National Academy of Science, U.S.A.* 81, 935-939.

Evans, H.M. and Simpson, M.E. (1931) Hormones of the anterior hypophysis. *American Journal of Physiology* 98, 511-523.

Florini, J.R. (1987) Hormonal control of muscle growth. *Muscle and Nerve* 10, 577-598.

Gaskins, H.R., Kin, J.W., Wright, J.T., Rund, L.A. and Hausman, G.J. (1990) Regulation of insulin-like growth factor-I, ribonucleic acid expression, polypeptide secretion, and binding activity by growth hormone in porcine preadipocyte cultures. *Endocrinology* 126, 622-630.

Gordon, J.W. (1989) Transgenic animals. *International Review of Cytology* 115, 171-229.

Gordon, J.W., Scangos, G.A., Plotkin, D.J., Barbosa, J.A. and Ruddle, F.H. (1980) Genetic transformation of mouse embryos by microinjection of purified DNA. *Proceedings of the National Academy of Science, U.S.A.* 77, 7380-7384.

Green, H., Morikawa, M. and Nixon, T. (1985) A dual effector theory of growth-hormone action. *Differentiation* 29, 195-198.

Hamby, P.L., Stouffer, J.R. and Smith, S.B. (1986) Muscle metabolism and real-time ultrasound measurement of muscle and subcutaneous adipose tissue growth in lambs fed diets containing a beta-agonist. *Journal of Animal icience* 63, 1410-1417.

Hammer, R.E., Pursel, V.G., Rexroad Jr, C.E., Wall, R.J., Bolt, D.J., Ebert, K.M., Palmiter, R.D. and Brinster, R.L. (1985) Production of transgenic rabbits, sheep and pigs by microinjection. *Nature* 315, 680-683.

Hanrahan, J.P., Fitzsimons, J.M., McEwan, J.C., Allen, P. and Quirke, J.F. (1987) Effects of the beta-agonist cimaterol on growth, food efficiency and carcass quality in sheep. In: Hanrahan, P.J. (ed.) *Beta-Agonists and Their Effects on Animal Growth and Carcass Quality*. Elsevier, London pp.106-117.

Hanrahan, J.P., Quirke, J.R., Bomann, W., Allen, P., McEwan, J., Fitzsimons, J., Kotzian, J. and Roche, J.F. (1986) β-agonists and their effects on growth and carcass quality. In: Haresign, W. (ed.) *Recent Advances in Animal Nutrition*. Butterworth, London pp.125-138.

Hinrichs, B.D. and Hallford, D.M. (1987) Growth response, reproductive activity, and serum growth hormone in fine-wool ewe lambs treated with triiodothyronine. *Theriogenology* 28, 205-212.

Hodate, K., Johke, T., Kawabata, A., Fuse, H., Ohashi, S., Shiraki, M. and Sawano, S. (1985) Influences of dose, age and sex on plasma growth hormone response in goats and sheep to synthetic human growth hormone releasing-factor. *Japanese Journal of Zootechnical Science* 56, 41-48.

Hoskinson, R.M., Djura, P., Welch, R.J., Harrison, B.E., Brown, G.H., Donnelly, J.B. and Jones, M.R. (1988) Failure of antisomatostatin antibodies to stimulate the growth of crossbred lambs. *Australian Journal of Experimental Agriculture* 28, 161-165.

Isaksson, O.G.P., Lindahl, A., Nilsson, A. and Isgaard, J. (1987) Mechanism of the stimulatory effect of growth hormone on longitudinal bone growth. *Endocrine Reviews* 8, 426-438.

James, S. and Cottingham, J.D. (1989) Active immunization using short hormone fragments can enhance the biological activity of growth hormone. *Immunobiology* (Suppl.) 4, 87 (abstr.).

Johnsson, I.D., Hart, I.C. and Butler-Hogg, B.W. (1985) The effects of exogenous bovine growth hormone and bromocriptine on growth, body development, fleece weight and plasma concentrations of growth hormone, insulin and prolactin in female lambs. *Animal Production* 41, 207-217.

Johnsson, I.D., Hathorn, D.J., Wilde, R.M., Treacher, T.T. and Butler-Hogg, B.W. (1987) The effects of dose and method of administration of biosynthetic bovine somatotropin on live-weight gain, carcass composition and wool growth in young lambs. *Animal Production* 44, 405-414.

Jones, R.W., Easter, R.A., McKeith, F.K., Dalrymple, R.H., Maddock, J.J. and Bechtel, P.J. (1985) Effects of the beta-adrenergic agonist cimaterol on the growth and carcass characteristics of finishing swine. *Journal of Animal Science* 61, 905-913.

Kann, G and Martinet, J. (1989) Procedure for increasing milk yield in mammals, and also their birthweight and post natal growth. *European Patent Application*, 13 pp.

Kempster, A.J. (1989) Carcass and meat quality research to meet market needs. *Animal Production* 48, 483-496.

Kim, Y.S., Lee, Y.B. and Dalrymple, R.H. (1987) Effect of the repartitioning agent cimaterol on growth, carcass and skeletal muscle characteristics in lambs. *Journal of Animal Science* 65, 1392-1399.

Koppel, J., Kuchar, S., Rynikova, A., Mozes, S., Noskovic, P. and Boda, K. (1988) Effects of bovine growth hormone in suckling lambs. *Experimental and Clinical Endocrinology* 91, 223-226.

Lee, M.O. and Schaffer, N.K. (1934) Anterior pituitary growth hormone and composition of growth. *Journal of Nutrition* 7, 337-363.

Machlin, L.J. (1972) Effect of porcine growth hormone on growth and carcass composition of the pig. *Journal of Animal Science* 35, 794-800.

Mersmann, H.J. (1989) Potential mechanisms for repartitioning of growth by β-adrenergic agonists. In: Campion, D.R., Hausman, G.J. and Martin, R.J. (eds.) *Animal Growth Regulation*. Plenum Press, New York, pp. 337-357.

Muir, L.A., Wein, S., Duquette, P.F., Rickes, E.L. and Cordes, E.H. (1983) Effects of exogenous growth hormone and diethylstilbestrol on growth and carcass composition of growing lambs. *Journal of Animal Science* 56, 1315-1323.

Murphy, L.J., Bell, G.I. and Friesen, H.G. (1987) Tissue distribution of insulin-like growth factor I and II messenger ribonucleic acid in the adult rat. *Endocrinology* 120, 1279-1282.

Murray, J.D., Nancarrow, C.D., Marshall, J.T., Hazelton, I.G. and Ward, K.A. (1989) Production of transgenic merino sheep by microinjection of ovine metallothionein-ovine growth hormone fusion genes. *Reproduction, Fertility & Development* 1, 147-155.

Nancarrow, C.D., Marshall, J.T.A., Clarkson, J.L., Murray, J.D., Millard, R.M., Shanahan, C.M., Wynn, P.C. and Ward, K.A. (1991) Expression and physiology of performance regulating genes in transgenic sheep. *Journal of Reproduction and Fertility* Suppl. 43, 277-291.

Nilsson, A., Carlsson, B., Isgaard, J., Isaksson, O.G.P. and Rymo, L. (1990) Regulation by GH of insulin-like growth factor-I mRNA expression in rat epiphyseal growth plate as studied with in-situ hybridization. *Journal of Endocrinology* 125, 67-74.

Novakofski, J., McKeith, F.K., Grebner, G.L., McLaren, D.G., Brenner, K., Easter, R.A., Bechtel, P.J., Jones, R.W. and Ingle, D.L. (1988) Pig carcass and pork sensory characteristics for animals given daily injections of natural or recombinant porcine somatotropin from 57 to 103 kg. *Animal Production* 46, 487 (abstr.).

O'Connor, R.M., Butler, W.R., Finnerty, K.D., Hogue, D.E. and Beermann, D.H. (1992) Acute and chronic hormone and metabolite changes in lambs fed the beta-agonist, cimaterol. *Domestic Animal Endocrinology* (In press).

Palmiter, R.D., Brinster, R.L., Hammer, R.E., Trumbauer, M.E., Rosenfeld, M.G., Birnberg, N.C. and Evans, R.M. (1982) Dramatic growth of mice that develop from eggs microinjected with metallothionein-growth hormone fusion genes. *Nature* 300, 611-615.

Palmiter, R.D., Norstedt, G., Gelinas, R.E., Hammer, R.E. and Brinster, R.L. (1983) Metallothionein-human GH fusion genes stimulate growth in mice. *Science* 222, 809-814.

Pastoureau, P., Charrier, J., Blanchard, M.M., Boivin, G., Dulor, J.P., Theriez, M. and Barenton, B. (1989) Effect of chronic GRF treatment on lambs having low or normal birth weight. *Domestic Animal Endocrinology* 6, 321-329.

Pell, J.M., Elcock, C., Harding, R.L., Morrell, D.J., Simmonds, A.D. and Wallis, M. (1990) Growth, body composition, hormonal and metabolic status in lambs treated long-term with growth hormone. *British Journal of Nutrition* 63, 431-435.

Pell, J.M., Flint, D.J., James, S. and Aston, R. (1991) Immunomodulation of hormones of the somatotropin axis. In: van der Wal, P. and van der Wilt, F. (eds.) *Biotechnology for the Control of Growth and Product Quality in Meat Production: Implications and Acceptability*. Wageningen Agricultural University, Wagenigen, Netherlands. PUDOC. pp. 51-66.

Pomp, D., Nancarrow, C.D., Ward, K.A. and Murray, J.D. (1992) Growth, feed efficiency and body composition of transgenic mice expressing a sheep metallothionein 1a-sheep growth hormone fusion gene. *Livestock Production Science* (in press).

Pullar, R.A., Johnsson, I.D., Chadwick, P.M.E. and Hart, I.C. (1986) Recombinant bovine somatotropin is growth promoting and lipolytic in fattening lambs. *Animal Production* 42, 433 (Abstr.).

Pursel, V.G., Pinkert, C.A., Miller, K.F., Bolt, D.J., Campbell, R.G., Palmiter, R.D., Brinster, R.L. and Hammer, R.E. (1989) Genetic engineering of livestock. *Science* 244, 1281-1288.

Pursel, V.G., Hammer, R.E., Bolt, D.J., Palmiter, R.D. and Brinster, R.L. (1990) Integration, expression and germ-line transmission of growth-related genes in pigs. *Journal of Reproduction and Fertility* Suppl. 41, 77-87.

Reeds, P.J., Hay, S.M., Dorwood, P.M. and Palmer, R.M. (1986) Stimulation of muscle growth by clenbuterol: lack of effect on muscle protein biosynthesis. *British Journal of Nutrition* 56, 249-258.

Rexroad, C.E., Jr. and Pursel, V.G. (1988) Status of gene transfer in domestic animals. *Proceedings 11th International Congress on Animal Reproduction and Artificial Insemination, Dublin* 5, 28-35.

Rexroad, C.E., Jr., Hammer, R.E., Bolt, D.J., Mayo, K.E., Frohman, L.A., Palmiter, R.D. and Brinster, R.L. (1989) Production of transgenic sheep with growth-regulating genes. *Molecular Reproduction and Development* 1, 164-169.

Rexroad, C.E., Jr., Hammer, R.E., Behringer, R.R., Palmiter, R.D. and Brinster, R.L. (1990) Insertion, expression and physiology of growth-regulating genes in ruminants. *Journal of Reproduction and Fertility* Suppl. 41, 119-24.

Roche, J.F. and Quirke, J.F. (1986) The effects of steroid hormones and xenobiotics on growth of farm animals. In: Buttery, P.J., Haynes, N.B. and Lindsay, D.B. (eds.) *Control and Manipulation of Animal Growth*. Butterworths, London, pp. 39-51.

Rosemberg, E., Thonnay, M.L. and Butler, W.R. (1989) The effects of bovine growth hormone and thyroxine on growth rate and carcass measurements in lambs. *Journal of Animal Science* 67, 3300-3312.

Salmon, W.D. Jr. and Daughaday, W.J. (1957) A hormonally controlled serum factor which stimulates sulfate incorporation by cartilage *in vitro*. *Journal of Laboratory and Clinical Medicine* 49, 825-836.

Simm, G., Dingwall, W.S., Murphy, S.V. and FitzSimons, J. (1990) Selection for improved carcass composition in Suffolk sheep. *Proceedings 4th World Congress Genetics Applied to Livestock Production, Edinburgh* 15, 100-103.

Sinnett-Smith, P.A., Woolliams, J.A., Warriss, P.D. and Enser, M. (1989) Effects of recombinant DNA-derived bovine somatotropin on growth, carcass characteristics and meat quality in lambs from three breeds. *Animal Production* 49, 281-289.

Spencer, G.S.G. (1984) Effect of immunization against somatostatin on growth rate of lambs. In: Roche, J.F. and O'Callaghan, D. (eds.) *Manipulation of growth in farm animals.* Martinus Nijhoff Publishers, Boston, pp. 122-133.

Spencer, G.S.G. and Williamson, E.D. (1981) Increased growth in lambs following immunization against somatostatin: preliminary observations. *Animal Production* 32, 376 (Abstr.)

Spencer, G.S.G., Hallett, K.G. and Fadlalla, A.M. (1985) A novel approach to growth promotion using autoimmunization against somatostatin. III. Effects in a commercial breed of sheep. *Livestock Production Science* 13, 43-52.

Spoon, R.A. and Hallford, D.M. (1989) Growth response, endocrine profiles and reproductive performance of fine-wool ewe lambs treated with ovine prolactin before breeding. *Theriogenology* 32, 45-53.

Sun, Y.X., Drane, G.L., Currey, S.D., Lehner, N.D., Gooden, J.M. and Hoskinson, R.M. (1990) Immunization against somatotropin release inhibiting factor improves digestibility of food, growth rate and wool production of cross-bred ewes. *Australian Journal of Agricultural Research* 41, 401-411.

Thompson, J.M. (1990) Correlated responses to selection for growth and leanness in sheep. *Proceedings 4th World Congress Genetics Applied to Livestock Production, Edinburgh* 16, 266-275.

Thornton, R.F., Tume, R.K., Payne, G., Larsen, T.W., Johnson, G.W. and Hohenhaus, M.A. (1985) The influence of the β_2-adrenergic agonist, clenbuterol, on lipid metabolism and carcass composition of sheep. *Proceedings of the New Zealand Society Animal Production* 45, 97-101.

Timmerman, H. (1987) β-adrenergics; physiology, pharmacology, applications, structures and structure-activity relationships. In: Hanrahan, P.J. (ed.) *Beta-Agonists and Their Effects on Animal Growth and Carcass Quality.* Elsevier, London, pp. 13-28.

Trout, W.E. and Schanbacher, B.D. (1990) Growth hormone and insulin-like growth factor-I responses in steers actively immunized against somatostatin or growth hormone-releasing factor. *Journal of Endocrinology* 125, 123-129.

Turner, J.D., Rotwein, P., Novakofski, J. and Bechtel, P.J. (1988) Induction of mRNA for IGF-I and -II during growth hormone-stimulated hypertrophy. *American Journal of Physiology* 255, E513-E517.

Van Kassel, A.G., Korchinski, R.S., Hampton, C.H. and Laarveld, B. (1990) Effect of immunization against somatostatin in the pregnant ewe on growth and endocrine status of the neonatal lamb. *Domestic Animal Endocrinology* 7, 217-227.

Vernon, R.G. and Flint, D.J. (1989) Role of growth hormone in the regulation of adipocyte growth and function. In: Heap, R.B., Prosser, C.G. and Lamming, G.E. (Eds.) *Biotechnology in Growth Regulation.* Butterworths, London, pp. 57-71.

Wagner, J.F. and Veenhuizen, E.L. (1978) Growth performance, carcass composition and plasma growth hormone levels in wether lambs when treated with growth hormone and thyrotropin. *Journal of Animal Science* 47 (Suppl. 1), 397 (Abstr.)

Ward, K.A., Nancarrow, C.D., Murray, J.D., Shanahan, C.M., Byrne, C.R., Rigby, N.W., Townrow, C.A., Leish, Z., Wilson, B.W., Graham, N.M., Wynn, P.C., Hunt, C.L. and Speck, P.A. (1990) The current status of genetic engineering in domestic animals. *Journal of Dairy Science* 73, 2586-2592.

Williams, P.E.V., Pagliani, L., Innes, G.M., Pennie, K., Harris, C.I. and Garthwaite, P. (1987) The effect of a β-agonist (clenbuterol) on growth, carcass composition, protein and energy metabolism of veal calves. *British Journal of Nutrition* 57, 417-428.

Wolf, B.T., Smith, C., King, J.W.B. and Nicholson, D. (1981) Genetic para-
    meters of growth and carcass composition in crossbred lambs. *Animal
    Production* 32, 1-7.
Wrutniak, C., Cabello, G., Charrier, J., Dulor, J.P., Blamchard, M. and Baren-
    ton, B. (1987) Effects of TRH and GRF administration on GH, TSH, T4 and
    T3 secretion in the lamb. *Reproduction, Nutrition, Development* 27, 501-510.
Zainur, A.S., Tassell, R., Kellaway, C. and Dodemaide, W.R. (1989) Recomb-
    inant growth hormone in growing lambs: effects on growth, feed utilization,
    body and carcass characteristics and on wool growth. *Australian Journal of
    Agricultural Research* 40, 195-206.

Chapter 11

Fibre Production from Sheep and Goats

A.J.F. Russel

The Macaulay Land Use Research Institute, Pentlandfield, Roslin, Midlothian, EH26 6RF, UK

Introduction

This chapter deals with recent findings in the field of wool production and the production of the two principal types of goat fibre: mohair and cashmere. The fibre produced by mohair x cashmere goats, which has recently been named "cashgora" and described as a new animal fibre but which has in fact been in existence as long as the Angora goat, is not specifically considered in this review. Little research has been carried out on cashgora *per se*; experience in its production indicates that its growth and response to nutrition is as would be expected from a mohair-cashmere intermediate.

Many distinct breeds of sheep, most of which produce commercial quantities of wool, are recognized throughout the world. Only one breed of goat, the Angora, produces mohair, although South African, Texan, Australasian and other strains which each have their own characteristics, exist. In contrast, there is no single breed of cashmere goat. Indeed, of the more than 300 goat breeds identified, almost all but the Angora carry an undercoat of short, fine, non-medullated fibres which in the majority of cases would comply with the commercial criteria of cashmere. Only a few breeds, however, produce this undercoat in sufficient quantity to make cashmere production economically viable.

Skin Follicles and Fibre Growth

Some elementary appreciation of the structure and activity of skin follicles is required to understand the relevance of recent research to wool and goat fibre production.

Both sheep and goats have primary and secondary skin follicles. Both types of follicles are formed prenatally by a downwards growth of the epidermis into the underlying dermis. Within this depression hair is formed from the mitotic division of cells in the follicle bulb. Primary follicles are distinguished by their close association with an apocrine sweat gland, a bilobed sebaceous gland and an arrector pili muscle; secondary follicles are smaller and have only an acinar sebaceous gland. In both sheep and goats primary follicles are typically, but not invariably, arranged in groups of three with from four to ten or more times that number of secondaries. The subject of wool follicle morphology has been comprehensively reviewed by Orwin (1989).

In sheep, primary follicles produce three types of fibre: wool, hair and kemp, and in goats they produce mohair, hair and kemp. Secondary follicles produce wool in sheep, mohair in Angora goats and cashmere, sometimes referred to as "down", in cashmere goats.

Three main phases of follicle activity are recognized: anagen, which is the phase of active cell division and fibre growth; catagen, in which the follicle regresses, fibre growth stops and the "brush end" (sometimes referred to as the "root" or "bulb") is formed; and telogen or resting phase, in which the fibre is often, but not always shed. Sometimes the fibre may be retained throughout the telogen phase and shed in the early part of the subsequent anagen phase when, for a short period, two fibres, the old and the new, can be seen within the one follicle.

There are also different types of pattern in which the fibres covering most of the body surface of sheep and goats are replaced. Some sheep breeds, e.g. the Wiltshire Horn, exhibit a marked seasonal pattern of fibre growth with all follicles showing a well defined telogen phase. Most domesticated sheep breeds, however, do not have a synchronized moult; their fleece grows continuously throughout the year, although its rate of growth varies during the course of the year and individual fibres are shed from time to time.

The pattern of fibre growth in Angora goats is similar to that in most sheep breeds, i.e. it grows continuously throughout the year, although the rate of growth can be much reduced in winter and a substantial proportion of follicles may be in the telogen phase at that time.

Cashmere goats exhibit a markedly seasonal pattern of fibre growth as described, for example by Mitchell *et al.* (1989). In general the secondary follicles are reputed to be active from the summer solstice to the winter solstice and to be in the resting phase during the remaining six months. The undercoat of cashmere is shed in the spring. In practice the anagen phase of follicular activity can extend from significantly earlier than the longest day to substantially later than the shortest day. Large differences between breeds are evident in this characteristic.

Fuller descriptions of the development and anatomy of skin follicles, of the phases of follicular activity and of patterns of fibre growth and replacement are contained in the authoritative writings of Wildman (1932), Carter (1955), Ryder and Stephenson (1968), Shelton (1968), Dry (1975),

Chapman and Ward (1979) and Reis (1982).

Fibre growth is ultimately under endocrine control, although the mechanisms through which the many hormones which influence follicular activity act are poorly understood. These hormones are in turn controlled by many environmental factors including photoperiod, temperature and nutrition as well as by physiological state and genotype. The following sections consider some of the recent research findings in these varied fields.

Follicular Physiology and Endocrinology

The Wiltshire Horn is one of the few breeds of sheep showing marked annual cycles of wool growth and shedding, as is the norm in cashmere goats. Although these sheep have a limited potential in commercial wool production enterprises, they are a useful model for studies on the seasonality of fibre growth. Maxwell *et al.* (1988) noted that in sheep of this breed subjected to continuous long days (18 hours light, 6 hours dark) the frequency of the seasonal rhythm increased to a cycle of about 8 months. If a similar seasonality exists in breeds exhibiting continuous wool growth, this finding indicates a possible means of manipulating follicular activity to increase annual production.

In cashmere goats kept under continuous light McDonald and Hoey (1987) and McDonald *et al.* (1987) noted reductions in the cycle period of cashmere fibre growth to 252-261 days. After 2 years, however, an extended cycle emerged which the authors attribute to photodesensitization, recurring presumably as a result of the animals not experiencing any darkness, as opposed to an unvarying light:dark ratio as used by Maxwell *et al.* (1988) in sheep. This again indicates a potential for the manipulation of seasonal follicular activity and fibre growth.

Evidence of seasonality in fibre growth in a major wool producing breed was found by Woods and Orwin (1988) working with New Zealand Romney sheep. Using an elegant ^{3}H-cystine radiolabelling technique in sheep on a maintenance level of feeding they showed that fibre diameter and growth rate (in length and weight) varied seasonally, with maximum and minimum values occurring in summer and winter respectively. The amplitude and timing of the cycles of growth differed between sheep, perhaps indicating opportunities for genetic selection to increase wool growth rate; however, differences also occurred between fibres from the same sheep and the authors conclude that this is evidence of the complexity of the physiological mechanisms determining follicular activity.

Seasonal effects in sheep and goats are caused by changes in photoperiod which determine the secretion of melatonin from the pineal gland. In sheep, most of the work on the use of exogenous melatonin has been directed at the manipulation of the breeding season. One study which specifically examined effects on wool growth was carried out by Harris *et al.* (1989).

Although they observed significant effects on testis diameter in Romney rams, the sustained release of a pharmacological dose of melatonin for one month had no significant effect on wool production.

In cashmere goats the studies involving melatonin have been largely directed at influencing fibre production. These have shown that melatonin administered in late winter or early spring either orally (2 or 4 mg/d) (Scheurman *et al.*, 1987), as a slow-release capsule (18 mg) given on three occasions at eight-week intervals (Moore *et al.*, 1989a) or on a variable number of occasions at six-week intervals (Litherland *et al.* 1990) or incorporated in a silicon rubber implant (Betteridge *et al.*, 1987; Welch *et al.*, 1990) resulted in significant increases in annual cashmere production. In some cases the relative increases have been extremely large - production has been more than doubled - although the absolute increases have been in the order of 10-20 g. These studies have shown that the administration of melatonin to cashmere goats in the spring brings about an earlier initiation of down growth. Although some workers have not recorded subsequent events, it appears that this growth of cashmere is shed in the autumn and followed by a second growth which is shed about one month later than the normally-grown undercoat in untreated animals.

Studies of the seasonality of cashmere growth by workers in New Zealand have revealed the presence of very short or vellus hairs in late spring or early summer (Nixon *et al.*, 1991). These are considered to be a vestige of the summer coat grown by many mammals. It is probable that the administration of melatonin in the spring causes these fibres to elongate and produce an additional crop of cashmere. If this is indeed the case, further research is required to determine with more precision the protocol for melatonin administration and the timing of the harvesting of the two cashmere crops to ensure that fibre lengths are adequate for manufacturing purposes and that no fibre is lost through premature shedding.

The effects of giving exogenous melatonin to cashmere goats in the winter have been studied by Lynch and Russel (1989). They found that continuous administration of melatonin (18 mg capsules on three occasions at six-week intervals) beginning prior to the winter solstice advanced the initiation of cashmere shedding by some four to six weeks. The treatment also resulted in an advance by a similar period of the normal seasonal increase in circulating prolactin concentration. The same treatment given some three to four months after the winter solstice significantly lowered circulating prolactin concentrations.

It thus appears that, while melatonin can be used to advance the initiation of cashmere growth, it cannot be used to extend the period of growth beyond the end of the normal growing season. Lynch and Russel (1989) have suggested that this is because the animals become refractory to melatonin after a period of six months of decreasing daylength. Their work also suggests that the effect of photoperiod on skin follicle activity is mediated through circulating prolactin concentrations.

Over the years the effects on wool production of hormones known to influence the growth and metabolism of other body tissues have been studied extensively. Thyroxine has probably been the prime example in earlier decades; more recently attention has turned to growth hormone. Studies in the USA in which ewe lambs were injected with 2.5 or 5.0 mg ovine growth hormone daily for 10 days and then on alternate days for 20 days (Holcombe *et al.*, 1988) or with 2.5 mg ovine growth hormone on alternate days for 98 days (Heird *et al.* 1989) failed to show any significant effects of the treatment on greasy or clean fleece weights, staple length, fibre diameter or any other component of wool production. A similar lack of response in wool growth to the administration to lambs to recombinant growth hormone which produced significant effects on rates of live weight gain, feed conversion efficiency and carcass weight has been reported from Australia by Zainur *et al.* (1989).

Positive responses in wool production and live weight gain in lambs given somatostatin antigen at 25 day intervals have been reported by Sun *et al.* (1988). In later work the same group (Sun *et al.*, 1990) observed increases in live-weight gain, wool growth and fibre diameter following immunization against somatotrophin release inhibiting factor.

This contrasts with other Australian work by Wynn *et al.* (1988) with Merino ewes fed at either 1 or 1.6 times maintenance energy requirements and injected with 10 mg somatotrophin daily for 28 days and which demonstrated a decrease of 20% in wool production during the treatment period in the animals on the higher level of energy intake. This was due mainly to a reduction in wool fibre diameter. After the end of the treatment wool growth increased by 20% for 12 weeks, due to increases in fibre length and diameter. The authors attribute the changes in wool growth induced by the growth hormone treatment to a change in the partitioning of amino acids between the muscle mass and the skin.

The different, and in some cases opposite, responses in fibre growth to treatment of animals with growth hormone may perhaps be reconciled through a better understanding of the precise nutritional limitations to body tissue and fibre growth prevailing in different situations.

Another area of recent work in Australia has been on epidermal growth factor. Studies on this substance have shown that its use induces a break, or at least a zone of weakness, in wool fibres such that the fleece can be peeled off by hand about seven days after administration (Panaretto *et al.*, 1989). This is like earlier methods of "chemical shearing" which have been advocated from time to time as an alternative to the traditional method of shearing. The use of such techniques which deny the sheep even the protective stubble of fibre left in conventional shearing leaves the animal extremely vulnerable to sunburn, sunstroke and hypothermia and is likely to prove unacceptable on welfare grounds.

Recently published results of 20 years of genetic selection for high and low fibre diameter and high and low staple length in medium-wooled

Merinos provide an interesting and tantalizing insight into the complexity of skin follicle physiology. Moore *et al.* (1989b) record that over the course of the experiment each line responded in the desired direction producing fleeces composed of thick or thin fibres and long or short wool staples. However, the variation in the amounts of wool grown which might have been expected from these responses was compensated by changes in other unselected traits. For example, the anticipated difference in fleece weight between the high and low staple length lines was reduced by an increase in fibre crimp frequency in the low length line. Similarly, differences in fleece weight in the fibre diameter lines were smaller than expected because of large and inverse changes in follicle numbers; towards the end of the selection regime the mean follicle density in the low diameter line was twice that in the high diameter sheep. Analysis of data from all four lines revealed a high significant negative linear correlation (r = -0.89) between follicle density and fibre diameter.

The implication of these results is that the number of follicles initiated in the skin during the fetal stage has a direct influence on the characteristics of the wool fibres produced in the adult sheep. The authors conclude that both features (i.e. follicle number and fibre characteristics) must be under the control of a single developmental mechanism. What is of especial significance is that since the expression of each of these traits is separated in time, the mechanism must be activated during the earlier event, i.e. prenatally, at or before the phase of follicle initiation.

Nutrition and Fibre Growth

Although seasonal changes in wool growth are due in part to the circannual changes in photoperiod discussed above, they are also a result of the quantitative and qualitative changes in the nutrition available to the grazing animal in the course of the year and to changes in physiological state which tend to be associated with season. The combined effects of nutrient intake and physiological state can be considerable. Hall (1987) recorded clean wool growth rates in grazing crossbred ewes in New South Wales ranging from 3.6 g per day for twin-bearing ewes in winter to 15.3 g per day for dry ewes in spring. In New Zealand, Sumner and McCall (1989) observed clean wool growth rates of dry sheep ranging from 7.2 g per day in winter to 12.5 g per day in summer; wool growth rates of single- and twin-rearing ewes averaged 63 and 55% of that of dry ewes in winter and 90 and 82% of dry ewe production in summer.

An example of the magnitude of the effect of physiological state on wool growth is contained in a report by Hawker and Thompson (1987) working with New Zealand Romney ewes. These workers estimated that pregnancy reduced wool growth rate by 18% by the third month of gestation and that in late pregnancy ewes carrying one fetus grew 21% less wool than non-pregnant ewes and 17% more wool than ewes with twin fetuses. Williams and Butt (1989) demonstrated that such effects are wholly attributable to nutrition. In an experiment in which non-pregnant and single- and twin-bearing ewes were fed to maintain fleece-free maternal live weight (i.e. exclusive of the weights of wool and conceptus) there were no differences between ewe classes in wool growth and fibre diameter. These results show conclusively that loss of wool production is not an obligate consequence of pregnancy and can be prevented by appropriate nutritional management.

The effect of nutrition in one season can affect wool production in the next. Rowe *et al.* (1989), working with young Merino sheep observed a carry-over effect of supplementary feeding during summer and autumn on wool growth during winter. For each additional 1 g wool grown during the period of grain feeding, about 1.4 g were measured in the final fleece weight. This work also gives an indication of the magnitude of the effects of nutrition on wool growth. Supplements of 750 g grain increased clean fleece weight from 2.65 to 3.59 kg, wool fibre diameter from 18.6 to 21.1 μm, staple length from 73.5 to 86.7 mm and staple breaking force from 17.4 to 26.1 N per ktex.

Energy intake can affect wool production, particularly at high levels of protein intake, as has been demonstrated by Reis *et al.* (1988). The major nutrient determining fibre production in all species is, however, protein, or more specifically the sulphur-containing amino acids, methionine and cystine. Unlike monogastric species, sheep and goats can synthesize methionine in the rumen. The amino acids are absorbed in the small intestine and subsequently methionine undergoes transsulphuration to cystine. This process does not occur in the skin, but most probably takes place in the liver; it is likely, however, that metabolism at the level of the follicle determines the rate at which cysteine, synthesized from cystine, is assimilated and fibre production activated (Cobon *et al.*, 1988).

Since methionine plays such a fundamental role in the synthesis of keratin, it is scarcely surprising that there has been a great deal of interest in investigating ways of increasing the supply of cystine and methionine to the sites of absorption in the small intenstine. Reis *et al.* (1989) studied the incorporation into wool of ^{35}S-labelled cystine and methionine administered either intravenously or via the abomasum. At optimal dose rates D-methionine, L-methionine and L-cystine, given by the abomasal routine, were all equally effective as supplements for wool growth. However, a large dose of methionine (10 g) given by the same route resulted in a high proportion being excreted in urine and a low rate of incorporation into wool. The results from the intravenous administration of both amino acids showed that at moderate levels of methionine availability (2-3 g per day), methionine is utilized in sheep for synthesis of wool with about 80% of the efficiency of cystine.

Further work from the same group (Reis *et al.*, 1990) indicates a specific role for methionine in the control of wool growth, other than the provision of cysteine. This role was postulated to be related to some function of S-adenosylmethionine. In this work mixtures of up to ten amino acids, and containing 3 g methionine per daily dose, increased the volume of wool grown by up to 86%. When cysteine completely replaced methionine wool growth was markedly reduced.

In trials involving the feeding of methionine, hydroxymethyl-methione, methionine hydroxy analogue and "protected" (i.e. fatty-acid coated) methionine, Cottle (1988b, 1988c, 1988d) observed increases in wool growth of up to 23%. In some trials the best results were obtained when the supplement was fed with a high roughage diet, and in others when it was given with a high grain diet. It was concluded that the wool growth response to hydroxymethyl-methionine was uneconomic and that, although methionine hydroxy analogue appeared to be a relatively inefficient way of supplying methionine to the intestines, it could be used economically in supplements fed to housed superfine Merino wethers (the "Sharlea" system of quality wool production). In general, the responses to this type of supplementation have been substantially less than those recorded in post-ruminal infusion studies (Cottle 1988e).

Responses in wool growth to protected methionine have also been recorded in Sardinian ewes which are kept principally for milk production (Floris *et al.*, 1988). These was a small (8%) increase in milk production but no effect on milk composition.

Another means of increasing the supply of amino acids to the small intestine is by manipulation of the populations of rumen micro-organisms. Fenn and Leng (1989) observed that in sheep given a roughage-based diet supplemented with bentonite (30 g per day as a dry powder or 60 g per day as a suspension in the drinking water) wool growth rates were increased by 19 or 20% respectively. These compared with no effect from supplementary cysteine and a 16% increase from the addition of methione to the drinking water. In subsequent work Fenn and Leng (1990) showed that supplementation with bentonite (30, 50 or 60 g per day) consistently increased the density of rumen protozoa. They noted a 17% increase in wool growth following the addition of 15 g bentonite per day to the drinking water. These workers conclude that the increased protozoal populations in the rumen fluid of bentonite-treated sheep allow a greater flow of protozoal protein to the intestines, which in turn results in an increase in wool growth.

Cottle (1988a) and Fenn and Leng (1990) also observed increases of 6.5% and 25% respectively in wool growth in defaunated sheep (i.e. in sheep in which the rumen protozoal populations had been eliminated). This presumably is a response to changes in microbial protein supply to the intestine.

The effect of bentonite on rumen fermentation and protozoal populations has been studied *in vitro* by Wallace and Newbould (1991). Using a rumen simulation technique (Rusitec) they observed a reduction, not an increase, in protozoal numbers. This was accompanied by an increase in bacterial numbers and the authors conclude that the net protein yield resulting from rumen fermentation might be expected to increase in animals treated with bentonite.

Although an adequate supply of dietary protein, or at least of the nitrogenous substrates from which microbial protein may be synthesized, is essential for wool growth, it does not follow that the provision of additional protein in the diet will invariably result in increased wool production. Denney and Hogan (1987) point out that in sheep eating high protein diets there can be a substantial wastage of crude protein associated with high rumen ammonia concentrations. This can appreciably reduce the quantity of amino acids absorbed and may limit wool growth.

The influence on wool production of the supply of amino acids at the sites of absorption in the small intestine is such that Neutze (1990) considered that wool growth response might serve as a means of estimating the amount of protein escaping degradation in the rumen. It was, however, concluded that the technique was not sufficiently sensitive for this purpose.

Leng (1989) has considered the potential for the manipulation of fermentative digestion in the rumen as a means of improving ruminant production, including wool. He argues convincingly that in pastoral situations the main challenges are to improve digestibility and increase protein relative to volatile fatty acids in the products of digestion. Without correction of dietary deficiencies, manipulations of the rumen or of the animal's genome are unlikely to improve production and may even be detrimental.

The means by which an increased provision of nutrient substrates is translated into increased wool growth have been examined in a study on wool follicle kinetics by Hynd (1989). The principal findings were that a nutritionally induced response in clean fleece production of 33% reflected increases in fibre diameter of 8% and in rate of length growth of 26%. The volume of the germinative region of the follicle bulb increased by 30% and the rate of bulb cell division by 35%. Williams and Winston (1987) calculated that a nutritionally induced response in wool production resulted from a 17% increase in the rate of incorporation of cortical cells into fibre, compared with a difference of 20% in rate of incorporation due to the results of genetic selection.

The effects on wool growth of a variety of feed additives, notably avoparcin and flavomycin, have been studied by workers in Western Australia (Aitchison *et al.*, 1988, 1989a, 1989b). In general the results indicate that the use of such additives can increase both live-weight gain and wool growth in rapidly growing sheep eating a high-protein diet, but have little or no consistent effects in sheep eating lower quality feedstuffs.

Most of the research work conducted on the effects of nutrition on wool growth has been aimed at increasing production; any detrimental effects have usually been concerned with increases in fibre diameter (i.e. a lowering of quality) as a result of improved nutrition. Wool quality can, of course, also be adversely affected by poor nutrition. The effect of poor nutritional regimes, imposed over either short periods or continuously, on wool fibres and their processing characteristics has been extensively studied by workers in South Africa (Hunter et al., 1990). They observed differences of up to $10\mu m$ between the mean diameters of fibre segments grown during periods of nutritional stress and normal feeding. Effects of this order, which is equivalent to a reduction in fibre diameter of up to 40%, clearly have profound consequences as regards the tensile properties and processing performance of wool from sheep subjected to even very short-term nutritional stress.

The general effects of nutrition on mohair production are analogous to those on wool growth. This is well illustrated by the report of Weshuysen et al. (1985) that supplementary feeding of six-month-old Angora kids during drought conditions more than doubled fibre production over a six-month period. In work with grazing adult animals McGregor (1986) showed that improved nutrition increased both live weight and mean fibre diameter, the rate of increase in the latter being 0.26 μm per kg increase in live weight. In a second experiment with pen-fed animals the increase in fibre diameter was 0.40 μm per kg increase in live weight (McGregor, 1986). Similar effects of improved nutrition on mohair production and fibre diameter have also been reported by McGregor and Hodge (1989).

Mohair production and quality have also been shown to respond to both the quantity and degradability of dietary protein. Throckmorton et al. (1982) reported increased mohair production following the feeding of formaldehyde-treated casein, and Sahlu et al. (1988) demonstrated an increase of 23% in clean fleece yield and of 5.2% in mean fibre diameter associated with an increase in dietary protein from 12 to 18%. The replacement of solvent-extracted soyabean meal by heat-treated (protected) soyabean meal also increased both clean fleece yield and fibre diameter.

With mohair, quality is determined by a number of factors including freedom from medullated fibres and kemp, as well as by mean fibre diameter. McGregor (1984) reported that a lower level of nutrition was associated with an increased medullated fibre content of the fleece, although this is not borne out by later studies by Bigham et al. (1990). In a recent review of factors influencing the degree of medullation in mohair, Lupton et al. (1991) conclude that nutrition has little if any effect on the incidence of this undesirable characteristic.

The subject of nutritional effects on cashmere production is one on which the evidence is less clear than in the cases of wool and mohair. McGregor (1988), working with Australian cashmere goats, found that those which consumed 0.68 of maintenance energy requirements produced less and finer cashmere (146 g, 16.7 μm) than those with greater than maintenance levels of intake (221 g, 17.7 μm). There was no differences in cashmere growth or diameter between goats fed 1.25 or 1.5 maintenance or *ad libitum*.

Norton and his co-workers in Queensland have investigated the effects of many different feeding regimes and forms of supplementation on cashmere production in Australian feral goats. Their nutritional treatments have ranged from maintenance levels of feeding to the *ad libitum* feeding of lucerne pellets, and have incorporated a wide range of protein intakes, including various "protected" protein supplements (Ash and Norton, 1984; Norton *et al.*, 1990; Norton, 1991). In none of these experiments was there any evidence of a response in cashmere production or in fibre diameter when either the level or quality of feed was above that needed for the maintenance of live weight.

Working with Scottish feral goats, Russel (1990) could demonstrate no significant effect on either cashmere growth or fibre diameter of a wide range of levels of nutrition including a diet containing a high concentration of undegradable dietary protein and fed at 1.7 maintenance. This confirms the results of similar work by Johnson and Rowe (1984).

The difference in responsiveness of wool and cashmere growth to nutritional manipulation is highlighted by the results of Ash and Norton (1987a) who showed that methionine supplementation, which would have been expected to greatly increase wool growth, had no effect on cashmere production. A lack of response in cashmere growth to supplementation with formaldelhyde-treated protein meals has also been reported by Ash and Norton (1987b) and to supplementation with protected casein by McGregor (1988).

The evidence thus indicates that while cashmere growth may be reduced by relatively severe undernutrition, such as that imposed by McGregor (1988), it is unresponsive in terms of increased weight or fibre diameter to the provision of any nutrients beyond those required for maintenance. This statement does not, however, apply to the other component of the cashmere goat fleece - the guard hair. Many experiments (e.g. Ash and Norton, 1987b; McGregor, 1988), which have shown no effect of nutrition on cashmere growth, have demonstrated substantial effects on guard hair production. McGregor (1988) concludes from this that energy deprivation favours a diversion of nutrients to cashmere growth, while better-fed goats partition nutrients towards hair growth. This conclusion is improbable. The substrates for the growth of both types of fibre are the same, and alterations in the relative growth rates of cashmere and hair fibres are more likely to be due to a differences between follicle type in the production of substrate, and most likely in the transsulphuration of methionine to cystine. Ash and Norton (1987a) suggest that in goats this process does not occur in the liver or kidneys as it does in sheep.

It therefore appears at the present time that mohair production from Angora goats will respond to nutritional manipulation in much the same way as will wool production in sheep. Improvements in nutrition, and particularly in undegradable dietary protein and the sulphur-containing amino acids, will generally result in increases in the weight and diameter of the fibre grown. The magnitude of responses to supplementation with such nutrients will depend on the quantity and nutrient balance of the basal diet; the more closely that approaches the optimum specification, although that has yet to be defined, the less will be the production advantages to be gained from supplementation.

The chemical composition of cashmere may be essentially the same as that of mohair and wool, but its growth is quite different, as is its response to nutritional manipulation. The opportunities for increasing cashmere production by supplementation appear to be very limited, if indeed they exist, in cashmere goats which can be unequivocally stated to have no Angora influence in their ancestry. Whether this unresponsiveness persists as cashmere production increases as a result of genetic selection remains to be established.

The mechanism by which fibres from the primary follicles respond to nutritional factors while those from the secondary follicles do not has yet to be elucidated. An understanding of this fundamental difference might conceivably indicate means whereby cashmere production could be manipulated by nutritional or endocrine treatment.

Genetics of Fibre Production

The scope for modification of wool characteristics by genetic selection has been studied in a simulated breeding programme by Sumner and Clarke (1990). Their results, based on New Zealand Romney data, indicate that the greatest monetary return would be achieved by direct selection for either greasy or clean fleece weight, which they estimate would increase by about 1.5% per year. Loose wool bulk, a characteristic which has assumed increasing importance in recent years, was the next most financially worthwhile characteristic, with an expected increase in gross returns being a third of those from fleece weight. It is, however, negatively correlated with fleece weight and selection for increased bulk would be expected to reduce fleece weight.

Sumner and Clark (1990) put the estimated genetic response in wool characteristics into perspective by comparing the magnitude of changes expected from a 10-year period of selection with those which might be expected from the manipulation of non-genetic factors. They compare the results of a nutritional study in which the best fed sheep grew 45% more clean wool, with 15% of the increase over a 10-year period due to genetic selection. The corresponding nutritional and genetic effects on other characteristics were increases of 3.7 and 4.5 μm in fibre diameter and 18 and 27 μm in staple length respectively. Yellowness was increased by the same amount by both nutritional manipulation and genetic selection. They also demonstrate that substantial effects on wool characteristics can be achieved by the farm and post-farm production practices involved in shearing, and wool handling and processing. While these effects are relatively large in relation to the responses obtained through genetic selection, the latter are cumulative and set the biological base against which the non-genetic manipulations can be made.

The results of a selection programme reported by Hynd *et al.* (1989) and combining the characteristics of clean fleece weight, fibre diameter and hogget live weight showed that fleece weight and hogget weight increased by more than 23% over 11 years, indicating that Sumner and Clark's (1990) estimate of a theoretical rate of progress of 1.5% per year can be readily achieved in practice. The substantial response in fleece weight to Hynd's *et al.* (1989) index was achieved with virtually no change in mean fibre diameter and was associated with an increase in S/P ratio and a reduction rather than an increase in variance in fibre size.

The parallel increases in fleece weight and live weight in Hynd's *et al.* (1989) selection work confirm differences reported by Wiener *et al.* (1988). They observed that inbred sheep were substantially smaller and lighter in live weight than outbred sheep and that these differences were mirrored in wool growth. Inbreeding had no effect on the growth of wool per unit area of skin.

In unselected sheep, however, Mortimer and Atkins (1989) found that genetic correlations between body weight and a number of fleece characteristics including weight, fibre diameter and staple length were all essentially zero. Their report also contains estimates of heritabilities of a number of wool production characteristics derived from data on Merino sheep. Davis *et al.* (1990) have recently reported genetic parameters for the Gansu Alpine Finewool sheep in China; their estimates of heritabilities of wool production traits are generally lower than those found in finewool breeds in other countries, including those of Mortimer and Atkins (1989).

An indication of the increasing attention being devoted by research workers serving the wool producers to the needs and requirements of the processors is the establishment of a programme of selective breeding for high staple strength wool (Rogers *et al.*, 1990).

There have been a number of recent reports of attempts to increase wool production from sheep by genetic engineering. This topic has been considered by Rogers (1990) who suggests three main ways in which wool production might be improved by such means. The first is the production of transgenic sheep to which bacterial genes controlling cysteine synthesis or involved in the glyoxalate metabolic pathway had been transferred. Accounts of work designed to produce such gene constructs has been given by D'Andrea *et al.* (1989) and Ward *et al.* (1989). The latter group report that the results of their initial experiments suggest that wool growth was increased in the transgenic sheep.

The second possible approach outlined by Rogers (1990) is the manipulation of rumen micro-organisms to improve the efficiency of cellolase digestion or the availability of essential amino acids. The third involves the manipulation of forage plants to improve their amino acid balance and thus the wool production of the sheep by which they are grazed. Future reviews will doubtless contain reports of progress in these areas.

There are fewer reports on the breeding of Angora and cashmere goats for fibre production than there are for wool production from sheep. The most comprehensive studies on the genetic parameters for mohair traits are those of Nicoll (1985), based on more than 1000 young goats, and of Nicoll et al. (1989), comprising three data sets derived from a total of more than 5000 individual records. Both reports refer to Angora goats in New Zealand. Estimated heritabilities of 6-month, 12-month and yearling fleece weight from the earlier study were 0.23, 0.40 and 0.53 respectively; corresponding estimates from the later and larger study were 0.24, 0.25 and 0.36 respectively. Fibre diameter had a high heritability of 0.51 while those of percentage medullation (0.16) and kemp (0.02) were low. These estimates imply that selection for reduced levels of medullation and kemp is likely to result in only very small responses. These undesirable traits are, however, highly and positively correlated with fleece weight (Nicoll et al., 1989) indicating that selection for increased mohair production is likely to lead to higher levels of both medullation and kemp.

Bigham et al. (1990) consider that these problems of the New Zealand, and presumably also the Australian, mohair industries could be alleviated to a large extent through a backcrossing programme with Texan or South African Angoras, by exploiting the additive genetic breed differences. They believe that this would rapidly reduce the levels of medullation and kemp.

Although there are a number of strains of the one breed of goat producing mohair - the Angora - the breeds of goat which produce cashmere probably number hundreds (see Millar, 1986, 1987). Goats bearing undercoats of fibre which comform to the generally accepted specifications of cashmere, i.e. non-medullated goat fibre showing both ortho- and para-cortical structures and with a mean fibre diameter of less than $19\mu m$, include breeds not usually considered as cashmere goats. These include the Boer (Couchman, 1988), the German Improved Fawn, the West African Dwarf goat, and various crosses between these breeds (Pfaff et al., 1988).

In recent years there have been a number of reports in the Soviet literature of crossing programmes with a variety of fibre-producing goat breeds. These have generally been designed either to introduce white colour to a cashmere breed such as the Altai Mountain, by crossing with Soviet Mohair and Don goats (Al'kov and Kraskova, 1984) or to improve the productivity of a local breed such as the Dagestan, by crossing with the Altai Mountain (Musalaev, 1984). Other studies have sought to increase the weight of cashmere produced by the use of Soviet Mohair, Don and Orgenburg sires (Dauletbaev, 1985) or by crossing with the Angora or Angora-type breeds (Mamadaliev, 1985; Akynbaev, 1985; Al'meev and Makhmudkhodzhaev, 1985). Many of these breeds and crosses are reputed to produce very high weights of down, but it is difficult to be confident of actual levels of production as there is often doubt, because of differences in terminology, as to whether figures reported refer to weights of combed fibre or actual undercoat. There is also doubt as to whether the undercoat from many of these breeds and crosses can be classed as cashmere, at least in the western world, as mean fibre diameters are frequently greater than 19 μm. Despite these reservations, these breeds and crosses clearly include some individual animals which produce very large quantities of down (between 500 and 1000 g) with mean fibre diameters well within the cashmere range.

The most comprehensive reports of the inheritance of cashmere characteristics, genetic parameters and responses to selection relate to Australian goats which have been extensively studied by Restall and Pattie (1989) and Pattie and Restall (1989, 1990, 1991a, 1991b). The heritabilities of down weight (0.61), diameter (0.47) and length (0.70) were high and greater than would generally be found for wool, although those for S/P ratio (0.29) and for primary and secondary follicle densities (0.16 and 0.17 respectively) were low. These high heritabilities for down weight, diameter and length suggest that these characteristics would respond rapidly to genetic selection. However, the strong positive genetic correlation between down weight and diameter (0.62) indicates that selection for increased weight of cashmere on its own would result in a rapid coarsening of the fibre. If fibre diameter is to be kept within acceptable limits, the rate of increase in down weight will inevitably be reduced.

These authors also report a strong positive genetic correlation (0.88) between down weight and down length. This can be used to advantage in a breeding programme designed to increase down weight (which is not easily estimated) by using down length (which can be readily measured on the animal) as an indirect estimate of down weight.

Relationships between down weight and length in different ages and sexes of cashmere goats, and which provide the basis of a simple and inexpensive method of estimating cashmere production, have also been presented by McDonald (1988).

There have also been reports from the UK of a breeding programme designed to improve the cashmere production of Scottish feral goats. Russel (1989) and Russel and Bishop (1990) have described the programme in which rapid initial improvements are being sought by crossing the native animals with superior stock imported from Iceland, Tasmania, New Zealand and Siberia to give 16 crossbred genotypes. This also serves to give an exceptionally broad genetic base from which selection for increased down weight and for quality (i.e. lower fibre diameter) cashmere can be practised.

Conclusion

There is, at the present time, an increasing interest in animal fibre production, of which wool, mohair and cashmere are the most important, but which also includes angora and fibre from the South American camelids. This stems at least in part from the increasing degree of self-sufficiency in foodstuffs, and particularly those derived from animal products, being achieved by developed countries. This in turn is being reflected in the research interest and endeavour being devoted to animal fibre.

Much remains to be elucidated about the mechanisms of fibre growth and especially about the control of endocrine and metabolic events in the follicle. The array of techniques now being brought to bear on this area of work is impressive and can be expected to lead to a more comprehensive understanding of the process of fibre growth and hence to means whereby it may be manipulated to advantage.

References

Aitchison, E.M., Ralph, I.G. and Rowe, J.B. (1989a) Evaluation of feed additives for increasing wool production from Merino sheep. 1. Lasalocid, avoparcin and flavomycin including in lucerne-based pellets or oaten chaff fed maintenance. *Australian Journal of Experimental Agriculture* 29, 321-325.

Aitchison, E.M., Tanaka, K. and Rowe, J.B. (1988) Evaluation of the feed additives flavomycin and M139603 for increasing liveweight gains and wool production. *Proceedings of the Australian Society of Animal Production* 17, 373.

Aitchison, E.M. Tanaka, K. and Rowe, J.B. (1989b) Evaluation of feed additives for increasing wool production from Merino sheep. 2. Flavomycin and tetro-nasin included in lucerne-based pellets and wheatenchaff fed *ad libitum*. *Australian Journal of Experimental Agriculture* 29, 327-332.

Akynbaev, M. (1985) The results of specialisation. *Ovtseveodstvo*, No. 4, 13-14.

Al'kov, G.V. and Kraskova, Z.K. (1984) The formation of a new type of white goats within the Altai Mountain breed. *Rezvedenie Ovets i Koz. Sherstovedenie* 43-47.

Al'meev, I.A. and Makhmudkhodzhaev, A. Sh. (1985) A new breeding region for cashmere goats. *Ovtsevodstvo* 4, 14-15.

Ash, A.J. and Norton, B.W. (1984) The effect of protein and energy intake on cashmere and body growth of Australian cashmere goats. *Proceedings of the Australian Society of Animal Production* 15, 247.

Ash, A.J. and Norton, B.W. (1987a) Effect of dl-methionine supplementation on fleece growth by Australian cashmere goats. *Journal of agricultural Science, Cambridge* 109, 197-199.

Ash, A.J. and Norton, B.W. (1987b) Productivity of Australian cashmere goats grazing Pangola grass pastures and supplemented with untreated and formaldehyde treated protein meals. *Australian Journal of Experimental Agriculture* 27, 779-784.

Betteridge, K., Welch, R., Pomroy, W., Lapwood, K. and Devantier, B. (1987) Out of season cashmere growth in feral goats. *Proceedings of 2nd*

International Cashmere Conference, New Zealand pp. 137-143.

Bigham, M.L., Bown, M. and Nicoll, G.B. (1990) The manipulation of kemp and medullation in the mohair fleece by breeding and management. *Proceedings 8th Research Conference, New Zealand* II pp. 277-283.

Carter, H.B. (1955) The hair follicle group in sheep. *Animal Breeding Abstracts* 28, 101-116.

Chapman, R.E. and Ward, K.A. (1979) Histological and biochemical features of the wool fibre and follicle. In: *Physiological and Environmental Limitations to Wool Growth.* J.L. Black and P.J. Reis (eds.). The University of New England, Armidale, Australia.

Cobon, D.H., Suter, G.R., Connelly, P.T., Shepherd, R.K. and Hopkins, P.S. (1988) The residual effects of methionine supplementation on the wool growth performance of grazing sheep. *Proceedings of the Australian Society of Animal Production* 17, 383.

Cottle, D.J. (1988a) Effects of defaunation of the rumen and supplementation with amino acids on the wool production of housed Saxon Merinos. 1. Lupins and extruded lupins. *Australian Journal of Experimental Agriculture* 28, 173-178.

Cottle, D.J. (1988b) Effects of defaunation of the rumen and supplementation with amino acids on the wool production of housed Saxon Merinos. 2. Methionine and protected methionine. *Australian Journal of Experimental Agriculture* 28, 179-185.

Cottle, D.J. (1988c) Effects of defaunation of the rumen and supplementation with amino acids on the wool production of housed Saxon Merinos. 3. Cottonseed meal and hydroxymethyl-methionine. *Australian Journal of Experimental Agriculture* 28, 699-706.

Cottle, D.J. (1988d) Effects of defaunation of the rumen and supplementation with amino acids on the wool production of housed Saxon Merinos. 4. Cottonseed meal, analogues of methionine and avoparcin. *Australian Journal of Experimental Agriculture* 28, 707-711.

Cottle, D.J. (1988e) Effects of cottonseed meal, methionine and analogues and avoparcin on the wool production of young, grazing wethers. *Australian Journal of Experimental Agriculture* 28, 713-718.

Couchman, R.C. (1988) Recognition of cashmere down on the South African Boer goat. *Small Ruminant Research* 1, 123-126.

D'Andrea, R.J., Sivaprasad, A.V., Bawden, S., Kuczek, E.S., Whitbread, L.A. and Rogers, G.E. (1989) Isolation of microbial genes for cysteine synthesis and prospects for their use in increasing wool growth. *The biology of wool and hair.* G.E. Rogers, P.J. Reis, K.A. Ward, R.C. Marshall (eds.), London, Chapman and Hall.

Dauletbaev, B.S. (1985) To mobilise reserves in production. *Ovtsevodstvo* 4, 8-9.

Davis, G.P., Ma, F.Z., Piper, L.R., Song, S.Z. and Whiteley, K.J. (1990) Genetic parameters for wool quality and production traits of Gansu Alpine Finewool sheep in China. *Proceedings of the 4th World Congress on Genetics Applied to Livestock Production, Edinburgh 23-27 July 1990. XV. Beef cattle, sheep and pig genetics and breeding, fibre, fur and meat quality,* pp. 185-188.

Denney, G.D. and Hogan, J.P. (1987) Digestion and wool production of sheep grazing grass/medic pasture in the low rainfall wheatbelt. *Temperate pastures: their production, use and management.* J.L. Wheeler, C.J. Pearson, G.E. Robards (eds.), East Melbourne, Victoria, Australia; Commonwealth Scientific and Industrial Research Organisation, pp. 356-358.

Dry, F.W. (1975) *The Architecture of Lambs' Coats.* Massey University, Palmerston North, New Zealand.

Fenn, P.D. and Leng, R.A. (1989) Wool growth and sulfur amino acid entry rate in sheep fed roughage based diets supplemented with bentonite and sulfur amino acids. *Australian Journal of Agricultural Research* 40, 889-896.

Fenn, P.D. and Leng, R.A. (1990) The effect of bentonite supplementation on ruminal protozoa density and wool growth in sheep either fed roughage based diets or grazing. *Australian Journal of Agricultural Research* 41, 167-174.

Floris, B., Bomboi, G. and Sau, F. (1988) Protected methionine in Sarda sheep: effect on lactation and wool growth. La Metionina protetta nella pecora sarda: effetti sulla lattazione e sulla crescita della lana. *Bollettino Societa Italiana Biologia Sperimentale* 64(12), 1143-1149.

Hall. D.G. (1987) Seasonality of growth and quality of wool from crossbred ewes on improved pastures in southern New South Wales. In: *Temperate Pastures: their production, uses and management*. J.L. Wheeler, C.J. Pearson, G.E. Robards (eds.), East Melbourne, Victoria, Australia; Commmonwealth Scientific and Industrial Research Organisation, 486-488.

Harris, P.M., Xu, Z.Z., Blair, H.T., Dellow, D.W., McCutcheon, S.N. and Cockrem, J. (1989) The effect of exogenous melatonin, administered in summer, on wool growth and testis diameter of Romneys. *Proceedings of the New Zealand Society of Animal Production* 49, 35-38.

Hawker, H. and Thompson, K.F. (1987) Effects of pasture allowance in winter on liveweight, wool growth, and wool characteristics of Romney ewes. *New Zealand Journal of Experimental Agriculture* 15, 295-302.

Heird, C.E., Hallford, D.M., Spoon, R.A., Holcombe, D.W., Pope, T.C., Olivares, V.H., Herring, M.A. and Lupton, C.J. (1989) Reproductive responses and wool characteristics of ewe lambs after long-term treatment with ovine growth hormone. *Proceedings, Western Section, American Society of Animal Science and Western Branch Canadian Society of Animal Science* 40, 299-301.

Holcombe, D.W., Hallford, D.M., Hoefler, W.C., Ross, T. and Lupton, C.J. (1988) Wool quality and yield in fine-wool ewe lambs after short-term administration of ovine growth hormone. *SID Research Journal, Sheep Industry Development* 6, 21-25.

Hunter, L., van Wyk, J.B., de Wet, P.J., Grobbelaar, P.D., Pretorius, P.S., Morris, J. de V. and Leeuwner, W. (1990) The effects of nutritional and lambing stress on wool fibre and processing characteristics. *Proceedings of the 8th International Wool Textile Research Conference, New Zealand* II, pp.145-156.

Hynd, P.I. (1989) Effects of nutrition on wool follicle cell kinetics in sheep differing in efficiency of wool production. *Australian Journal of Agricultural Research* 40, 409-417.

Hynd, P.I., Ponzoni, R.W. and Wayte, C.L. (1989) The effect of a WOOLPLAN-type selection programme on primary and secondary fibre size and shape. *Wool Technology and Sheep Breeding* 37, 101-106.

Johnson, T.J. and Rowe, J.B. (1984) Growth and cashmere production by goats in relation to dietary protein supply. *Proceedings of the Australian Society of Animal Production* 15, 400.

Leng, R.A. (1989) The scope for manipulation of fermentative digestion in the rumen to improve ruminant production. In: *The biology of wool and hair*, G.E. Rogers, P.J. Reis, K.A. Ward, R.C. Marshall (eds.), London, Chapman and Hall, pp. 205-215.

Litherland, A.J., Paterson, D.J., Parry, A.L., Dick, H.B. and Staples, L.D. (1990) Melatonin for cashmere production. *Proceedings of New Zealand Society of Animal Production* 50, 339-343.

Lupton, C.J., Pfeiffer, F.A. and Blakeman, N.E. (1991) Medullation in mohair. *Small Ruminant Research* 5, 357-365.

Lynch, P. and Russel, A.J.F. (1989) The endocrine control of fibre growth and shedding in cashmere goats. *Animal Production*, 48, 632-633.

McDonald, B.J. (1988) Estimation of cashmere production from cashmere fibre length in goats. *Australian Journal of Experimental Agriculture* 28, 37-39.

McDonald, B.J. and Hoey, W.A. (1987) Effect of photo-translation on fleece growth in cashmere goats. *Australian Journal of Agricultural Research* 38, 765-773.

McDonald, B.J., Hoey, W.A. and Hopkins, P.S. (1987) Cyclical fleece growth in cashmere growth. *Australian Journal of Agricultural Research* 38, 597-609.

McGregor, B.A. (1984) Growth and fleece production of Angora wethers grazing annual pastures. *Proceedings of the Australian Society of Animal Production* 15, 715.

McGregor, B.A. (1986) Liveweight and nutritional influences on fibre diameter of mohair. *Proceedings of the Australian Society of Animal Production* 16, 420.

McGregor, B.A. (1988) Effects of different nutritional regimes on the productivity of Australian cashmere goats and the partitioning of nutrients between cashmere and hair growth. *Australian Journal of Experimental Agriculture* 28, 459-467.

McGregor, B.A. and Hodge, R.W. (1989) Influence of energy and polymer-encapsulated methionine supplements on mohair growth and fibre diameter of Angora goats fed at maintenance. *Australian Journal of Experimental Agriculture* 29, 179-181.

Mamadaliev, F. Kh. (1985) It is important to improve the breed. *Ovtsevedstvo* 4, 9-10.

Maxwell, C.A., Scaramuzzi, R.J., Foldes, A. and Carter, N.B. (1988) The effects of long term exposure to continuous long or short days on conditioned clean wool weight in Wiltshire Horn x Merino ewes. *Proceedings of the Australian Society of Animal Production* 17, 246-249.

Millar, P. (1986) The performance of cashmere goats. *Animal Breeding Abstracts* 54, 181-199.

Millar, P. (1987) Breeds and types of cashmere-producing goats. In: *Scottish Cashmere*. A.J.F. Russel and T.J. Maxwell (eds.), Scottish Cashmere Producers Association, pp. 26-31.

Mitchell, R.J., Betteridge, K., Welch, R.A.S., Gurnsey, M.P. and Nixon, A.J. (1989) Fibre growth cycles of unselected, reproducing cashmere does discussed in relation to winter shearing. *Proceedings of the New Zealand Society of Animal Production* 49, 163-164.

Moore, R.W., Bigham, M.L. and Staples, L.D. (1989a) Effect of Regulin implants on spring fertility, lactation and down growth of cashmere does. *Proceedings of the New Zealand Society of Animal Production* 49, 39-41.

Moore, G.P.M., Jackson, N. and Lax, J. (1989b) Evidence of a unique developmental mechanism specifying both wool follicle density and fibre size in sheep selected for single skin and fleece characters. *Genetical Research* 53, 57-62.

Mortimer, S.I. and Atkins, K.D. (1989) Genetic evaluation of production traits between and within flocks of Merino sheep. I. Hogget fleece weights, body weight and wool quality. *Australian Journal of Agricultural Research* 40, 433-443.

Musalaev, Kh. Kh. (1984) The possibility of using crossbred white cashmere-type bucks in the improvement of the performance of local goats in Dagestan. *Razvedenie Ovets i Koz. Sherstovedenie* 29-32.

Neutze, S.A. (1990) Use of wool growth response to estimate escape of protein supplements from the rumen. *Australian Journal of Agricultural Research* 41, 761-767.

Nicoll, G.B. (1985) Estimates of environmental effects and some genetic parameters for weaning weight and fleece weights of young Angora goats.

Proceedings of the New Zealand Society of Animal Production 45, 217-219.

Nicoll, G.B., Bigham, M.L. and Alderton, M.J. (1989) Estimates of environmental effects and genetic parameters for live weights and fleece traits of Angora goats. *Proceedings of the New Zealand Society of Animal Production* 49, 183-189.

Nixon, A.J., Gursey, M.P., Betteridge, K., Mitchell, R.J. and Welch, R.A.S. (1991) Seasonal hair follicle activity and fibre growth in some New Zealand Cashmere-bearing goats. *Journal of Zoology, London* 224, 589-598.

Norton, B.W. (1991) Management of Australian cashmere goats for maximum fleece growth. *Proceedings Cashmere Research Seminar, Ballina.* New South Wales Department of Agriculture and Fisheries, pp. 28-41.

Norton, B.W., Wilde, C.A. and Hales, J.W. (1990) Grazing management studies with Australian cashmere goats. 1. Effects of stocking rate on the growth and fleece production of weaned goats grazing tropical pastures. *Australian Journal of Experimental Agriculture* 30, 769-775.

Orwin, D.F.G. (1989) Variations in wool follicle morphology. In: *The biology of wool and hair*. G.E. Rogers, P.J. Reis, K.A. Ward, R.C. Marshall (eds.), London, Chapman and Hall, pp. 227-241.

Panaretto, B.A., Moore, G.P.M., Robertson, D.M., Bennett, J.W., Tunks, D.A., Chapman, R.E., Hollis, D.E., Dooren, P.H. van, Sharry, L.F., Radford, H.M., Stockwell, P.R., Hazelton, I.G., Mattner, P.E. and Brown, B.W. (1989) Biological wool harvesting. *Australian Journal of Biotechnology* 3, 258-263.

Pattie, W.A. and Restall, B.J. (1989) The inheritance of cashmere in Australian goats. 2. Genetic parameters and breeding values. *Livestock Production Science* 21, 251-261.

Pattie, W.A. and Restall, B.J. (1990) Breeding for cashmere. In: *Scottish Cashmere - the Viable Alternative*. A.J.F. Russel (ed.), Scottish Cashmere Producers Association, Edinburgh, UK, pp. 13-31.

Pattie, W.A. and Restall, B.J. (1991a) The genetics of cashmere. *Proceedings of the Cashmere Research Seminar, Ballina*. New South Wales Department of Agriculture and Fisheries, pp. 44-48.

Pattie, W.A. and Restall, B.J. (1991b) Breeding for cashmere. *Proceedings of the Cashmere Research Seminar, Ballina*. New South Wales Department of Agriculture and Fisheries, pp. 80-88.

Pfaff, G., Matter, H.E. and Steinbach, J. (1988) Dynamics of undercoat growth in goats. Untersuchungen zur Dynamik es Wollwachstums bei Ziegen. *Deutsche Tierärztliche Wochenschrift* 95, 178-181.

Reis, P.J. (1982) Growth and characteristics of wool and hair. In: *Sheep and Goat Production*. I.E. Coop (ed.), World Animal Science, Elsevier, Oxford, C1, pp. 205-223.

Reis, P.J., Tunks, D.A. and Munro, S.G. (1988) Relative importance of amino acids and energy for wool growth. *Proceedings of the Nutrition Society of Australia* 13, 122.

Reis, P.J., Tunks, D.A. and Sharry, L.F. (1989) Incorporation of abomasal and intravenous doses of cystine and methionine into wool. *Journal of Agricultural Science* 112, 313-319.

Reis, P.J., Tunks, D.A. and Munro, S.G. (1990) Effects of the infusion of amino acids into the abomasum of sheep, with emphasis on the relative value of methionine, cysteine and homocysteine for wool growth. *Journal of Agricultural Science* 114, 59-68.

Restall, B.J. and Pattie, W.A. (1989) The inheritance of cashmere in Australian goats. 1. Characteristics of the base population and the effects of environmental factors. *Livestock Production Science* 21, 157-172.

Rogers, G.E. (1990) Improvement of wool production through genetic engineering. *Trends in Biotechnology* 8, 6-11.

Rogers, G.R., Orwin, D.G.F., Fraser, M.C., Bray, A.C., Smith, M.C., Baird, D.B., Clarke, J.N. and Bickerstaff, R. (1990) Selective breeding for high and low strength Romney wools. *Proceedings 8th International Wool Textile Research Conference, New Zealand* II. pp. 180-188.

Rowe, J.B., Brown, G., Ralph, I.G., Ferguson, J. and Wallace, J.F. (1989) Supplementary feeding of young Merino sheep, grazing wheat stubble, with different amounts of lupin, oat or barley grain. *Australian Journal of Experimental Agriculture* 29, 29-35.

Russel, A.J.F. (1989) Goats and the international breeding scene. *Proceedings of the 3rd International Cashmere Conference,* Adelaide, Australia.

Russel, A.J.F. (1990) Nutrition of Cashmere Goats. In: *Scottish Cashmere - the viable alternative.* A.J.F. Russel (ed.), Scottish Cashmere Producers Association, Edinburgh, pp. 32-46.

Russel, A.J.F. and Bishop, S.C. (1990) Breeding for cashmere in feral and imported goats in Scotland. *Proceedings of the 4th World Congress on Genetics Applied to Livestock Production, Edinburgh.* XV, 204-208.

Ryder, M.L. and Stephenson, S.K. (1968) *Wool growth.* Academic Press, London.

Sahlu, T., Fernandez, J.H. and Manning, R. (1988) Dietary protein degradability and mohair production. *Proceedings of the Third Annual Field Day of the American Institute for Goat Research, Langston University, Oklahoma* pp. 81-84.

Scheurman, E., Staples, L., McPhee, S. and Galloway, D. (1987) The effect of exogenous melatonin on reproductive and fleece parameters when administered to Australian goats prior to the natural breeding season. *Proceedings 2nd International Cashmere Conference, New Zealand* pp. 145-157.

Shelton, M. (1968) Fibre Production. In: *Goat Production.* C. Gall (ed.), Institute for Animal Breeding and Genetics, Germany, pp. 381-409.

Sumner, R.M.W. and Clarke, J.N. (1990) Scope for modification of Romcross wool characteristics through selection. *Proceedings 8th Research Conference, New Zealand* II pp. 157-168.

Sumner, R.M.W. and McCall, D.G. (1989) Relative wool production of wethers and ewes of different rearing status. *Proceedings of the New Zealand Society of Animal Production* 49, 209-213,

Sun, Y.X., Drane, G.L., Currey, S.D., Lehner, N.D., Gooden, J.M., McDowell, G.H. and Wynn, P.C. (1988) Digestibility of food, growth and wool production in crossbred lambs fed to support moderate and high growth rates and immunised with somatostatin. *Proceedings of the Nutrition Society of Australia* 13, 155.

Sun, Y.X., Drane, G.L., Currey, S.D., Lehner, N.D., Gooden, J.M., Hoskinson, R.M., Wynn, P.C. and McDowell, G.H. (1990) Immunization against somatotrophin release inhibiting factor improves digestibility of food, growth and wool production of crossbred lambs. *Australian Journal of Agricultural Research* 41, 401-411.

Throckmorton, J.C., Ffoulkes, D., Leng, R.A. and Evans, J.V. (1982) Response to bypass protein and starch in Merino sheep and Angora goats. *Proceedings of the Australian Society of Animal Production* 14, 661.

Wallace, R.J. and Newbould, C.J. (1991) Effects of bentonite on fermentation in the rumen simulation techniques (Rusitec) and on rumen ciliate protozoa. *Journal of agricultural Science, Cambridge* 116, 163-168.

Ward, K.A., Murray, J.D., Shanahan, C.M., Rigby, N.W. and Nancarrow, C.D. (1989) The creation of transgenic sheep for increased wool productivity. In: *The biology of wool and hair.* G.E. Rogers, P.J. Reis, K.A. Ward, R.C. Marshall (eds.), London, Chapman and Hall.

Watson, M.J. and Bogdanovic, B. (1988) Avoparcin supplements for wool growth of sheep fed roughage diets. *Proceedings of the Australian Society of Animal Production* 17, 354-357.

Welch, R.A.S., Gurnsey, M.P., Betteridge, K. and Mitchell, R.J. (1990) Goat fibre response to melatonin given in spring in two consecutive years. *Proceedings of New Zealand Society of Animal Production* 50, 335-338.

Westhuysen, J.M. van der, Wentzel, D. and Grobler, M.C. (1985) *Angora goats and mohair in South Africa.* South African Mohair Growers' Association, Port Elizabeth.

Wiener, G., Woolliams, C. and Slee, J. (1988) A comparison of inbred and out-bred sheep on two planes of nutrition. 1. Growth, food intake and wool growth. *Animal Production* 46, 213-220.

Wildman, A.B. (1932) Coat and fibre development in some British sheep. *Proceedings of the Zoological Society of London* 2, 257-285.

Williams, A.J. and Butt, J. (1989) Wool growth of pregnant Merino ewes fed to maintain maternal liveweight. *Australian Journal of Experimental Agriculture* 29, 503-507.

Williams, A.J. and Winston, R.J. (1987) A study of the characteristics of wool follicle and fibre in Merino sheep genetically different in wool production. *Australian Journal of Agricultural Research* 38, 743-755.

Woods, J.L. and Orwin, D.F.G. (1988) Seasonal variations in the dimensions of individual Romney wool fibres determined by a rapid autoradiographic technique. *New Zealand Journal of Agricultural Research* 31, 311-323.

Wynn, P.C., Wallace, A.L.C., Kirby, A.C. and Annison, E.F. (1988) Effects of growth hormone administration on wool growth in Merino sheep. *Australian Journal of Biological Sciences* 41, 177-187.

Zainur, A.S., Tassell, R., Kellaway, R.C. and Dodemaide, W.R. (1989) Recombinant growth hormone in growing lambs: effects on growth, feed utilization, body and carcass characteristics and on wool growth. *Australian Journal of Agricultural Research* 40, 195-206.

Chapter 12

The Production of Transgenic Sheep for Improved Wool Production

K.A. Ward and C.D. Nancarrow

CSIRO Division of Animal Production, PO Box 239, Blacktown, NSW 2148, Australia

Introduction

The amount of wool produced by modern breeds of sheep is dramatically increased compared with that of their primitive ancestors. This results largely from the successful application of selective breeding techniques, in which animals of high wool production are mated to produce superior progeny. There are, however, significant limitations to the procedure, imposed by the fact that the unit of genetic exchange is the intact chromosome. Thus, large numbers of genes which have no direct bearing on wool production must be transferred along with the few genes encoding the genetic information for the sought-for improved phenotype. In addition, conventional breeding makes it impossible to transfer genetic information from one species of animal to another. Recently, a new technology has emerged that offers a solution to these obstacles by providing a method for the direct transfer of isolated pieces of DNA into animal embryos.

The 50 year period from 1940 to 1990 has been one of great discovery in biology. During this time we have seen the elucidation of the genetic code, the confirmation of the one-gene one-enzyme concept, the establishment of differential gene activity as the source of developmental regulation and a preliminary understanding of the detailed molecular events involved in gene regulation. As is the case with most periods of rapid discovery, these advances have been made possible by the development of new technology, which, for biology, consisted of a range of sophisticated techniques allowing the examination and manipulation of biological macromolecules. During the last decade, these techniques have been refined to the point where they now provide the ability to manipulate the genetic properties of animals. Commonly called "genetic engineering", this powerful technology allows purified DNA to be transferred between species, without regard to the constraints imposed by conventional breeding, and makes it possible to consider altering the productivity of domestic animals in ways that would be quite impossible by normal selective breeding.

The application of genetic engineering to the sheep and goat industry is particularly attractive because the textile fibres (e.g. wool, cashmere and mohair) are potential targets for genetic manipulation. In addition, it should eventually be possible to alter milk production and quality, carcass composition, fecundity and even disease resistance in transgenic sheep and goats. However, the field is still in its infancy, with active research needed to increase the efficiency of gene transfer and to understand the nature of the regulation of the genes being transferred. In this chapter, the progress that has been made towards the goal of increasing wool production by gene transfer to sheep will be reviewed.

The Methodology for the Production of Transgenic Sheep

One of the most significant technical advances in modern biology was the demonstration that the growth properties of mice could be altered by the successful transfer of recombinant DNA to mouse embryos by the technique of pronuclear microinjection of single-cell embryos (Costantini and Lacy, 1981; Gordon, 1981; Palmiter *et al.*, 1982; Palmiter *et al.*, 1983; Palmiter and Brinster, 1986; Stewart *et al.*, 1982; Wagner *et al.*, 1981). The application of the methodology to sheep was attractive as a way of improving productivity and several laboratories have committed significant resources towards the adaptation of the techniques to this species (Hammer *et al.*, 1985; Murray *et al.*, 1989; Rexroad *et al.*, 1988; Rexroad *et al.*, 1989; Simons *et al.*, 1988). These efforts have resulted in the clear demonstration that transgenic sheep can be produced in numbers sufficient for the evaluation of novel gene sequences.

The bulk of the experiments involved in the initial establishment of gene transfer technology for sheep utilized growth hormone genes placed under the control of metallothionein promoters. While these experiments were successful in establishing the technique in the species, they were a failure from a practical farming sense, because all transgenic sheep which expressed such transgenes were acromegalic and suffered impaired health as a result. These experiments served to highlight the need to tightly regulate the expression of the foreign "transgenes" when they encode molecules with a critical role in the maintenance of animal homeostasis.

The conclusion to be drawn from this early work is that a detailed understanding of the physiology of important production traits is vital if transgenic technology is to be of value in a practical sense for improving the productivity of sheep. It has become apparent over the past few years that very few such traits actually fall into this category. In addition, it becomes important for the appropriate molecular modifications to be made to the chosen gene so that it has the desired effect on the animal.

Wool growth is one area where these criteria are largely met and hence the application of genetic engineering might be expected to have a substantial impact on productivity. There are two areas where the technology can be readily applied to directly influence fibre production. The first of these is to manipulate the biochemistry of the animals in order to remove substrate limitations to keratin protein synthesis, while the second area is to manipulate the keratin proteins themselves in order to improve the efficiency of production of the fibre and also to change the actual properties of wool.

The Modification of Sheep Biochemistry for Increased Wool Growth

One way in which genetic engineering is likely to have a large impact on sheep productivity is through the introduction of genes that alter the species' inherent biochemical capability. A large number of biochemical pathways have been lost during evolution and the compounds produced by these pathways must be supplied as essential nutrients in the diet. Genetic engineering now provides the ability to restore some of these lost biochemical functions by the transfer of genes from organisms where the pathways are functional. In order for this to be of practical use in increasing wool production however, it is necessary to identify substrate or nutrient limitations to important production traits.

The cysteine biosynthetic pathway

The amino acid cysteine represents such a substrate. As shown by Reis *et al.* (Reis and Schinkel, 1963; Reis, 1967; Reis *et al.*, 1973a; Reis *et al.*, 1973b) the amino acid is rate-limiting for wool growth when sheep are fed a diet that simulates that of the grazing animal. While cysteine itself is not an essential amino acid for mammals, it can only be synthesized from methionine and therefore the supply of either methionine or cysteine is essential in the sheep diet. If the circulating plasma levels of one or other of these two compounds is increased, the rate of wool growth is also increased. However, direct dietary supplementation is not effective because the added amino acid is degraded in the rumen of the sheep by the resident microflora. Since this problem was identified, several different approaches have been taken to overcome the degradation that occurs. For example, proteins with low ruminal degradability have been included in the diet. These proteins can be of natural origin, or can be "protected" by treatment with formaldehyde (Ferguson, 1975), which prevents ruminal degradation but allows digestion in the distal part of the digestive tract. Alternatively, methionine itself can be encapsulated in a medium that survives the rumen but not the lower digestive tract (Ferguson, 1975). However, while these methods have all proven effective in increasing the amount of available cysteine, they have not been economically viable for the farmer.

An alternative approach is to introduce into the sheep genome the genes that encode the cysteine biosynthetic pathway by the techniques of genetic engineering. The appropriate genes are functional in prokaryotes and the auxotrophic eukaryotes, and are thus available for isolation, modification for expression in the digestive tract of sheep, and transfer to the sheep genome by embryo microinjection. In Australia, two laboratories are currently pursuing this approach, using the genes from *Escherichia coli* (Ward *et al.*, 1984, 1990, 1991) or *Salmonella typhimurium* (Sivaprasad *et al.*, 1989; D'Andrea *et al.*, 1989; Rogers, 1990).

The pathway for the biosynthesis of cysteine in *E. coli* and *S. typhimurium* is complex and consists of two discrete parts, namely a simple two step pathway that combines sulphide with the amino acid serine to produce cysteine and a pathway for the reduction of a sulphur source to sulphide. In *E. coli* only two genes are required to encode the first part of the process. These are the *cysE* and *cysK* genes, encoding the enzymes serine transacetylase and O-acetylserine sulph-hydrylase. These catalyse the following reaction:

serine + acetyl-CoA --------------> O-acetylserine + CoA-SH
                    serine transacetylase

O-acetylserine + H_2S ---------------> cysteine + acetate
                    O-acetylserine sulph-hydrylase

The *cysE* and *cysK* genes from *E. coli* and the *cysE, cysK* and *cysM* genes from *S. typhimurium* have all been isolated and fully characterized (Denk, and Bock, 1987; Byrne *et al.*, 1988a; D'Andrea *et al.*, 1989), thus providing the necessary coding sequences for the gene transfer experiments.

The research involved in this type of genetic engineering project can be divided into four parts:

1. Isolation and characterization of the appropriate DNA coding and promoter sequences.
2. Construction and testing of various fusion genes in cell culture to determine the optimum configuration of the various components.
3. Testing of the preferred gene in transgenic mice to determine the efficiency and tissue-specificity of expression *in vivo*.
4. Transfer of the preferred gene to transgenic sheep.

Figure 12.1 Diagrammatic representation of the modifications made to bacterial coding sequences to provide expression in eukaryotic cells. Gene 1 contains exon 5 of the sheep growth hormone gene 3' to the bacterial sequence. Gene 2 is similar but contains the entire sheep growth hormone gene. Gene 3 consists of a fusion of two gene 1 sequences such that a single piece of DNA encodes the enzymes necessary for the cysteine synthesis or glyoxylate cycle biochemical pathways.

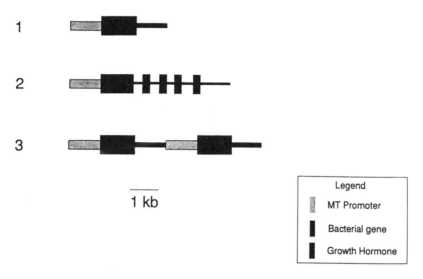

1 kb

Legend	
	MT Promoter
	Bacterial gene
	Growth Hormone

The general design of the modifications made to the bacterial genes in our laboratory is shown in Figure 12.1. The bacterial coding sequence has been inserted downstream from the sheep metallothionein-Ia promoter and the sheep growth hormone gene used for stabilization of the mRNA transcribed from the fusion gene product (Ward *et al.*, 1990; Ward *et al.*, 1991). A number of different combinations of the various components of the fusion genes were constructed in order to determine the configuration conferring optimal production of the encoded enzymes in eukaryotic cells. The ability of the modified genes to be transcribed and translated was tested initially in mouse L-cells in culture. The genes all produced mRNA transcripts of the predicted sizes (Ward *et al.*, 1991), and extracts prepared from the transformed cells contained the readily detectable levels of the enzymes SAT and/or OAS (Table 12.1). Substantial differences in expression were observed with the different genes, demonstrating the importance of investigating a number of different combinations of the various components used to construct the fusion genes. Genes containing only exon 5 of the sheep growth hormone gene (MTCE10, MTCK7 and MTCEK1) were expressed at much higher levels than genes containing the entire growth hormone sequence.

Table 12.1 Activity of serine transacetylase (SAT) and O-acetylserine sulph-hydrylase in extracts from cells transformed with various fusion genes containing either the *cysE* (CE) or *cysK* (CK) gene of *Escherichia coli*. Genes were constructed as shown in Figure 8.1. pMTCE10, pMTCK7 and pMTCEK1 contain only exon 5 of the sheep growth hormone gene. pMTCE11 and pMTCK11 contain the complete sheep growth hormone gene. Enzyme activity is expressed as umoles substrate utilized (SAT) or product formed (OAS) /30 min /mg protein.

Construct	Zn - induced		Uninduced	
	SAT	OAS	SAT	OAS
pMTCE10	2796	-	777	-
pMTCE11	255	-	68	-
pMTCK7	-	1350	-	-
pMTCK11	-	162	-	-
pMTCEK1	268	6960	86	1242
pMTCE10 + pMTCK7	654	490	196	162

The various *cys*-encoding fusion genes were transferred to mice by embryo microinjection and their expression measured in intestinal, liver and kidney tissues of the resulting transgenic animals. As predicted from the results in tissue culture, only those genes containing exon 5 of the sheep growth hormone gene located 3' to the bacterial coding sequence were expressed at detectable levels in zinc-induced transgenic mice. This information is of general relevance to the expression of genes in transgenic animals, since the genes containing only exon 5 do not possess any exon/intron structure. Introns therefore are not an obligatory requirement for the expression of foreign genes in transgenic mice.

Having established the fact that both genes can be independently transcribed and translated in transgenic mice, a combination gene which contains both the *cysE* and *cysK* sequences in a single piece of DNA (Figure 12.1) was transferred. Both *E. coli*-derived sequences were transcribed and translated from this DNA sequence (Table 12.2). Expression of the *cysE* and *cysK* genes was found at high levels in the intestinal epithelium and at lower levels in the kidney and liver. Expression was dependent on induction by zinc. These animals thus possess in these tissues the enzymes necessary for the synthesis of cysteine from H_2S. Clearly, the actual biosynthesis of cysteine in these animals requires the presence of the necessary substrates in the tissues that express the transgenes and current experiments are directed towards establishing whether such biosynthetic activity can be demonstrated in the mice.

Table 12.2 Activity of serine transacetylase (SAT) and O-acetylserine sulph-hydrylase (OAS) in tissue extracts prepared from transgenic mice. CK7-26 contains the gene pMTCK7, CE10-29 contains pMTCE10 and EK1-28 and EK1-08 contain pMTCEK1. Specific activity is measured as nmoles substrate utilized (SAT) or product formed (OAS) / 30 min /mg protein.

Mouse line	Organ	SAT	OAS
CK7-26	Intestine	-	206
	Kidney	-	352
	Liver	-	13
CE10-29	Intestine	6546	-
	Kidney	0	-
	Liver	0	-
EK1-28	Intestine	1161	2797
	Kidney	0	24
	Liver	0	3
	Brain	16	85
EK1-08	Intestine	4522	12778
	Kidney	105	128
	Liver	9	3
	Brain	0	245

The results obtained from transgenic mice indicate that a combination of genes appears to have been established that is suitable for expression in animals, since high levels of zinc-inducible expression of both genes has been obtained in the intestinal epithelium. Earlier studies in our laboratory have shown that there is close correspondence between the expression in transgenic mice and transgenic sheep, although the zinc status of sheep can result in genes that are regulated by metallothionein promoters being more highly expressed in sheep than in mice (Murray *et al.*, 1989; Shanahan *et al.*, 1989). We have therefore commenced the transfer of these genes to sheep.

A conceptually similar approach has been taken by Rogers and colleagues (D'Andrea *et al.*, 1989; Sivaprasad *et al.*, 1989; Rogers, 1990) to achieve the same goal of enabling sheep to synthesize cysteine from H_2S. The source of genes to provide the appropriate coding sequences are again a bacterium, in this case *S. typhimurium*. The two genes being used are the *cysE* gene, which is essentially the same as the *E. coli* gene, and the *cysM* gene, which encodes an OAS enzyme significantly different from that encoded by the *cysK* gene both in *E. coli* and *S. typhimurium*. These two bacterial genes have been expressed in sheep cells in culture after being fused to the SV40 late promoter and SV40 polyadenylation signal sequences

(Sivaprasad *et al.*, 1989; Rogers, 1990), where they have given rise to active SAT and OAS enzyme activities. More recently, the two genes have been joined together in a single piece of DNA and each coding sequence has been placed under the control of a promoter derived from the long terminal repeat of the Rous sarcoma virus. When the whole construct was transferred to mice by pronuclear microinjection, constitutive expression was obtained (Sivaprasad *et al.*, 1989). The same gene has also been transferred to a number of transgenic sheep and constitutive expression of SAT and OAS observed in tail tissue (Rogers, 1990). These experiments are significant because they demonstrate that the transcriptional and translational products of the genes, in spite of their bacterial origins, are stable in at least some sheep tissues. The research now in progress aims to provide the genes with a promoter more suited to directing their expression in tissues where the appropriate substrates for cysteine biosynthesis might be available (Rogers, 1990).

The glyoxylate cycle

Wool production, in addition to its dependence on an adequate supply of sulphur amino acids, also requires substantial supplies of glucose as a source of metabolizable energy (Chapman and Ward, 1979). In sheep, as in all ruminants, glucose is not available from the digestive tract. This results from the ruminant digestive process, which is largely dependent on the bacteria which populate the rumen. These microorganisms consume essentially all available carbohydrate in the ingested feed and produce a range of fermentation products, the most important of which, from an energy viewpoint, are the volatile fatty acids. These are absorbed by the sheep and are used directly for energy or, if gluconeogenic, are converted to glucose to provide the carbohydrate that is essential for the proper function of several key tissues, including the wool follicles (Chapman and Ward, 1979). On some pastures, the predominant volatile fatty acid produced in the rumen is acetate, which is not gluconeogenic, and on such pastures sub-optimal growth of wool is predicted (Armstrong *et al.*, 1957; Lindsay, 1979; Van Soest, 1982). By utilizing the abundant supply of acetate in these animals, it has been suggested that wool growth might be increased.

The ability to transfer functional biochemical pathways to animals provides a possible way to achieve this objective, because acetate can serve as a source of glucose in organisms that possess the enzymes necessary to catalyse the reactions of the glyoxylate cycle (Cioni *et al.*, 1981) (Figure 12.2). Acetate enters the tricarboxylic acid cycle in normal fashion to produce isocitric acid, but when the glyoxylate cycle is operational, the isocitrate is cleaved to succinate and glyoxylate by the action of the enzyme isocitrate lyase. Succinate is a gluconeogenic substrate and the glyoxylate produced in the reaction is combined with another molecule of acetate to produce malate, thus providing the necessary substrate for continuation of the cycle. This second reaction is catalysed by the enzyme malate synthase.

Figure 12.2 The biochemical reactions of the glyoxylate cycle.

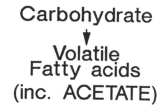

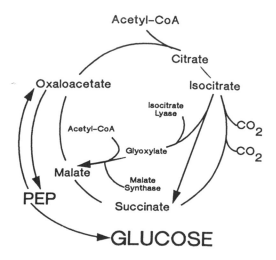

In *E. coli*, the enzyme isocitrate lyase is encoded by the gene *aceA*, while malate synthase is encoded by the *aceB* gene. These have both recently been isolated and sequenced (Byrne *et al.*, 1988b; Matsuokae and McFadden, 1988; Rieul *et al.*, 1988). In order to transfer this genetic information in functional form to sheep, a protocol similar to that described earlier for the cysteine biosynthesis genes is being followed. The same general construction of the genes has been used (Figure 12.1), utilizing the sheep MT-Ia promoter and exon 5 of the sheep growth hormone gene. Five genes utilizing different combinations of the various sequence components have been tested for expression in mouse L-cells and transgenic mice.

In L-cells, all genes were transcribed giving RNA transcripts of the predicted sizes (Byrne, 1990; Ward *et al.*, 1990). Upon translation, active isocitrate lyase and malate synthase were produced in extracts prepared from the transformed cells (Table 12.3). The genes were then transferred to mice and their expression examined in liver, kidney and intestinal tissues. As was found for the genes encoding cysteine biosynthesis, those genes containing only exon 5 of the sheep growth hormone gene were the most actively transcribed and translated in transgenic animals. In all three tissues examined, the genes produced mRNAs which hybridized with appropriate probes for the bacterial coding sequences and were of the predicted sizes (Byrne, 1990). Cell-free extracts from the same tissues showed active isocitrate lyase and malate synthase activities, indicating that the animals have the enzymic potential for the operation of the cycle (Byrne, 1990). The detailed physiology of these animals is currently under investigation.

Table 12.3 Activity of isocitrate lyase and malate synthase in extracts of zinc-induced L-cells transformed with fusion genes encoding the enzymes of the glyoxylate cycle. pMTaceA1 and pMTaceB1 contain the complete sheep growth hormone sequence, while pMTaceA2 and pMTaceB2 contain only exon 5, as shown in Figure 12.1. Specific activities are expressed as nmoles of product formed /20 min /mg protein and are corrected for a low level of malate synthase activity in L-cell control extracts. No isocitrate lyase activity was detected in untransformed L-cells.

Construct	Isocitrate lyase	Malate synthase
pMTaceA1	76	-
pMTaceB1	-	1.7
pMTaceA2	68	-
pMTaceB2	-	34.3

Structural Protein Modification

The second major area where genetic engineering of sheep should prove useful in increasing wool production is in the manipulation of the keratin proteins that comprise the wool fibre itself. This is an attractive target area because their modification would not be expected to affect the viability of the animal. The potential for such modification remains speculative at present, for very little experimentation is currently directed towards this goal. However, the growing acceptance of the relevance of transgenic animals in domestic animal production has resulted in the approach receiving serious consideration.

The important textile fibres, wool, cashmere and angora, are composed of a unique group of sulphur-rich proteins, the keratins (Fraser *et al.*, 1972), which are combined in a precise arrangement of microfibrils and surrounding matrix. The geometry of this combination is essentially the same in all fibres, so the different properties of the various fibres arise from differences in keratin protein composition and from differences in the structure and arrangement of the keratinized cells. Thus, it is probable that manipulation of one or more of the keratins may be able to alter the textile properties of the fibre. However, changing specific keratin proteins by gene modification requires a detailed knowledge of the genes involved. While this information is still incomplete, it is nevertheless sufficient for some experiments to be considered in this area.

The keratins can be divided into sub-groups, each of which has an important role in the final epidermal structure. Each sub-group of keratins is encoded in the genome by a multi-gene family, and gene probes for each of these families have now been isolated (Ward *et al.*, 1982). Many of the genes encoding the wool keratins have been isolated (Powell *et al.*, 1983, 1986; Wilson *et al.*, 1988; Frenkel *et al.*, 1989) and their sequences are being examined for conserved regions that might indicate potential control and regulatory regions.

The critical nature of the interaction between these various proteins in the wool fibre has recently been shown by an experiment in which a complete keratin gene encoding a type II wool protein was expressed in transgenic mice (Powell and Rogers, 1990). The gene was expressed at different levels in various transgenic animals, and when the expression was high, the wool protein caused significant abnormalities in the structure of the mouse hair. This is probably the result of incompatible pairing of the wool protein with the mouse keratins and hence provides some guide to the limits for keratin manipulation. The experiment demonstrates that modifications can be made to the keratin protein sequences while maintaining tissue-specific expression in the hair or wool follicle, and that when the keratin sequences are altered, significant effects can be observed on fibre properties. The challenge ahead is to determine what sequence manipulations can be made that confer more desirable properties on the wool fibre.

Since wool biosynthesis is limited by the supply of the sulphur amino acids because of the high cysteine content of the wool keratin proteins, an alternative approach to increasing the supply of cysteine is to reduce the cysteine requirement of the fibre. This might be achieved by modifying one or more of the genes encoding matrix keratins so that the number of cysteine residues is reduced, and then transferring the modified genes to animals such that they are expressed at high levels. Since the cysteine content of wool can vary considerably without apparent effect on its textile properties (P.J. Reis, unpublished), the expression of low-cysteine matrix keratins could increase wool production substantially.

It would be of significant advantage to the wool industry if the fibre produced by sheep could be given some of the desirable textile attributes of the goat cashmere fibre. The desirable textile qualities associated with cashmere fibre arise partly from the fine diameter of the fibre ($< 16 \ \mu$m) and partly from a cuticle structure different from that of wool (Maddocks and Jackson, 1988). The factors that are responsible for the structure of the cashmere cuticle are not known, but good candidates are the cuticle-specific proteins that fill the cells. The genes for several cuticle-specific proteins have recently been identified (McKinnon *et al.*, 1990), thus making it possible to consider altering the cuticle proteins of the wool fibre so that they approximate the composition of those of the cashmere fibre.

Other modifications to the structural proteins of the textile fibres are possible. For example, proteins with increased numbers of amino acids possessing side-chains important for dye-retention might be introduced into the fibre. Inclusion of proteins with specific toxicity to moths might prevent the destructive action of the keratinases secreted by these insects. It might also be possible to reduce or eliminate the shrinkage that wool fabrics can undergo during normal washing. These examples remain speculation at this time but serve to demonstrate the ever-widening range of possibilities that is opening up in the area of textile fibre protein modification.

Community Acceptance of Genetically-Engineered Animals

The introduction of transgenic sheep into the farming community is probably some years away. Nevertheless, it is probable that some genetically-engineered livestock will eventually be used by farmers to improve wool production. There exist at present widespread community concerns about the safety and long-term utility of such animals. However, transgenic sheep modified for increased wool production efficiency are less likely to cause major concern compared with animals in which fundamental hormone levels are altered. Thus, such wool-modified transgenic sheep represent a possible route for scientists to initiate the processes necessary to allay the community fears. This can only be achieved by a combination of community discussion and effective legislation, organized in such a way that the science necessary for the development and testing of the animals is not impeded but providing at the same time a safe and open regulatory framework for release of the animals.

A wide range of regulatory regimes currently operates in different countries, from a very liberal approach in Italy to extremely strict requirements in Germany and Denmark. Canada, the UK, the USA and Australia all use similar methods for the review and controlled release of genetically-engineered organisms and in the Australian situation, a Government enquiry has been under way since mid-1990 to formalize the procedures that must be followed in this country. There are many difficulties associated with the formulation of appropriate legislation because of the conflicting interests of the many parties involved. Thus, the general public need an assurance that their health and environment are adequately protected, interest groups need a mechanism for expressing their concerns, scientists need clear guidance on the requirements for release so that they can design appropriate experiments and politicians must consider the economic needs of the nation and yet at the same time properly represent the concerns of their local constituents.

It is essential that major public interest groups become involved at an early stage in the planning of experiments which have as their ultimate goal the release of genetically-engineered sheep to the farming community. In this way, many of the environmental concerns of the public could be addressed and allayed well in advance of the time when release of the modified animals is required. While this inevitably means the loss of some commercial confidentiality, it is not impossible to consider mechanisms whereby the patentable details of the work are held in confidence while allowing full and open discussion of the general concepts. It is important that the question of community acceptance be addressed as a matter of priority, because it is likely that the public perception of genetic engineering may become as great a hurdle to the practical application of the research as the actual production of useful animals.

The Future Direction for Genetic Engineering of Sheep for Increased Wool Production

It is clear that the actual technology for the production of transgenic sheep is now in place, although substantial increases in efficiency of the animals would be of considerable practical value. It still remains to be demonstrated that the technology can in fact alter the productivity of sheep in a positive way for the farmer. The work on the manipulation of biochemical pathways that has been described here will serve as a useful guide to the extent to which the techniques can be used to alter the fundamental reactions involved in animal metabolism. Should these experiments prove successful in sheep over the next few years, the field will be opened up for the manipulation of a number of pathways involved in the production of rate-limiting substrates and cellular metabolites. Also in the next few years, there is likely to be more emphasis placed on the production of sheep which produce altered keratin proteins in their wool follicles. This may result in novel useful changes to the properties of wool itself.

It is certain that considerable emphasis will be placed on understanding the mechanisms whereby transgenes might be regulated. A number of original approaches to this difficult problem are currently being investigated, and considering its importance for the future application of the technology, such investigations will continue to be supported. An area of major research during the next five years is likely to be the alteration of the activity of existing genes. Powerful new approaches to this technology are now available, including the use of genes encoding anti-sense RNA molecules and RNA molecules that contain catalytic cleavage sites (ribozymes) (Cameron and Jennings, 1989; Cotten and Birnstiel, 1989) which can destroy specific mRNA molecules. In addition, the replacement of existing genes with modified sequences may soon be possible in domestic animals. This will involve the use of embryonic stem (ES) cells, the production of which in livestock remains in its infancy.

It is now a decade since the first transgenic animal was produced and, since that time, much has been promised to the farming community concerning the potential gains to be expected in animal productivity by the utilization of the technology. However, it is clear that such gains will require considerable effort and a very clear understanding of the physiology of the species under investigation. Because the physiology and biochemistry of wool growth has been studied for some time, many of the details of its production are known and it therefore has some advantage as a potential target area. It is possible that the first genuinely practical application of transgenic technology for the farming community will be achieved in this area. However, considerable research remains to be done and it is to be hoped that the wool growers' patience will endure.

Acknowledgments

The authors would like to express their sincere thanks to their colleagues Dr. Jim Murray, Dr. Carolyn Byrne, Dr. Zdenka Leish, Dr. Cathy Shanahan, Dr. Cheryl Hunt, Mr. Bruce Wilson and Dr. John Bonsing for their contributions and dedication to much of the research described in this review. The expert help of Ms. Cathy Townrow, Mrs. Nola Rigby, Mrs. Astrid Dafter, Mrs. Gina Hardwicke and Ms. Kathy Lewis in the microinjection of sheep embryos and gene construction work, and of Mr. Ian Hazelton, Mr. Jim Marshall, Mr. Tony Redden and Mr. Peter Mitchell in the transfer of embryos to transgenic sheep is gratefully acknowledged.

References

Armstrong, D.G., Blaxter, K.L. and Graham, N.McC. (1957) Utilization of the end products of ruminant digestion. *Proceedings of the British Society of Animal Production* 3-15.

Byrne, C.R. (1990) *The isolation of bacterial acetate metabolism genes and their expression in a eukaryote.* Ph.D. thesis. Sydney, Australia, Macquarie University.

Byrne, C.R., Monroe, R.S., Ward, K.A. and Kredich, N.M. (1988a) DNA sequences of the *csyK* regions of *S. typhimurium* and *E. coli* and linkage of the *cysK* regions to *ptsH*. *Journal of Bacteriology* 170, 3150-3157.

Byrne, C.R., Stokes, H.W. and Ward, K.A. (1988b) Nucleotide sequence of the *aceB* gene encoding malate synthase A in *Escherichia coli*. *Nucleic Acids Research* 16, 9342.

Cameron, F.H. and Jennings, P.A. (1989) Specific gene suppression by engineered ribozymes in monkey cells. *Proceedings of the National Academy of Science, U.S.A.* 86, 9139-9143.

Chapman, R.E. and Ward, K.A. (1979) Histological and biochemical features of the wool fibre and follicle. In *Physiological and Environmental Limitations to Wool Growth*, J.L. Black and P.J. Reis (eds.) Armidale, University of New England Publishing Unit, pp. 193-208.

Cioni, M., Pinzauti, G. and Vanni, P. (1981) Comparative biochemistry of the glyoxylate cycle. *Comparative Biochemistry and Physiology* 70B, 1-26.

Costantini, F. and Lacy, E. (1981) Introduction of a rabbit beta-globin gene into the mouse germ line. *Nature* 294, 92-94.

Cotten, M. and Birnstiel, M.L. (1989) Ribozyme mediated destruction of RNA *in vivo*. *Embo Journal* 8, 3861-3866.

D'Andrea, R.J., Sivaprasad, A.V., Bawden, C.S., Kuczek, E.S., Whitbread, L.A. and Rogers, G.E. (1989) Isolation of microbial genes for cysteine synthesis and prospects for their use in increasing wool growth. In *The Biology of Wool and Hair*, G.E. Rogers, P.J. Reis, K.A. Ward and R.G. Marshall (eds.) New York and London, Chapman and Hall, pp. 447-464.

Denk, D. and Bock, A. (1987) L-cysteine biosynthesis in *Escherichia coli*: nucleotide sequence and expression of the serine acetyltransferase (*cysE*) gene from the wild-type and a cysteine-excreting mutant. *Journal of General Microbiology* 133, 515-525.

Ferguson, K.A. (1975) The protection of dietary proteins and amino acids against microbial fermentation in the rumen. In *Digestion and Metabolism in the Ruminant*, I.W. McDonald and A.C.I. Warner (eds.) Armidale, University of New England Publishing Unit, pp. 448-464.

Fraser, R.D.B., MacRae, T.P. and Rogers, G.E. (1972) *Keratins, their composition, structure and biosynthesis*. Springfield, Charles C. Thomas.

Frenkel, M.J., Powell, B.C., Ward, K.A., Sleigh, M.J. and Rogers, G.E. (1989) The keratin BIIIB family: isolation of cDNA clones and structure of a gene and a related pseudogene. *Genomics* 4, 182-191.

Gordon, J.W. (1981) Integration and stable germ line transmission of genes injected into mouse pronuclei. *Science* 214, 1244-1246.

Hammer, R.E., Pursel, V.G., Rexroad, C.E., Jr., Wall, R.J., Bolt, D.J., Ebert, K.M., Palmiter, R.D. and Brinster, R.L. (1985) Production of transgenic rabbits, sheep and pigs by microinjection. *Nature* 315, 680-683.

Lindsay, D.B. (1979) Metabolism in the whole animal. *Proceedings of the Nutrition Society* 38, 295-301.

Maddocks, I.G. and Jackson, N. (1988) *Structural studies of sheep, cattle and goat skin*. Blacktown, CSIRO Division of Animal Production.

McKinnon, P.J., Powell, B.C. and Rogers, G.E. (1990) Structure and expression of genes for a class of cysteine-rich proteins of the cuticle layers of differentiating wool and hair follicles. *Journal of Cell Biology* 111, 2587-2600.

Matsuokae, M. and McFadden, B.A. (1988) Isolation, hyperexpression and sequencing of the *aceA* gene encoding isocitrate lyase in *Escherichia coli*. *Journal of Bacteriology* 170, 4528-4563.

Murray, J.D., Nancarrow, C.D., Marshall, J.T., Hazelton, I.G. and Ward, K.A. (1989) Production of transgenic sheep by microinjection of ovine metallothionein-ovine growth hormone fusion genes. *Reproduction, Fertility and Development* 1, 147-155.

Palmiter, R.D., Brinster, R.L., Hammer, R.E., Trumbauer, M.E., Rosenfeld, M.G., Birnberg, N.C. and Evans, R.M. (1982) Dramatic growth of mice that develop from eggs microinjected with metallothionein-growth hormone

GH fusion genes stimulate growth in mice. *Nature* 300, 611-615.

Palmiter, R.D., Norstedt, R.E., Gelinas, R.E., Hammer, R.E. and Brinster, R.L. (1983) Metallothionein-human GH fusion genes stimulate growth in mice. *Science* 222, 809-814.

Palmiter, R.D. and Brinster, R.L. (1986) Germline transformation of mice. *Annual Review of Genetics* 20, 465-500.

Powell, B.C., Sleigh, M.J., Ward, K.A. and Rogers, G.E. (1983) Mammalian keratin gene families: organisation of genes coding for the B2 high-sulphur proteins of sheep wool. *Nucleic Acids Research* 11, 5327-5346.

Powell, B.C., Cam, G.R., Feitz, M.J. and Rogers, G.E. (1986) Clustered arrangement of keratin intermediate filament genes. *Proceedings of the National Academy of Science U.S.A.* 83, 5048-5052.

Powell, B.C. and Rogers, G.E. (1990) Cyclic hair-loss and regrowth in trans-genic mice overexpressing an intermediate filament gene. *Embo Journal* 9, 1485-1493.

Reis, P.J. and Schinkel, P.G. (1963) Some effects of sulphur-containing amino acids on the growth and composition of wool. *Australian Journal of Biological Science* 16, 218-230.

Reis, P.J. (1967) The growth and composition of wool. *Australian Journal of Biological Science* 20, 809-825.

Reis, P.J., Tunks, D.A. and Downes, A.M. (1973a) The influence of abomasal and intra-venous supplements of sulphur-containing amino acids on wool growth rate. *Australian Journal of Biological Science* 26, 249-258.

Reis, P.J., Tunks, D.A. and Sharry, L.F. (1973b) Plasma amino acid patterns in sheep receiving abomasal infusions of methionine and cystine. *Australian Journal of Biological Science* 26, 635-644.

Rexroad, C.E. Jr., Behringer, R.R., Bolt, D.J., Miller, K.F., Palmiter, R.D., and Brinster, R.L. (1988) Insertion and expression of a growth hormone fusion gene in sheep. *Journal of Animal Science* 6, 267.

Rexroad, C.E. Jr., Hammer, R.E., Bolt, D.J., Mayo, K.M., Frohman, L.A., Palmiter, R.D. and Brinster, R.L. (1989) Production of transgenic sheep with growth regulating genes. *Molecular Reproduction and Development* 1, 164-169.

Rieul, C., Bleicher, F., Duclos, B., Cortay, J.C. and Cozzone, A.J. (1988) Nucleotide sequence of the *aceA* gene coding for isocitrate lyase in *Escherichia coli*. *Nucleic Acids Research* 16, 5689.

Rogers, G.E. (1990) Improvement of wool production through genetic engineer-ing. *Trends in Biotechnology* 8, 6-11.

Shanahan, C.M., Rigby, N. W., Murray, J.D., Marshall, J., Townrow, C., Nancarrow, C.D. and Ward, K.A. (1989) Regulation of expression of a sheep growth hormone fusion gene in transgenic mice. *Molecular and Cell Biology* 9, 5473-5479.

Simons, J.P., Wilmut, I., Clark, A.J., Archibald, A.L., Bishop, J.O. and Lathe, R. (1988) Gene transfer into sheep. *Biotechnology* 6, 179-183.

Sivaprasad, A.V., D'Andrea, R.J., Bawden, C.S., Kuczek, E.S. and Rogers, G.E. (1989) Towards a new sheep genotype with increased wool growth by transgenesis with microbial genes for cysteine synthesis. *Journal of Cellular Biochemistry* 13B (suppl.), 183.

Stewart, T., Wagner, E.F. and Mintz, B. (1982) Human beta-globin gene sequ-ences injected into mouse eggs, retained in adults and transmitted to progeny. *Science* 217, 1046-1048.

Van Soest, P.J. (1982) *Nutritional Ecology of the Ruminant*. Portland, Durham and Downey.

Wagner, E.F., Stewart, T.A. and Mintz, B. (1981) The human beta-globin gene and a functional viral thymidine kinase gene in developing mice. *Proceedings of the National Academy Science, U.S.A.* 78, 5016-5020.

Ward, K.A., Sleigh, M.J., Powell, B.C. and Rogers, G.E. (1982) The isolation and analysis of the major wool keratin gene families. In *Proceedings of the 2nd World Congress on Genetics Applied to Livestock production*, 6, pp. 146-156.

Ward, K.A., Murray, J.D., Nancarrow, C.D., Boland, M.P. and Sutton, R. (1984) The role of embryo gene transfer in sheep breeding programmes. In *Reproduction in sheep*, D.R. Lindsay and D.T. Pearce (eds.), pp. 279-285.

Ward, K.A., Nancarrow, C.D., Byrne, C.R., Shanahan, C.M., Murray, J.D., Leish, Z., Townrow, C., Rigby, N.W., Wilson, B.W. and Hunt, C.L. (1990) The potential of transgenic animals for improved agricultural productivity. *Rev. sci. tech. Off. int. Epiz.* 9, 847-864.

Ward, K.A., Byrne, C.R., Wilson, B.W., Leish, Z., Rigby, N.W., Townrow, C.R., Hunt, C.L., Murray, J.D. and Nancarrow, C.D. (1991) The regulation of wool growth in transgenic animals. *Advances in Dermatology* 1, 70-76.

Wilson, B.W., Edwards, K.J., Sleigh, M.J., Byrne, C.R. and Ward, K.A. (1988) Complete sequence of a type-1 microfibrillar wool keratin gene. *Gene* 73, 21-31.

Index